SOCIOBIOLOGY AND CONFLICT

SOCIOBIOLOGY AND CONFLICT

Evolutionary perspectives on competition, cooperation, violence and warfare

Edited by

J. VAN DER DENNEN

Polemological Institute,
University of Groningen, The Netherlands

and

V. FALGER

Department of International Relations,
University of Utrecht, The Netherlands

CHAPMAN AND HALL

London • New York • Tokyo • Melbourne • Madras

UK	Chapman and Hall, 2–6 Boundary Row, London SE1 8HN
USA	Chapman and Hall, 29 West 35th Street, New York NY10001
JAPAN	Chapman and Hall Japan, Thomson Publishing Japan, Hirakawacho Nemoto Building, 7F, 1-7-11 Hirakawa-cho, Chiyoda-ku, Tokyo 102
AUSTRALIA	Chapman and Hall Australia, Thomas Nelson Australia, 480 La Trobe Street, PO Box 4725, Melbourne 3000
INDIA	Chapman and Hall India, R. Seshadri, 32 Second Main Road, CIT East, Madras 600 035

First edition 1990

Softcover reprint of the hardcover 1st edition 1990

Typeset in 10/12 pt Plantin Light by
Leaper and Gard Ltd, Bristol
Printed in Great Britain by
St Edmondsbury Press, Bury St Edmunds,
Suffolk

ISBN-13: 978-94-010-7316-5 e-ISBN-13: 978-94-009-1830-6
DOI: 10.1007/978-94-009-1830-6

British Library Cataloguing in Publication Data

A CIP catalogue record for this book
is available from the British Library.

Library of Congress Cataloging-in-Publication Data

Available

Contents

PART THREE

'Primitive' Warfare

PART FOUR

The Conflict about Sociobiology

Contributors

Dan Cohen	Professor of Ecology and Evolutionary Biology at the Department of Botany, the Hebrew University, Jerusalem, Israel
Johan M.G. van der Dennen	Researcher at the Polemological Institute of the State University of Groningen, The Netherlands
Vincent S.E. Falger	Lecturer in International Relations at the University of Utrecht, The Netherlands
Penny Anthon Green	Teaches at the Department of Sociology at the University of Texas at Austin
Michael Hopp	Member of the Hebrew University School of Nutrition and has a private applied social research institute
Colin Irwin	Held a Killam Post-Doctoral Fellowship at Dalhousie University at Halifax, Nova Scotia, Canada, while preparing his contribution to this volume
Joseph Lopreato	Professor of Sociology at the University of Texas at Austin and a former Chairperson of his Department
Umberto Melotti	Professor of Political Sociology and Cultural Anthropology at the universities of Rome and Pavia, Italy
Peter Meyer	Lecturer of Sociology at the Economics and Social Sciences Department at the University of Augsburg, FRG

Popko P. van der Molen	Conducts research at the University of Groningen, The Netherlands
Uzi Motro	Senior Lecturer at the Department of Genetics and the Department of Statistics of the Hebrew University of Jerusalem, Israel
Yochanan Peres	Associate Professor of Sociology at Tel Aviv University, Israel
Olwen Rasa	Associate Professor of Animal Behaviour at the University of Pretoria, South Africa
Ullica Segerstråle	Assistant Professor of Sociology at Illinois Institute of Technology, Chicago
Lionel Tiger	Professor of Anthropology at Rutgers University in New Brunswick, New Jersey

Acknowledgements

The Sociobiology of Conflict was the topic of the ninth meeting of the European Sociobiological Society, held on January 10 and 11, 1987. It was Michael Hopp's initiative to organize this conference in Jerusalem, Israel, a symbolic place in many respects. Thanks to the scientific and personal qualities of Professor Amotz Zahavi, from Tel-Aviv University, many non-Israeli participants were able to experience the naturalistic, geographical and political history of the country in an impressive guided tour which influenced clearly the presentations and discussion in the conference.

Without the hospitality and financial support of the Van Leer Jerusalem Institute the meeting would not have been possible. The Institute's director, Professor Yehuda Elkana, and Mrs Rivka Ra'am, member of the Executive Board of the Van Leer Jerusalem Institute, in close cooperation with local organizer Michael Hopp, contributed very much to the success of the meeting itself. The Board of the European Sociobiological Society expresses its gratitude for this vital support.

In the conference itself Vincent Falger, Lea Gavish, Johan Goudsblom, Anne Rasa, Avi Shmida, Jan Wind and Amotz Zahavi presented papers next to those elaborated and collected in this volume. ESS Board members Jan Wind, Hans van der Dennen and Vincent Falger organized those aspects of the conference not immediately connected with the meeting in Jerusalem.

Finally, the patience and trust of Tim Hardwick, former senior editor with Chapman and Hall, and his successor Bob Carling, were indispensable for this book to be published. It is fortunate that in human society cooperation is not less essential than conflict. This volume combines both, not surprisingly.

CHAPTER ONE

Introduction

1.1 THE STUDY OF CONFLICT

Polemos Pantoon Pater

Heraclitus

Conflict on all levels of organic existence is pervasive, persistent, ubiquitous. Conflict is the universal experience of all life forms. Organisms are bound in multiple conflict-configurations and -coalitions, which have their own dynamic and their own logic. This does not mean, however, that the more paroxysmal forms of conflict behaviour, naked violence and destruction, are also universal. Conflict and cooperation are always intertwined. Conflicts do, however, have a propensity to gravitate towards violence.

There is, as Pettman (1975) pointed out, no accepted or agreed list of the social units by which conflicts might be classified. To talk of conflict in intra-personal, inter-personal, familial, group, class, ethnic, religious, intra-state or inter-state terms is to assume, perhaps erroneously, that 'each kind of social unit, having its own range of size, structure, and institutions, will also have its own modes of interaction and thus its own patterns of conflict with other social units' (Fink, 1968) like and unlike itself. Such an assumption merits scrutiny on its own, since, despite the plausibility of some sort of analytical link between the parties to a conflict and the nature of the confrontation that ensues, the link should be demonstrated and not allowed to stand by assertion alone.

This volume is devoted to one type of analysis of conflict, the socio-biological one. In *The Sociobiology of Ethnocentrism*, a book closely related to many of the ideas and some authors of this volume, sociobiology was defined as 'the branch of biology that concerns itself with the explanation of social behaviour in all species, including our own. It is thus, essentially, evolutionary biology, and relies on Darwinian, Mendelian and Hamiltonian ideas – concepts such as natural selection, genetics, and, especially the individual's inclusive fitness – for its underlying explanatory schema' (Reynolds, Falger and Vine, 1987). Inclusive fitness theory, first proposed by Hamilton in 1964 and repeatedly referred to in this book, shows that genes will spread if their carriers act to increase not only their own fitness or reproductive success but also that of other individuals carrying copies of the same genes. A person's inclusive fitness is his or her personal fitness plus the increased fitness of relatives that he or she has in some way caused by his or her actions.

It may sound very deterministic or even 'geneticistic' to try to explain social phenomena on the basis of some supposed underlying biological process, but that would be too restricted an interpretation of the effort to draw attention to biology as one place to look. What is encouraging about a recourse to biology is that there is a body of theory and a wealth of empirical data relating to other species. This provides an excellent background against which to compare and contrast human processes and situations, the main focus of attention here. None of the authors in the present volume is proposing to reduce the study of conflict to biology, to account for it simply as any simple instinct. But all are alert to the existence of similar processes in animals and are trying to use the theory of inclusive fitness to explain the evolution of these processes.

However, a sociobiology of human behaviour would not be taken seriously if it explains in terms of fitness something that can be better explained in strictly cultural terms or in terms of the market. That is why sociobiology of any human behaviour can and must seek links with social psychology, social anthropology, sociology, economics, political science and even history. Beyond that – and this is very important – it would be quite pretentious to assert that sociobiology could solve all questions that have arisen since the phenomena of conflict were studied in a systematic way. To know what these questions are and why it makes sense to add a biological–evolutionary oriented approach to the study of conflict and competition, it is worthwhile to draw some broad historical lines. Therefore, this introduction aims literally at introducing the reader to the traditional scientific discourse on conflict, which usually means *human* conflict. Then a condensed overview on the (socio)biology of conflict and competition tries to make the non-specialist familiar with the most relevant theoretical and conceptual problems in this field. Those readers, however, who want to cut short and demand an immediate answer on the question why a sociobiological perspective on conflict and competition is a valuable contribution, should skip the next three main parts of this introduction and turn straight to its fifth part.

1.2 THE STUDY OF CONFLICT IN PERSPECTIVE

Conflict may be defined as: incompatibility of interests, goals, values, needs, expectations, and/or social cosmologies (or ideologies). Ideological conflicts especially have a tendency to become malicious (cf. Berger and Luckman, 1966). Webster's Dictionary defines conflict as 'Clash, competition or mutual interference of opposing or incompatible forces or qualities (as ideas, interest, wills)'. Coser (1956) defined social conflict as 'a struggle over values and claims to scarce status, power and resources in which the aims of the opponents are to neutralize, injure or eliminate their rivals'.

McEnery (1985) suggests as a new definition of conflict: 'the interaction of any two or more value systems'. Conflict is either malignant or benign depending on whether the particular interaction of the value systems tends to destructive disruption or creative progress.

According to Galtung (1965), an action-system is said to be in conflict if the system has two or more incompatible goal states. In the case of one actor the conflict is called a dilemma, *l'embarras de choix*, or intra-individual conflict, consisting of incompatible motivational or behavioural tendencies (approach/approach, avoidance/avoidance, and approach/avoidance conflict, see Hinde, 1966 for a thorough review of this literature).

A distinction should be made between conflict, conflict attitude, and conflict behaviour, which may be depicted as a triangle. A conflict process may start in any corner of the triangle. One of the means of conflict resolution is to eliminate or incapacitate one or more of the actors in the conflict. This may be done either nondestructively or destructively. The latter we call violent conflict. 'Thus we may distinguish between destructive and non-destructive behaviour, although this is, of course, a continuum and not a dichotomy. Two of the most celebrated propositions about conflict can now be made use of: conflict behaviour tends to become destructive behaviour (because of the frustration–aggression cycle) and destructive behaviour tends to become self-reinforcing' (Galtung, 1965). The same author lists the following conflict-resolution mechanisms: chance mechanisms, oracles, ordeals, regulated warfare, fights, private duels, judicial duels, verbal duels, debates, mediation, arbitration, courts and voting (Galtung, 1965).

Fink (1968) defines social conflict as: 'any social situation or process in which two or more social entities are linked by at least one form of antagonistic psychological relation or at least one form of antagonistic interaction. This emphasizes that while *antagonism* (which for the moment remains undefined) is the common element in all conflicts, there are a number of different kinds of psychological antagonisms (for example, incompatible goals, mutually exclusive interests, emotional hostility, factual or value dissensus, traditional enmities etc.) and a number of different kinds of antagonistic interaction (ranging from the most direct, violent and unregulated struggle to the most subtle, indirect, and highly regulated forms of mutual interference), none of which is necessarily present in all instances of conflict. This is a disjunctive definition which subsumes any form of social antagonism, thus making the theory of conflict equivalent to a theory of antagonistic social relations in general.'

There is, furthermore, no agreed classification for conflict itself. Rapoport (1960), for example, talks in terms of an 'ideal' division between fights, games and debates. Bernard (1951) outlines three competing approaches – social-psychological, sociological and semanticist – which place their main emphasis in turn upon personal frustrations, mutually incompatible interests, and mutual misunderstanding. There is also the familiar distinction by discipline between biological, economic, political, psychological and social conflict (Pettman, 1975).

Taxonomy and dimensions of conflict: objective vs. subjective; horizontal vs. assymetrical; fractionating vs. cross-cutting; zero-sum vs. variable-sum; absolute vs. relative; realistic vs. projected; violent vs. nonviolent; perceived vs. nonperceived; regulated vs. unregulated; indirect (parallel striving) vs. direct (mutual interference); unconscious vs. conscious; impersonal vs. personal; continuous vs. intermittent; communication absent vs. communication present; large admixture of cooperation vs. relatively pure antagonism; object-centered vs. opponent-centered; based on scarcity vs. based on incompatibility, inconsistency etc.; economic goals vs. non-economic goals; nondisruptive vs. disruptive; etc.

Conflicts may also be distinguished according to the number of actors, the number of goals, and the level of complexity of actors (individual, group, state); etc.

Stagner (1967) distinguishes size, duration, evaluation, intensity, polarization and regulation and Rapoport (1966) distinguishes two conflicting paradigms of conflict: cataclysmic vs. strategic (or Tolstoyian vs. Clausewitzian).

The two basic and polar types of conflict structures outlined by Freeman (1972) are fractionating conflict and cross-cutting conflict.

Fractionating conflict structures exist when opponent groups are cleaved apart by differences on all significant value fronts – economic, political educational, religious, ethnic, racial. Adversaries are opponents on all. There are no cross-cutting attachments to common values. There is no common ground upon which to compromise, no incentive to negotiate. Opponents ascribe to each other less than human qualities reflecting their lack of shared values. Violence is condoned by both groups in order to protect against the extreme threat represented by the other.

Cross-cutting conflict structures exist when opponent groups are in opposition over a limited number of cleavage fronts, but are allied in common cause in other significant conflicts. Actors in disagreement over one or more value preferences find shared attachments when they approach other issue areas. Here lay the roots of social cohesion. Cross-cutting cleavages over values stitch society together by facilitating constantly renewed willingness to negotiate disputes and seek ground for compromise. Total involvement of an actor in any one conflict against any single opponent is precluded. Roles and statuses include interaction with a range of opponents on some issues who are allies on other conflict fronts. Multiple involvement in cross-cutting social cleavages precludes polarization on any one axis and keeps social groups open to ideas and innovations from each other. Cross-cutting cleavage patterns make for low propensities to engage in violence and for high propensities to tolerate change, deviance, and innovation.

Conflict can also be seen as a pervasive or marginal feature of human society, as an objective phenomenon or one that is subjectively derived, as functional or dysfunctional in its effect, and each of these views and combinations of them will predispose certain preferred modes of conflict termination (Pettman, 1975).

1.3 GENERAL THEORIES OF SOCIAL CONFLICT

It must be recognized that general theories of social conflict (or at least conceptual frameworks, assumptions, and hypotheses oriented to the analysis of conflict in general) have been around for a long time, both in the social sciences and in the general culture. General orientations toward conflict are present in all cultures and appear not only in social science but also in religious, ethical, political, and philosophical systems from Heraclitus to Hegel, Macchiavelli to Hobbes, and Locke to Mill (Sorokin, 1928, 1947, 1966: Singer, 1949a, b; Bernard, 1957).

Several writers trace the beginnings of the modern sociology of conflict back to Marx and to Social Darwinists like Bagehot, Gumplowicz, Ratzenhofer, and Oppenheimer (Sorokin, 1928; Bernard, 1950; Coser, 1956, 1967; Dahrendorf, 1959, 1967; Horowitz, 1962). As Angell (1965) has argued, these early sociological theories were not truly general since they did not deal with all forms of social conflict. Nevertheless, these theories (along with the theories of Pareto, Durkheim, Marshall, Weber, Freud, and many others) contain many concepts, assumptions, and hypotheses which greatly influenced later writers who did attempt to deal with conflict in general. By the late nineteenth century, highly general theories of conflict in physical, biological, and social systems were presented in such works as *Conflict in Nature and Life* (Patterson, 1883), *Les Luttes Entre Sociétés Humaines et Leurs Phases Successives* (Novikow, 1896), and *L'Opposition Universelle* (Tarde, 1897). General theories of social conflict appeared in the works of Tarde (1899), Simmel (1903), Carver (1908, 1915) and others. Various mixes of these and earlier theories, together with new contributions and applications of these theories to various kinds of social conflict, appeared in the general sociologies of such writers as Cooley (1918), Park and Burgess (1924), Ross (1930), Von Wiese and Becker (1932), MacIver (1937), Lundberg (1939), and Sorokin (1947), and in general treatments of conflict by Lasswell (1931), Simpson (1937), Lewin (1948), Singer (1949a, b), Wright (1942), Chase (1951), Lawner (1954), and Coser (1956), among others (Fink, 1968).

Again, various combinations of concepts, assumptions, and hypotheses from these earlier theories, together with game theory and other mathematical approaches, continue to dominate the general theories of conflict developed by such writers as Bernard (1951 *et seq.*), Mack and Snyder (1957), Boulding (1957, 1962), Dahrendorf (1958, 1961), Schelling (1958, 1960), Rapoport (1960, 1965, 1974), Galtung (1959), Rex (1961), DeKadt (1965), Thurlings (1965), Beals and Siegel (1966), Coser (1967), and Stagner (1967). (For further commentaries, please see: McNeil, 1965; Smith, C. 1971; Kriesberg, 1973; Oberschall, 1973; Brickman, 1974; Collins, 1975; Duke, 1976; Eldridge, 1979; Himes, 1980; Schellenberg, 1982.)

A landmark in the history of the study of conflict has been the so-called Simmel–Coser propositions on conflict (Simmel, 1903; Coser, 1956). Some of their more counterintuitive findings include:

1. Conflict serves to establish and maintain the identity and boundary lines of societies and groups.
2. Conflict with other groups contributes to the establishment and reaffirmation of the identity of the group and maintains its boundaries against the surrounding social world.
3. Patterned enmities and reciprocal antagonisms conserve social divisions and systems of stratification.
4. A distinction has to be made between conflict and hostile or antagonistic attitudes. Social conflict always denotes social interaction, whereas attitudes or sentiments are predispositions to engage in action.
5. Conflict is not always dysfunctional for the relationship within which it occurs.
6. Social systems provide for specific institutions which serve to drain off hostile and aggressive sentiments. These safety-valve (*Ventilsitten*) institutions help to maintain the system by preventing otherwise probable conflict or by reducing its disruptive effects.
7. Aggressive or hostile 'impulses' do not suffice to account for social conflict.
   It has often been pointed out (Bernard, 1951; DeVree, 1982; among others) that hostile stereotypes, prejudice, threat perception, general hostility, and aggression (however conceptualized) are more likely to be the result of conflict than its cause.
8. Antagonism is usually involved as an element in intimate relationships. A conflict is more passionate and more radical when it arises out of close relationships.
9. Conflict with another group leads to the mobilization of the energies of group members and hence to increased cohesion of the group.
10. Groups engaged in continued struggle with the outside tend to be intolerant within. Rigidly organized struggle groups may actually search

for enemies with the deliberate purpose or the unwitting result of maintaining unity and internal cohesion.

Dahrendorf (1958 *et seq.*) pictures contemporary sociology as split between two viable models of society. The consensus-and-equilibrium model is followed by many sociologists most of the time, but Dahrendorf prefers (as did Marx) a conflict-and-change model. He leaves the impression that there is no good way to bridge these two models. However, it can well be argued that the bridge has already been built by his fellow German sociologist, Max Weber. Weber's work provides a thorough analysis of diverse forms of social organization and how in them social norms come to be stabilized so that raw social conflict is rarely observed. But Weber's theory also is based upon assumptions about the fundamental role of power and conflict in society.

Duke (1976) is among those who recently have pointed out the bridge which Weber provides between conflict and consensus theories of society. In his able review of sociological theories of conflict, he shows a special appreciation for Weber both as a sociologist of conflict and as a sociologist of social order. The propositions Duke uses to paraphrase and summarize Weber's conflict theory can be used as well as a summary of the neo-Marxian position as developed by Dahrendorf:

1. Conflicts of interest are endemic in social life.
2. Power is differentially distributed among groups and individuals in any society.
3. Social order is achieved in any society through rules and commands issued by more powerful persons to less powerful persons and enforced through sanctions.
4. Both the social structure and the normative systems of a society are more extensively influenced by powerful persons than by weaker persons (true by definition), and come to represent the interests of these more powerful persons.
5. Social changes are often more disruptive to powerful persons than to less powerful persons. Powerful persons therefore generally favour the status quo and oppose changes that would reduce their power.
6. However, changes in a society occur as the result of actions by persons who stand to benefit from these changes and who accumulate power to bring them to pass.

Far from being always a negative factor or social pathology, social conflict may contribute in many ways to the maintenance and cohesion of groups and collectivities as well as to the cementing of interpersonal relations. Any relationship between individuals or groups necessarily involves conflictual as well as cooperative or integrative elements, however that relationship might start off (Coser, Gluckman, Blake and Mouton, North, Cooley, Deutsch,

McEnery etc.). The analogy of friction in physics illuminates the essential character of conflict as neither good nor evil. Friction impedes movement and has to be overcome if movement is to take place. But the initiation of movement is impossible without friction, which is therefore essential to movement.

Deutsch (1969) categorized conflict as to whether it is 'productive' or 'destructive'. This categorization was a major advance in removing the fog of misunderstanding of conflict as necessarily an undesirable force. However, the nomenclature rests on end results, 'production' or 'destruction'. It therefore does not fully reflect the fact that, because value systems are dynamic, any given conflict is not foreordained as to result. For these reasons, according to McEnery (1985), it is more accurate to categorize conflict as either 'malignant' (i.e. tending to produce destructive disruption) or 'benign' (i.e. tending to produce creative progress).

1.4 THE (SOCIO)BIOLOGY OF CONFLICT AND COMPETITION

Competition of two species for the same resources is, as Wilson (1970) explains, more fatal than a predator–prey relation. Competition eventually leads to the extermination of the species with the smaller growth capacity; a predatory–prey relation only leads to periodic oscillation around a mean value (Volterra, 1928; von Bertalanffy, 1968). Competition, as Miller (1967) modified the original Clements and Shelford (1939) definition, is 'the active demand by two or more individuals of the same species (intra-species competition) or members of two or more species at the same trophic level (interspecies competition) for a common resource or requirement that is actually or potentially limiting'. This definition is consistent with the assumptions of the Lotka–Volterra equations, which still form the basis of the mathematical theory of competition (Levins, 1968).

Intraspecific competition occurs when two or more individuals seek access to a resource that is somehow important to the fitness of each and that is restricted in abundance such that optimal utilization of the resource by one individual requires that another settle for suboptimal utilization. In other words, if there is enough to go around, then there is no reason for competition for example, few animals ever compete for air. However, severe competition may erupt over food, water, nesting sites, and/or appropriate mates (Barash, 1977).

According to Huntingford and Turner (1987), we can loosely divide the causes of such conflicts of interest between animals into two broad and overlapping categories, conflict over resources and conflict over other outcomes. Thus, conflicts of interest commonly arise when two or more individuals are competing for something which they both need but which is in short supply. Conflicts over outcomes include: killing or survival of prey;

occurrence of mating; occurrence of parental care; distribution of care between parents.

Nicholson (1954) distinguished between contest competition and scramble competition. Non-aggressive scramble competition occurs when each participant attempts to accumulate and/or utilize as much of the critical resource as it can, without regard to any particular social interaction with its competitors. If the resource is used up in the process, then the so-called winners of scramble competition are the individuals who have converted the largest part of that resource into copies of themselves, i.e., those that are most fit. Fitness in this case has been achieved by simply out-reproducing the competition, usually by being most efficient at garnering the resource in question. By definition, social interactions are excluded from this type of competition. If, on the other hand, competing individuals interact directly with each other and use the outcome of such interactions to determine access to resources, then contest competition is taking place. To the victor belongs the spoils in contest competition. In scramble competition, the victor is simply the one that scrambles the most spoils. This is an Easter egg hunt, in which every participant ignores every other and simply concentrates upon finding as many eggs as possible. In contrast, contest competition would be occurring if the participants first argued, fought, or somehow disputed among themselves, on the basis of which they decided who would look where, who would have first choice of the eggs collected, etc. (Barash, 1977).

Alternatively, competition may be sidestepped by mutual avoidance, either in space or in time. A more active response to a conflict of interest is to meddle with the activities of rivals. In some cases, a favourable outcome is gained by manipulating the behaviour of other animals. One potential response to a conflict of interest is to make active attempts to physically coerce other animals into giving up a disputed resource or into acquiescing to a particular outcome. Conversely, the victims of such attempts resist strenuously. In ecological terminology, responses to conflicting interests other than scramble competition are often referred to as interference competition. The use of physical coercion in response to a conflict of interest is often described as aggression (Huntingford and Turner, 1987).

Aggression is the proximate mechanism of contest competition. It takes place when individuals interact with each other such that one of them is induced to surrender access to some resource important to its fitness. The exact forms of aggression vary widely, from intimidating displays and threats to actual fights. Just as animals ought to exert themselves to acquire important resources or enlarge their supply, thereby enhancing their fitness, they also ought to resist the loss of important resources, thereby avoiding decrements to their fitness. Accordingly, animals may respond to aggression by threatening back, fighting back, and, occasionally, signalling their submission and/or running away (Barash, 1977).

Aggression can be recognized by a collection of features: by the forceful (and deliberate) attempt to inflict harm (either physical damage or exposure to an aggressive display) on another (reluctant) individual; this may be accompanied by strong physiological and emotional arousal and often functions to space animals out or determine status. Perhaps it is best to think of aggression as a special case of manipulation in which the desired outcome is brought about by intense displays, which can if required lead to direct physical conflict and injury.

It is generally accepted that in both human and animal conflict the initiation of an attack (offence) is not the same as protection against such an attack (defence). It is hard to draw a clear dividing line at any point of the continuum from offence or attack through offensive and defensive threat and submission to escape, yet it seems an abuse of language to include escape under the heading of aggression (Huntingford and Turner, 1987). Therefore this word is often replaced by the term agonistic behaviour (Scott and Fredericson, 1951), which refers to a 'system of behaviour patterns having the common function of adaptation to situations involving physical conflict'.

Another (and independent) distinction is between resource competition and sexual competition (Wilson, 1975a). Sexual competition involves access to receptive mates; it includes both contest and scramble forms. One form of sexual competition which is similar to scramble competition is unobtrusive mating, where a male sneaks up to one of a number of females which are being guarded by a male.

There are a number of indirect forms of sexual competition which fall into the category of 'contest' competition yet do not involve fighting. In males, competition may take the form of removing a rival's sperm prior to mating, sperm competition, or olfactorily induced pregnancy block. In females, it may take the form of suppression of the reproductive activity of other females (Archer, 1988).

The evolutionary rules underlying interspecific variations in competitive aggression for food resources have been well covered in a number of previous discussions, for example those of Brown (1964), Wilson (1975a), Clutton-Brock and Harvey (1976) and Geist (1978). When food is abundant, aggression will be unnecessary since the same benefits can be obtained without it. When food is scarce, it will often be advantageous for the animal to use its energy in foraging for food (i.e. scramble competition) rather than in aggressive competition. This will apply particularly when food is widely dispersed or difficult to find. In general, therefore, we might expect aggression to occur under conditions of intermediate food availability.

If it is advantageous for an animal to compete aggressively for food, it will be energetically more efficient for it to fight a relatively few times for an area which contains food than to compete for each item of food. Given such a broad generalization, we should expect all animals to seek to defend feeding territories either in groups, in pairs, or individually, unless it is not possible to defend their food supply in this way (Archer, 1988).

1.4.1 Game-theoretical (cost/benefit) analyses of animal conflict

Animals invest time and energy in agonistic behaviour and can run serious risks of injury or even death from fighting. Injury and death are obvious risks of fighting but displays and fights can also expose an animal to predators. Apart from the risk of attracting predators, males on lekking grounds run the risk of losing body condition or even starving because of the need to stay on the territory and keep displaying. Furthermore there are costs involved in 'aggressive neglect'. As well as the costs, however, there are also substantial benefits to being aggressive. Individuals can thereby gain exclusive use of a resource such as a food source, or may win exclusive mating rights. The more aggressive an animal is, the more benefits it may gain (such as extra food). But if an animal is too aggressive it might face unacceptably high costs (such as serious injury) so the animal must weigh up the relative costs and benefits of its action and choose an optimum level of aggression (i.e. maximize the net benefits). If the costs are too high and the benefits too low, avoiding a fight may be preferable to competing. In other cases it may be worth fighting vigorously for a valuable resource (Huntingford and Turner, 1987).

1.4.2 The hawk–dove game and the concept of ESS

When animals compete with one another the behaviour each adopts depends on what other individuals in the population do. The techniques of game theory, originally applied to human conflicts (notably Prisoners' Dilemma), have been used to investigate which behaviour is best for an individual to use in relation to what others are doing. Game theory treats evolution as a game in which the players use different patterns of behaviour. These behaviours are termed strategies but conscious decision-making is not necessarily implied.

Many games try to answer the question of how aggressive an animal should be when fighting over a disputed resource. For example, in the hawk–dove game a hawk always fights fiercely either until the opponent retreats or until one or other combatant is seriously injured. A dove, however, tries to settle the dispute amicably by displaying rather than fighting and will retreat if attacked. Hawks always beat doves but if they fight another hawk they stand an equal chance of winning or being injured. If two doves fight, each has an equal probability of winning; neither is injured but they both spend a lot of time displaying.

The best strategy to use depends on the resulting costs and benefits and the frequency of its use in a population. Thus if the benefits (of gaining the resource) exceed the costs (of injury), a hawk in a population will do better than a dove because the dove never reaps the benefits of winning. But if the costs are greater than the benefits, a dove in a population of hawks does better than a hawk because it never bears the costs of escalated fighting.

In the first case, where benefits exceed costs, the strategy of playing hawk is termed an evolutionary stable strategy (ESS). Such a strategy cannot be outcompeted by any other strategy defined in the model that might invade the population (dove in this case). Where costs exceed benefits, however, neither hawk nor dove is an ESS. The evolutionary stable strategy here is termed a mixed one: play hawk with probability p and dove with probability $p - 1$. At this evolutionary stable ratio the pay-offs in terms of costs and benefits of being a hawk or a dove are equal (Huntingford and Turner, 1987). Note that an ESS is not necessarily the best or even an optimal strategy for the individuals involved.

1.4.3 Asymmetries in animal conflicts

Early game theory models assumed that the combatants in a fight were equal in all respects. Clearly this is not so and more recent models consider contests in which individuals differ. The most studied questions have been whether animals use these differences (asymmetries) to settle fights and to decide how much to escalate, and whether the way they fight conveys information about the asymmetries. The asymmetries between the combatants are of three possible types (Maynard Smith and Parker, 1976):

1. **Resource-holding potential (RHP)**. One combatant may be better able to fight and defend a resource than the other. A large, obviously superior opponent might quickly win a dispute without any escalation to physical contact being necessary.
2. **Resource value**. One combatant may value a resource more highly than does its opponent. Food may be more valuable to a hungry animal than to a satiated one, for example.
3. A third type of asymmetry is not related to resource-holding potential or to a pay-off, being quite arbitrary. A convention such as 'if the opponents are otherwise equally matched, the owner of a territory wins, the intruder retreats' is of this type.

There is evidence of each type of asymmetry being used to settle a dispute. If the asymmetries are very weak or difficult to detect then the ownership convention alone may be used. Unnecessary and potential harmful escalation is thus avoided (Huntingford and Turner, 1987).

1.4.4 Assessment of asymmetries, intentions, and badging

In asymmetric contests, it would seem advantageous to lie (i.e. give false information about for example RHP), but there can be costs to doing so; and the more obvious the asymmetry involved, the harder and more costly (in terms of energy and time or risk of injury) it is likely to be to fake it. Conversely, cues will be reliable only when they cannot be faked (Zahavi, 1977). Where assessment is by trial of strength, faking may be too costly: roaring contests in red deer stags, for example, are physically exhausting (Clutton-Brock and Albion, 1979).

We would expect evenly matched opponents to conceal their intentions (to attack or to retreat) during an agonistic encounter – there is no point in saying you are going to give up right at the beginning of the fight because your opponent might also be prepared to give up later on. Whether or not an animal uses a display that might reveal its intentions about persisting or giving up the fight may be determined by the relative costs and benefits of that display. Displays can be time- or energy-consuming and an individual using them can face varying degrees of risk of retaliation so a combatant can afford to use high cost or high risk displays only if they are effective. Faking intentions may be possible within certain constraints (Maynard Smith, 1984), but the cost of lying may be high. A commitment to escalate may lead to serious injury if the cheat is physically unable to continue fighting and cheating will thus be limited.

Status signals or badges may be advantageous because they obviate the need for agonistic encounters. But if it is so advantageous, why do not all individuals signal dominance whatever their true status? There are several possible reasons why status signals can exist without cheats invading and taking over the population. In situations where individuals frequently encounter and recognize one another, cheats may be discovered. In addition, it may not pay some individuals to pretend to be dominant if (because of factors such as small size) they are not capable of living up to the demands of high status. Where encounters are most frequent between individuals of similar rank a cheat would often be involved in a contest with superior animals. Cheats may then be discovered or they may suffer costs as a result that outweigh the benefits of signalling dominance (Huntingford and Turner, 1987).

1.4.5 Dominance and territoriality

Similar cost/benefit calculations have been applied to explain the variety of animal dominance hierarchies and territory holding. In general, animals appear to adjust their rank and territorial behaviour according to both costs and benefits.

1.4.6 Game-theoretical analyses of specific conflicts

Conflict between the sexes:

1. conflict over the number of mates;
2. mate guarding, sneak copulations, rape;
3. conflict over the quality of mates.

Conflict over parental care:

1. desertion and paternal confidence.

Parent–offspring conflict:

1. infanticide and cannibalism.

Sibling rivalry:

1. fratricide.

If a comparison were made, it is likely, according to Johnson (1972), that intraspecific killing occurs with about the same statistical frequency in man and other animals. Even more outspoken is Wilson (1978), who states: 'Although markedly predisposed to aggressiveness, we are far from being the most violent animal. Recent studies of hyenas, lions, and langur monkeys, to take three familiar species, have disclosed that individuals engage in lethal fighting, infanticide, and even cannibalism at a rate far above that found in human societies. When a count is made of the number of murders committed per thousand individuals per year, human beings are well down on the list of violently aggressive creatures, and I am confident that this would still be the case even if our episodic wars were to be averaged in.'

1.5 WHY A SOCIOBIOLOGY OF CONFLICT AND COMPETITION?

Why this book? Why a sociobiological perspective on conflict, after all those other works on conflict? We will point to various merits of this approach. There are several reasons why a volume devoted to sociobiology and conflict and competition is neither redundant nor superfluous – besides the obvious need to study conflict from *every* point of view, in the hope that we might better understand and come to terms with the problems of human conflict:
(1) There is considerably *more* competitive and conflictuous (not necessarily agonistic) interaction going on than was ever dreamt of, on all levels of organismic existence, from intragenomic competition, through intraindividual conflict, sperm competition, interindividual contest competition, to the many intricacies of coalitional aggression in primates and man. The idea that competition may exist even at the intragenomic level is a relatively novel one. It is explained thus by Wind (1984, p. 13):

> Assuming that natural selection ultimately takes place at the level of the genes (or even at that of the nucleotides) there must be another mechanism that hitherto has hardly been taken into account. So far, one has focused only on the competition between their survival machines, the individual phenotypes. However, there is likely to be also a more direct competition by what I have called *non-interphenotypic, intragenomic gene control* (Wind, n.d.), and what Dawkins (1982) called the 'arms race' between the genes. Such competition is suggested by combining the Selfish Gene Theory with molecular-biological data.

Wind suggests that three kinds of the latter are of interest here. In the first place there is gonosomal and autosomal gene exchange through meiosis and mobility of DNA elements. This implies not only interallelic but, in general, intercoding-sequence competition. Secondly there is replication and expression of genes depending on the action of others. For instance, suppressor and regulator genes have been shown to control the expression of others. Lastly there are genes causing segregation distortion (or meiotic drive), i.e. causing themselves to be present in more than half of the gametes (see Crow (1979); Dawkins, (1982); for the idea of sperm competition see Parker (1970) and Trivers (1985)).

(2) There is also considerably more *violent conflict* than was formerly assumed (for example, by the German ethological school), in the form of infanticide in a wide range of species (Hausfater and Hrdy, 1984), involving pup-killing and/or cannibalism; infant mortality resulting from mate- and nest-desertion; siblicide; disproportional mortality of omega-individuals being forced into suboptimal habitats; lethal and injurious fighting between males over females; rape and courtship violence; predatory 'warfare' and slavery in the eusocial insects; to intergroup agonistic behaviour in primates, culminating in 'primitive warfare' in chimpanzees (Goodall, 1979 *et seq.*) and, of course, man. (For kin selection theory see Hamilton (1963 *et seq.*), Wilson (1975 a) and Trivers (1985); for sex ratio manipulation and parent–offspring conflict see Trivers (1974, 1985).)

These new data provide a falsification (or at least partial refutation) of ethological assumptions on inhibitory blocks against the killing of conspecifics. Thereby it has become clear that drive concepts of aggression, with their emphasis on the idea of action-specific energies (Lorenz, Eibl-Eibesfeldt), have become either entirely obsolete, or may only applyto a limited class of agonistic acts. Sociobiology can show that there are evolutionary 'optima' for behaviours such as aggression.

(3) *Game-theoretical models* of the evolution of agonistic behaviour are increasingly becoming more veridical and more robust in their predictive/postdictive power, thus more and more unravelling and revealing the 'cold calculus of evolution' in which reproductive success is the only currency.

These models not only include analyses of actual fighting behaviour, but also the evolutionary rationales behind the 'battle of the sexes', parent–offspring conflict, and sibling conflict (for example, Trivers, 1985; Stamps and Metcalf, 1980).

(4) Renewed interest is also shown by human sociobiologists in the evolutionary bases and vicissitudes of hominid and human primitive war, its causes, motives, dynamics etc. In this context it may be observed that many sociobiologists have been very reluctant to make propositions on the human species, while others have not hesitated to make sweeping statements (for example, on the alleged universality of human warlikeness). Several chapters in this volume testify to the renewed and critical interest in primitive war. Some sociobiologists – just like Freud, Lorenz, and others before them – try to explain war by recourse to the aggressive dispositions of people. This is a notion being criticized in this book. A merit of sociobiology is, however, to focus attention on the question of how war and human nature go together.

(5) On the other hand, it is not only becoming increasingly clear where and when conflict and (agonistic) competition are to be expected, but also, and perhaps more importantly, where and when not. For example, when limiting resources (of whatever kind) are abundant, an organism should not engage in agonistic contest competition (aggression) because the same benefits can be obtained without it. Similarly, when resources are scarce, it will often be energetically more efficient to engage in scramble competition (foraging) than aggressive competition (Archer, 1988). (For a review of optimal foraging theory see Barash (1982).)

(6) A more thorough analysis of altruistic behaviour, and the evolution of cooperative behaviour (Voorzanger, 1988; Axelrod and Hamilton, 1981; Axelrod, 1984) has revealed many intricacies and may constitute the necessary counterpoise to an overemphasis of conflict and competition in sociobiological thought. (For cultural evolution theories of human co-operation see Campbell (1975) and Boyd and Richerson (1985).)

There are two main reasons why sociobiology should not be expected to provide easy answers to the intricacies of human social behaviour. One reason is inherent in the discipline:

> While the basic paradigm of the sociobiology – the selfish-gene concept – is quite simple as well as scientifically quite valid, the difficulties in its application in behavioural analyses seem to increase exponentially when passing from viroids to viruses (in which genotype and phenotype are virtually identical) and unicellular organisms to simple multicellular ones and the higher vertebrates including man. In the same order the practical value of sociobiology decreases. (Wind, 1984, p. 18)

The other reason is more intricate and substantial. It has become increasingly clear that *Homo sapiens sapiens*, no longer the 'Crown of Creation' ever since

Darwin, is indeed an exceptional and odd species in the world of organisms. The time elapsed since our origin (some 40000 years or some 2000 generations ago) is – in evolutionary perspective – quite short. Therefore, many of our genes' frequencies and behaviours are still oscillating without having reached yet a less disequilibrized state as is usually found among other species. Stated in more traditional biological terms, adaptation still has to occur, or, in even less technical terms, we are still in the wake of our evolutionary origin. After all, man's genetic make-up was shaped when he or she was living in small family groups. Thus, at that time the overlap of group, kin, individual and gene selection was probably greater than nowadays. Yet, our behaviour (including altruism) is determined by largely the same genes interacting, however, with a totally different environment.

Enigmatically, man often seems to show sociobiologically odd properties such as celibacy and other nonreproductive behaviours. Some of these properties may, in fact, very well have, at the individual level, a negative selective value, and may be in the process of being selected against.

Because of the above reasons *Homo sapiens sapiens* is likely to show behaviours that can sociobiologically be qualified as an Evolutionary Quite Uncommon, Unstable Strategy (EQUUS), instead of an Evolutionary Stable Strategy (ESS) (Wind, 1984). In our view, however, this does *not* imply that it is superfluous or irrelevant to combine new biological theories and methodology with the study of human behaviour. We very well realize that sociobiology does not give definite answers to many fundamental questions. But since when is that enough reason not to try to expand the limits of understanding, in the last instance of ourselves?

That is why the interested reader is invited to study carefully the chapters that follow. The book is organized in four parts, but each chapter can also be read on its own. In Part I three chapters present theoretical and empirical studies on conflict from a biological perspective. The first chapter provides an introduction to the extensive ethological studies of conflict and reconciliation in primate groups. The human dimension is focused on in Part II where social scientists and biologists discuss the relevance of sociobiology for the explanation of the timeless and cross-cultural phenomenon of enmity. Part III highlights a supposed important origin of modern man: so-called 'primitive' warfare. Discussion of the existing literature, development of new hypotheses in this field and a detailed empirical case study make this part the most voluminous one of the book. The shortest is Part IV, containing one chapter on the conflict *about* sociobiology. Because we think, however, that the general debate about sociobiology is very important to everyone who is interested in – or disgusted with – sociobiology, we thought it appropriate to address the topic of controversy around it unambiguously in this volume.

Any study of conflict behaviour is also an exercise in human introspection. Animal conflict studies more often than not refer to problems which also exist in the human species. An evolutionary approach of conflict behaviour in general implies a preparedness to accept reductionistic argumentation to a certain degree. It also requires a relativization of the nature–nurture dichotomy, in particular relevant in the study of human conflict behaviour. Unfortunately, every effort to explain human behaviour with the help of (socio)biological axioms is risking to be categorized as an ideological justification of the social and political status quo. Political attitudes, such as a tendency to accept social inequality, racism and sexism, are seen by some as logically following from the premise that human behaviour and human biological evolution cannot be treated as completely separated compartments.

The political accusation that sociobiology is inevitably leading to conservative and even reactionary social views of the world, or supporting these, must be taken seriously, however incorrect the accusation may be. Of course, sociobiology can be abused to support political value judgements. Sociobiology, used in a very particular way, could provide the existing inequality between the sexes or Apartheid with a legitimation which *looks* scientific, especially for those who are not familiar with the generally accepted rules of the game called science. Nobody will deny that individual people, using sociobiology argumentation, sometimes have aired opinions which fall in the category of political abuse – although hard data on this abuse are quite scarce.

Is this situation really different from other disciplines, like genetics, archaeology, psychology, history or economics? Principally no, practically we think yes: it has happened far less than might be expected (and has been predicted by many radical anti-sociobiologists). Admitting that sociobiology can be abused politically is something very different from stating that inclusive-fitness theory, the essence of sociobiology, is a pseudo-scientific, reactionary political cover-up (Sociobiology Study Group, 1978; Rose (ed.), 1982a, b; Rose, Lewontin and Kamin, 1984; Levins and Lewontin, 1985. See for reactions against these charges Masters, 1982 and Falger, 1984). Personal values and motives do play a role in the research programmes of individual scientists, proponents and opponents of sociobiology. Ullica Segerstråle's contribution to this volume represents an interesting analysis of two well known Harvard based opponents. The question, however, is ultimately whether a contribution is made to the understanding of, in this case, human behaviour. Of course, we should not judge the results ourselves, but all contributors expressly want their work to be seen as an effort to create that understanding.

As far as reductionism is used here, it is not because individual human behaviour is considered only as a personal expression of the 'laws of human nature'. Nor does the use of inclusive-fitness theory in this volume imply teleology or genetic determinism. And finally, the cost-benefit calculus, which sociobiology shares with the hard core of economic theory, game theory and public choice theory, offers a model of explanation and prediction, not an accurate description of actual historical phenomena. So, whoever feels justified to draw political conclusions from any of the essays in this book, does so purposively contrary to the intentions of the authors and editors. The only extra-scientific commitment we have is the hope that the study of conflict from an evolutionary perspective adds to the understanding of a vital category of behaviour in animals and man.

PART ONE

Conflict and Biology

CHAPTER TWO

Intergroup competition and conflict in animals and man

J.A.R.A.M. Van Hooff

2.1 APPROACHES IN THE STUDY OF CONFLICT BEHAVIOUR

The use of frequently destructive and bloody violence in group encounters, which we know in its most extensive and organized form as war, seems to be as old as mankind itself, if not older. In any case, early hominid fossil finds indicate that death as a result of armed violence must not have been uncommon (Birdsell, 1972). In nearly all the cases the conduct of war and related heroics form an important subject for the arts of literature and design. It is almost a paradox for us, we who live under the apocalyptic threat of nuclear weapons, to ascertain that the tone of these expressions of artistry is seldom negative, on the contrary! Cultures in which violent confrontations between groups appear not to occur form startling but nevertheless marginal exceptions. Instead of disproving it, they confirm the rule that man is a belligerent being. They do, however, contradict any explanation saying that war is an inevitable result of a deep-rooted, insurmountable bellicosity in human nature. One idea which finds popular acceptance is that there exists in our species a kind of innate need for war, a trait in our make-up which manifests itself under particular circumstances with a certain necessity and inevitability. This idea has also found its advocates in scientific circles, where it rests on a 'drive model' of aggressive behaviour (Lorenz, 1963). There is, however, no evidence to support such a view (Hinde, 1960, 1973, 1974). Besides this, it is a defeatist and therefore dangerous point of view. It can become a self-fulfilling prophecy, because it can block efforts to search for constructive solutions. This does not detract from the fact that in the long history of our species characteristics which contributed to our successful engagement in tribal warfare could have gained an advantage in natural selection and thus have become incorporated into our genetic heritage, for example in the predisposition for suitable motivational emotional structures.

And so we arrive at the question what the causes of aggressively fought-out conflicts are, in particular of group conflicts. This question about the causes can be split up into a number of sub-sections.

1. At the macro-level: are there socio-ecological contexts which promote belligerence? Or is there usually question of accidental cultural developments? If the socio-ecological context gives a selective advantage to the victor in a group conflict under certain circumstances, is that then reflected in the causes which we can establish for the occurrence of a conflict, on the one hand, and in the motives which are given by the belligerent groups, on the other hand? (For these three categories are not necessarily identical, cf. Ferguson, 1984a, b.)
2. At the micro-level: how do the causal factors influence the social interactive processes so that individuals go on the war-path? Out of which motives does the decision to go to war arise, and what motivations drive the warrior into action? And what role do leaders and followers have here? To what extent can these processes be described in terms of aggresssive behaviour and aggressive motivations?
3. At the comparative level: No doubt the technologically, organizationally and logistically advanced professional activities which constitute modern warfare are completely different from primitive tribal battles, and from that which the boys from the Z-squad do to opponent fans on the football stands. To what extent is it possible to compare these processes on each of the levels mentioned in 1. and 2. above? And to what extent are the mechanisms involved here new, only coming into being with the rise of 'civilized' man? Or do motivational structures and behaviour dispositions play a part here, which had proven their usefulness already long before the dawn of mankind, and which have their equivalents in present day related species? And finally, animal species differ greatly in behaviour dispositions, such as those determining the nature and occurrence of intergroup conflicts. This raises the question, what explains these differences and what is their evolutionary *raison d'être*, and thus their adaptive significance? Can a comparative approach reveal general rules which will also further our understanding of the adaptive significance of human belligerence?

This last issue asks for some clarification. In biology there are two major levels of explanation for biological phenomena; so too in ethology, where behavioural structures are concerned.

First of all there is the question of how a process is regulated, of the mechanism responsible for its occurrence, or its proximate, that is, its immediate causation. As far as behaviour is concerned this means: what are the programmatic rules and the aims of a particular behavioural system by which the organism is made to display a particular behaviour pattern under certain circumstances? What effects does the organism seek to achieve, and to what influences and feed-backs is it responding in doing this? Behaviour systems are thus regarded as systems which fulfil a specific function because they have become programmed to maintain or achieve particular goals. This level is known as investigation into the proximate factors of behaviour. It is the level which occupies psychologists and ethologists interested in questions of motivation.

The second question concerns the evolutionary *raison d'être* or the ultimate causation. This is a typically biological question. Here one tries to ascertain the adaptive significance of a certain behavioural pattern. The starting point is the generally accepted neo-Darwinian evolutionary model. In this, natural selection plays a central role as a shaping factor, in that it tests characteristics on their adaptive significance. A characteristic is adaptive in as far as it imparts its bearer greater 'fitness', i.e. reproductive success, that is, if this bearer leaves more offspring in following generations than individuals possessing other variants of that characteristic. If the variation in the characteristic has a heritable component, the reproductively successful variant will by definition increase its numbers within the population! The relative contribution to the reproductive success of its carriers is therefore the ultimate gauge of biological functionality. An important contribution to this model was made by Hamilton (1964). He made it clear that hereditary predispositions which lead to behaviour damaging the survival and reproductive prospects of its bearers, can nevertheless spread in a population provided this damage is more than offset by an increased reproductive success of individuals which (partly) share this genetic predisposition. This can be the case if beneficiaries of this behaviour are relatives. Thus the overall 'fitness' of an animal could nevertheless be increased by this indirect contribution to its 'inclusive fitness'. This principle of so-called **kin selection** can explain the willingness, noted in a great many species, of animals to behave in an altruistic and even self-sacrificing way with respect to other animals of the same species, who are their fellows: in natural circumstances those 'fellows' are usually closer related than other individuals. One example of this is the loyalty towards relatives which is demonstrated so clearly in inter group conflicts (see, for example, Wilson (1975a) or Trivers (1985) for further elucidation).

2.2 BETWEEN-GROUP COMPETITION AS A SELECTION FACTOR IN THE EVOLUTION OF SOCIALITY

In practice the question of ultimate causation becomes particularly interesting when we put it as follows: what in fact are the consequences of a characteristic which lead to greater reproductive success, and under what circumstances is this the case? Let us use group-living in primates as an example because it immediately confronts us with the possible role of intergroup conflicts (van Schaik and van Hooff, 1983; van Schaik, 1983). Not all animals aspire to live in groups. In some species individuals prefer to operate alone. This could be, for example, because they can avoid the competition of their companions in this way. In other species the opposite is true. There an individual could profit from the fact that he is living together with his fellows in various ways, for example because each group member can:

1. share the knowledge acquired through experience concerning food, dangers, wander-routes and so on;
2. forage with a greater efficiency because he can profit from the discoveries of others, or because he can profit from cooperatively exploiting a resource;
3. enjoy a better protection against predators, for example, when these are signalled earlier or when a united defence is possible;
4. profit from a coordinated resistance against other groups competing for the same resources, to the mutual profit of members of the group, and so on.

This list is not complete; it merely sums up a few of the most important possible adaptive consequences of sociality. These could exert a selective pressure favouring dispositions to certain social inclinations, but then of course only if the sum of the advantages outweighs the disadvantages, such as:

1. increased competition when members of a group depend on the same resources;
2. increased chance of illness because infections are passed on more easily, etc.

Which of such factors really operate as the driving force in natural selection for a given species? At present only two hypotheses are seriously considered to be able to explain the evolution of sociality in primates (Dunbar, 1988). The first one supposes that a better protection against predation would constitute the most important advantage of group-living for diurnal primates. However, a heavy price might have to be paid for this, namely a decreased foraging efficiency due to intra-group competition (Alexander, 1974; van Schaik, 1983; van Schaik and van Hooff, 1983).

An alternative for this 'Predation–Within Group Competition' hypothesis (P–WGC) has been formulated by Wrangham (1979, 1980). His 'Between Group Competition' hypothesis (BGC) supposes that members of larger groups would each be better off than members of smaller groups because they could monopolize the access to better resources against strange conspecifics. This view implies that the advantage obtained by individual group members in the competition between groups amply offsets the disadvantage of a greater competition within the group (Wrangham, 1979, 1980).

There are two possible empirical approaches:

1. A broad comparison of species, and an inventory of the circumstances under which group-living does or does not take place.
2. Revealing the selection process at work by means of socio-ecological research. This entails measuring the success of individuals of a species under different circumstances (for example, the efficiency and success of foraging in large and small groups, and for animals who live in close proximity to each other and those who operate further apart, fecundity and survival of animals in larger and smaller groups; and so on).

Thus van Schaik (1985) demonstrated that the selective advantage for feral long-tailed macaques is primarily provided by the positive consequences presumed in 3.: in larger groups predators were discovered at greater distances; the most vulnerable set of animals, namely the newly independent juveniles, were shown to have a lower mortality rate in the larger groups. The disadvantage supposed under 1., that is, competition within the group, appeared to give an important counter-pressure however, which made any advantages gained by 2. dwindle into insignificance: animals in larger groups had a lower foraging efficiency; they also appeared to cover larger distances to obtain food and could devote less time to resting; moreover, female fecundity, which depends on physical condition, was lower in larger groups. Selection pressures are revealed which, within this species, must have determined their social system. These findings do not tally with the Between Group Competition hypothesis. For the majority of primates the P–WGC hypothesis appears to hold best; safety from predators appears to have been the most important factor promoting associative tendencies.

Competition within the group forms an important counterforce. The relative importance of both factors, which each pull in opposite directions, can vary from species to species, and their ratio is a determining factor in the extent of sociality. Among, for example, wolves and beavers, where cooperation in hunting and environmental control respectively are strongly developed, wholly different selective pressures must play a role.

I have gone into this matter in some detail, because outside the circle of behavioural biologists there is often an unjustified assumption that theories about the evolutionary background, and the adaptiveness of behaviour cannot be empirically tested. They can, however, because the activity of selective pressures can in principle be demonstrated.

As mentioned earlier, there are few indications that an advantage in competition between groups has provided an important stimulus for the development of group-living among primates (however, see Robinson, 1988, for an example of a possible exception). That is not to say, however, that groups, once they have formed, cannot become involved in conflicts between one another. Certainly in man, there are very good reasons to believe that between group competition and conflict are most important factors reinforcing the formation of larger and stronger units of social organization ('states')

(Cohen, 1984). Evidence concerning the conditions that favour the development of conflict behaviour between groups of non-human primates, has been obtained so far primarily by means of the comparative method (cf. Janson and van Schaik, 1988; Dunbar, 1988; van Hooff, 1988). This method may reveal some possible analogies about what can be said about war at a macro-level.

First of all though, let us pay some attention to the proximate factors of aggressive behaviour and intergroup conflicts.

2.3 AGGRESSIVE BEHAVIOUR: A HETEROGENEOUS CONCEPT

Attempts to influence other individuals, be they of the same species or not, by using or threatening to use violence is an almost universal phenomenon in the animal kingdom. This is the category of action we call aggressive behaviour. In previous decades the opinion has been common that the incidence of aggressive behaviour is determined by an aggressive drive or tendency, which fluctuates as a result of the performance of this behaviour (exhaustion as a result of frequent release, and accumulation if this behaviour was not used for long periods of time). The effectiveness of releasing stimuli varied depending on the aggressive drive. This 'drive model' corresponded with the motivation-models which were in fashion in ethology until well into the sixties, and it conformed with Freudian theories on anger. It was expressed most explicitly in Lorenz' book *On Aggression* (1963). At a fairly early stage, there were critics of this theory, on both the psychological side (Berkowitz, 1962) and the ethological side (for example, Hinde, 1960), who raised objections to the so-called 'energy-model' of motivation. Above all, aggression proved not to be conceivable as a unitary, homogeneous motivating factor (see for example Hinde, 1970, 1979).

It is more productive to see behaviour as a hierarchical structure of regulatory systems (for example, Baerends, 1979). Such regulatory systems seek to achieve or maintain certain norms or goals, such as liquid balance in the body tissues (by drinking or excretion), or a living area free of competitors, i.e. a territory (by defending this area against certain other members of the same species), and so on (Wiepkema, 1977b). The properties of such regulatory systems, their goal or norm values and the choice of behavioural routines will be adjusted in the long run, by natural selection of hereditary behavioural dispositions, and on short term within the individual by selection on the basis of experience (conditioning, cognitive evaluation). Thus the system will be tuned so as to bring about certain adaptive consequences most efficiently, with the least costs in terms of energy investment and risks. Behaviour which is aimed at inflicting damage to another individual, whether by hurting the creature itself or by afflicting structures on which it depends, can form an instrumental subroutine in achieving certain social goals in animals. Certainly in lower animals, this 'social punishment' is achieved by species-specific, sometimes 'ritualized', fixed-action type behaviours. In other words, there is an ethographical category of 'aggressive behaviours'. In higher animals, and certainly in man, the behaviour–function connection is more loose. Apart from the 'old' species-specific expressions of aggression, there is a variety of opportunistically available routines for 'aggressive functions' (cf. Tedeschi, Milburg and Rosenfeld, 1981).

An animal wanting to gain access to a resource in competition with other animals may use the threat of violence to force others into compliance. However, alternative instruments may be preferred, depending on the circumstances and the relationship between those involved. Thus the female pygmy chimpanzee may 'buy' compliance and priority of access from a male competitor by selectively offering her sexual favours (Kuroda, 1984).

In short, to achieve a desired goal, different subroutines may be used depending on the situation. The reverse can be true as well; a subroutine can also be activated in different functions of a higher hierarchical order. Female primates may present themselves sexually because they wish to make sexual contact in order to obtain the related satisfaction, that is, because of a sexual motive. They may do the same, however, to allay the threat of violence, and thus as an instrument of pacification, or, as mentioned above, to gain priority in the competition for food, that is, out of a motive to acquire food. The question whether one regards this as sexual behaviour can lead to fruitless discussions about definitions. The important thing to bear in mind is that this phenomenon can be regarded at different levels.

The same is true when one tries to define the limits of aggressive behaviour, for instance when one asks whether violent behaviour, in a group conflict or even in a war, should be described as aggressive. Violent behaviour can be woven into functional contexts in various ways, and the motivational and emotional mechanisms involved can differ.

Thus an individual can show different levels of arousal, for example in his expressive behaviour, and in physiological parameters, such as autonomous activities, muscle tension and the like. If harming behaviour is associated with a state of emotionality we speak of 'anger', 'rage' and 'fury'. When we are dealing with humans, we have the possibility to establish that these emotional states may be paired with the desire to humiliate the opponent, or to hurt him. Concerted use of violence which has as its immediate and primary aim to inflict harm on another, and which derives its satisfaction from achieving this, has been called 'pure' or 'hostile' aggression (for example, Hinde, 1973, and Berkowitz, 1981). This can be the case when hate or revenge are the driving impulses for the aggression. Even if the main aim is to hurt the other, another consequence, and one that could be advantageous to the aggressor, might be that some hindrance or claim which came from the opponent is removed in this way. However, the suspension of such infringement may also have been the primary aim, and then behaviour hurting the other is simply a means to an end. This is what Scott and Fuller (1965) called instrumental aggression. An example is the burglar who strikes down a guard in order to be able to work in peace, or the hired killer. In humans, and only in humans, one can discover whether one is dealing with coolly calculated instrumentality, which is not necessarily associated with the emotionally laden urge to inflict pain. One could maintain that such a person is not aggressive, despite the concerted use of violence, but simply doing his job. In doing so, however, one restricts the definition of aggressive behaviour, in that one makes the emotional urge to hurt an essential characteristic of the definition. It is an established fact that the destructive actions carried out by soldiers, certainly in modern technical warfare, are often not fed by aggressive motivations in this limited sense, and usually not even by the desire to see the war run a course favourable for the nationalistic or idealistic cause. Rather, the behaviour is governed by proximate motives which are very directly related to material and social consequences within the immediate surroundings of the soldier. We will return to this later.

It is clear that the term 'aggression' refers to a heterogeneous collection of phenomena with a number of more or less independent dimensions of morphological, functional and motivational variation. But as Hinde (1974) posited: it is not easy to make these categories have sharp outlines; they often contain a 'generally agreed nugget', but are 'shady at the edges'.

Apart from the dimensions of variation listed above different kinds of aggressive behaviour can be distinguished on the basis of the adaptive consequences which they entail, i.e. the functional context in which they are often shown. The question whether the animal is cognitively aware of these adaptive consequences is left unanswered. These consequences may or may not be represented as anticipated effects and thus may or may not play a role as motivational factors in the regulatory circuit of behaviour. We shall take a quick look at a categorization with a partly arbitrary choice (cf. Rosenzweig's, 1981, three-category system and Moyer's, 1968, six-category system) of five categories, which, anyway, is comprehensive.

Defensive aggression can be directed at something that an animal perceives as a threat to itself, its kin and in some cases also its fellows. It occurs especially when fleeing or hiding are not possible (anymore). This primarily reactive form of aggression is usually marked by a high level of fear. Often the new, the unknown or the strange can release a defensive attitude. Unfamiliar members of the own species can also cause a xenophobic reaction.

Territorial aggression against intruders is aimed at the exclusive use of a certain area for oneself or one's group. Other behaviour patterns occurring in the same functional context are patrolling or marking the limits of a territory (for example the 'song' of birds, gibbons and howler-monkeys; scent-marking etc.). Ritualized displays of strength are a particular form of this behaviour complex.

Competitive aggression. Violence or the threat of violence can be used to gain access to limited resources. As with the previous form of aggression, one can expect the choice of behaviour to be based on some form of weighing the 'estimated' costs in relation to the 'estimated' gains. That does not have to be a consciously thought out strategy; it may be an emotional evaluation of the goods in question, a 'feeling' of how easy they are to monopolize, the effort it will take to acquire it elsewhere, and the costs and risks of the fight (fear) taking into account the nature and resoluteness of the opponent.

A special form of this competitive aggression, put under a separate heading by some authors (for instance Moyer, 1969), is the competition for relationships with other conspecifics, notably for sexual partners. In most species the latter type of competition takes place primarily amongst the males.

Rank and status. Aggression can be seen, as stated above, as a means to safeguarding the prerequisites for a safe existence. In species which live within a stable group context, competition may lead not only to direct rivalry conflicts, but also to attempts to establish a general and robust dominance in social relationships. Many a conflict amongst primates has to do with the attainment and maintenance of status (see for example de Waal, 1978, 1982; Netto and van Hooff, 1986; Walters and Seyfarth, 1986). Characteristic behaviours in such circumstances are 'showing off' and 'bluffing', the ritualized show of the characteristics of high status (size, strength, stamina, the possesion of a specific position within an area). Thus Topi-antelope bucks in the savannah stand for hours on as high a termite-mound as possible. The opposite to this is behaviour to show submissiveness and recognition of the other's status (making oneself small and displaying harmlessness).

Frustration-aggression. If an animal experiences hinder or irritation because of a partner, for example, because it is frustrated in the attainment of certain goals, the 'supposed cause' of the hindrance may be attacked. Dollard *et al.* (1939) found much support for their opinion that aggression can always be related to some form or other of frustration. However, this view can only be made to hold water if all the other reasons given here for aggressive behaviour are subsumed under the heading 'frustration'. Then the concept clouds rather than clarifies the issue (cf. Sherif and Sherif, 1970). This does not detract from the fact that frustration and irritation in a narrower sense can be a cause for aggression. It is thought to be characteristic that frustration, deprivation and irritation can, if occurring repeatedly, lead to a gradual increase in aggressive tendency, which can be unleashed by comparatively minor and irrelevant stimuli, whether these be individuals or objects (the proverbial straw that broke the camel's back; redirection, Bastock, Morris and Moynihan, 1953). One particular form of frustration can arise when conspecifics display behaviour which deviates from expectations raised. Such 'violation of the rules' can lead to aggression, which in the case of humans we can also identify as 'moral aggression'.

Predatory aggression. Some authors have considered the behaviour with which predatory animals capture and kill their prey as a form of aggression (for example, Karli, 1956, and Myer and White, 1965). They used the mouse-killing behaviour of rats as a measure in research concerning the neurophysiological bases of aggression. Ethologists have repeatedly pointed out that this is extremely misleading. These behaviours are subroutines which are mobilized as parts of food acquisition strategies. Their motivational background is, therefore, completely different from that of violent behaviour intended to defend one's own interests and safety, and even of the victimization behaviour (vandalism, sadism) with which Rosenzweig (1981) classifies it. There is however some similarity between the behaviour patterns employed and the weapons used (for example, teeth and claws). That makes predatory behaviour an interesting phenomenon in connection with the origin of human warfare. One view is that the behavioural characteristics and techniques which primitive man had developed as a part of cooperative hunting appeared to be advantageous in group conflicts as well, and have promoted warefare as a response to between-group competition. We will elaborate on this later.

2.4 GOING INTO AGGRESSION TOGETHER, REGULATION OF THE PROCESS

The ethological literature contains no reports of antelopes, rabbits or cats joining in an attack against a rival group or a threatening predator. There are, however, reports of birds who attack birds of prey *en masse* (Curio, 1963), of chimpanzees who together beat up a leopard model with sticks (Kortlandt, 1965), of bee-eaters (Hegner, Emlen and Demong, 1982), hyenas (Kruuk, 1972), and wolves (Murie, 1944) who organize in clans and defend a common foraging territory against intruders from other clans; of jays and hoopoes who form breeding communes and unitedly defend these (Woolfenden and Fitzpatrick, 1974; Ligon and Ligon, 1982) etc.

Obviously there are vast differences between species in the extent to which they show group aggression. It is also obvious that the motivational mechanisms which bring about such group behaviour are of different kinds. In the most simple and least interesting cases, common action rests on the fact that each of the individuals is affected by the same behaviour-eliciting factors; the joint action is nothing but the sum of independent individual actions. More interesting are those cases where individuals influence each other so that a certain coordination of action results.

The simplest example of this is 'social sensitization'. The activity of a group member directs the attention of others towards the same source of stimuli, and thus has the effect of, perhaps inadvertently, synchronizing their activities. If this leads, let us say, to a more effective defence, and if the individual fitness of the animals involved is enhanced by so doing, then natural selection can be expected to promote the spreading within a population of all sorts of structures which contribute to such a synchronization and to even better coordination. In its simplest form, this is the case when animals warn each other about a threat with special alarm signals and the like. The evolution of breeding in colonies among birds such as gulls (for example, Tinbergen, 1963 and Gotmark and Anderson, 1984), and terns (Møller, 1982) must have been selectively promoted by the increased security resulting from the united call for defence.

A stage further in social influencing is the instigation or transfer of a particular attitude or mood. The receiver of an alarm call is then not merely made aware of a certain stimulus source, whereupon he can choose to react as he sees fit. The signal can also cause stimuli which were neutral for the animal to start with, now suddenly to acquire an 'emotional colour', for instance, that another animal is all at once seen as an 'enemy'. One example is the police dog. It can easily be set on a criminal, even though the criminal has never done anything against that dog. Try to achieve the same with cats or deer! In short, among some species of birds and mammals (including particular species of primates like macaques) aggressive display can have a contagious effect on members of the same group or species. Often these species know special forms of behaviour with which they can incite a fellow to turn against an opponent (see Fig. 2.1). (See for example van Hooff and de Waal, 1975; de Waal and van Hooff, 1981.)

2.5 THE CIRCUMSTANCES OF COALITION FORMATION

We just saw how in coalition formation the signals which come from a partner soliciting aid can be enough to make his opponent the object of aggressive behaviour. In this way participation in between-group conflict comes under the control of new mechanisms, in which the relationship with the conspecific(s) and group processes form an important element; one enters a new level of analysis.

One can distinguish between two main forms of support-giving by third parties in a conflict. In an analysis of interventions in a captive group of chimpanzees in the Arnhem Zoo in the Netherlands, de Waal (1978)

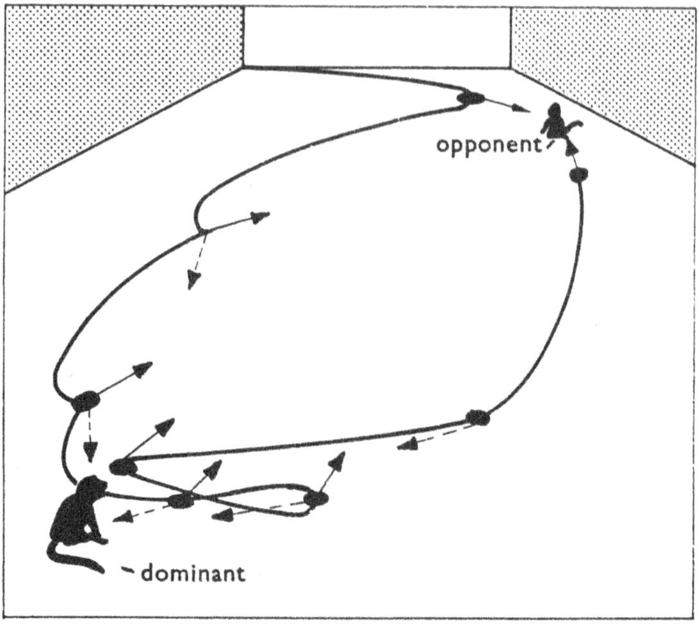

Fig. 2.1 Inciting behaviour in long-tailed macaques. The solid line shows the direction of walking of an aggressor. First he goes in the direction of a dominant group member, and tries to set him on his opponent. This he does by walking past the dominant animal, looking at him in quick repetitions (dotted arrows) and intermittently turning his head towards his opponent with sharp jerks (thin arrows). He points out his opponent with chin high and neck outstretched, while 'bark-growling'. From this position he may decide to attack his opponent (after de Waal, 1977).

identified a category of protective support, typically for the benefit of the weaker party in a conflict, and one of opportunistic support, typically for the benefit of the stronger party. The first form of support obviously involves greater strategic risk for the supporter, especially if the supporter in his turn is also weaker than the opponent. The intervention then has an altruistic character. In such cases the supporter chooses the side of the party in the conflict to whom he is most closely attached (in the literal sense). Support appears to be given above all to partners who belong to the same family groups, centred around one or more older females, which can be distinguished in this chimpanzee community. In the wild, the animals within these groups tend to be more closely related to each other than on average; in the originally artificially composed study group of the Arnhem Zoo this initially was the case to a much smaller extent. This indicates that this self-sacrificing inclination can contribute to what biologists call the inclusive fitness of the supporter, and that this inclination exists by the grace of kin selection. Of course this does not mean that a perception of genealogical relationship has to play a role in the choices of who supports whom. The Arnhem situation shows that the choice is determined by purely arbitrary historical associations (but, once again, under natural conditions such association is often determined historically by the relationships of common descent).

The second type, opportunistic support, concerns cases where an animal helps another who is stronger than his opponent, and who would have had a large chance of winning the conflict on his own anyway. In chimpanzees we encounter this exploitative and much less risky form of support particularly among males. They seek coalitions, within which they can attain a maximum of influence and freedom of movement. The choice of which side to support appears to be much less consistent over time (opportunistic) and to depend less on pre-existing relationships of interdependence. The support often seems to be mutual, or other 'favours' may be given in compensation, as a reward for the support given (for example, sexual permissiveness). Changes of allegiance which occurred after the dominance positions of individual animals changed, suggest that coalitions of this type are 'opportunistic agreements', which remain in being as long as both partners are better served by these than by other coalitions. Opportunistic support may reflect the principles of reciprocal altruism, as formulated by Trivers in 1971. It seems to be of a more calculated nature than protective support, which appears to be a more impulsive matter of the heart.

Indications about this kind of difference between the more altruistic form of support, which depends on a relationship of attachment and affiliation, and the more opportunistic kind, where intervention depends on the possible advantage gained by the supporter, have also been found in other primate species (see Netto and van Hooff, 1986).

Just as the giving of support can be facilitated by a relationship of attachment, so too can the reverse: joining forces against a common enemy may increase mutual attachment. A classic explanation for this rests on the so-called redirection of aggression, which is described in many species. If an animal receives aggression-stimulating signals from a conspecific, but at the same time also receives signals which thwart the expression of that aggression (for instance signals which stimulate its fleeing or sexual tendencies), the aggression may be redirected at an innocent third party. This redirection can be encouraged by one of the partners choosing a victim and picking on him. A typical example has been described for monogamous species such as ducks and geese. When there is tension or uncertainty in the pair bond the weaker party, often the female, may set its mate on an outsider. The bond-strengthening effect of joint action against such an outsider can now become the main causal factor of such behaviour, whereby fortification of the relationship is the primary goal. A good example is the triumph-ceremony of greylag geese, the origin of which can be traced back to the redirection incitation found in many duck species (Fig. 2.2). When pairs meet again after

Fig. 2.2 The 'triumph-ceremony' of the greylag goose. The gander makes an attack (1, 2, 3) on an imaginary enemy (F), and then returns (4) to the female who has followed him (5), after which they both 'pump' and 'chatter' (after Fischer, 1965).

some separation, they seem to be uncertain about the attitude of their partner. One of the two animals may now carry out an attack on an imaginary third party, in which he may be joined by his or her mate. After the attack both parties return to carry out synchronized, ecstatic pumping movements with their necks. Behaviour which was originally an aggressive redirection has here evolved into a declaration of attachment (Fischer, 1965). There are indications that in cognitively highly developed species such as primates, the redirection-victim is chosen not only because it is an enemy of the redirector, but also because the redirector knows that it has a difficult relationship with its potential ally. This kind of strategy demands a high level of appraisal of the interaction possibilities among the other group members, and it has been postulated that the cognitive demands necessary for such a process must have formed the most important selective pressure in the development of primate intelligence (Humphrey, 1976).

To sum up, joint aggression against a common enemy can be promoted by mutual attachment, but can also in turn be used as a means to strengthen attachment.

2.6 AGGRESSION BETWEEN GROUPS IN ANIMALS

Coalescing in aggression against third parties is something we observe in a rich diversity of forms and circumstances, whether it is within or between groups. In the simplest case, two partners join together, but the same mechanism can lead to the formation of vast coalitions. Studies on this subject have so far concentrated on conflicts and coalition formation within groups. And then it appears, for example among our closest relatives, the primates, that members of family-clans in particular engage in coalitions with one another (for example, Kurland, 1977; Datta, 1988; Netto and van Hooff, 1986; Walters and Seyfarth, 1986).

Hostile interactions between groups are rare in most species. Neighbouring groups tend in general to avoid each other. Should they happen to come in close range of one another, the group which is the most peripheral in its home range will tend to retreat. This is the case in a great many primate species, such as macaques and baboons (McKenna, 1982; Cheney, 1986).

Expressions of enmity between groups increase when population density increases and the areas in which the groups range become smaller. Under these circumstances, it also becomes easier for the animals to track down intruders on their own territory, and to chase them away (Mitani and Rodeman, 1979). This relation shows up not only when we compare different species, but also when we compare the same species, living in different environments. In rural areas in India, where rhesus monkey groups have large home ranges, these can overlap each other to a considerable extent. The groups avoid each other and display little aggressive behaviour. Around the temples and in the bazaars of Indian towns, however, these same monkeys live in small, non-overlapping territories, with sharply defined borders which they guard ferociously (Southwick, Beg and Siddiqi, 1965; Southwick, 1969).

Something similar has been observed in langurs, another Indian primate species (Yoshiba, 1968; Ripley 1967; Nagel and Kummer, 1974) and in Japanese macaques (Kawanaka, 1973).

Hyenas, social carnivores which hunt cooperatively, live in great densities in the animal-rich Ngorongoro crater in Tanzania and defend group territories there. This has not been observed elsewhere (Kruuk, 1972). Incidentally, not all attachment to a territory is the result of intergroup aggression. The advantages of such attachment are often that the animals are better able to exploit an area if they are familiar with it, and can better avoid the dangers which lurk there. This can remove every impulse to leave the area (McKenna, 1982).

On the other hand, not all agonistic group interactions concern territory. Hamadryas baboons differ from other baboons and macaques in that they live in clans, each one of which consists of a few 'harem units'. Within such units one adult male herds several females and maintains exclusive access to them. This species lives in arid steppes in Ethiopia, an area with little food, and so the clans disperse widely when they wander over large areas during the day. At night they seek safety on sheer cliff walls. These are not plentiful, however, with the result that many clans congregate to form large sleeping aggregations. This does not lead to skirmishes of any importance; the animals obviously behave according to accepted rules. The situation changed when Kummer (1968) offered the animals grain to eat at the base of the cliffs early in the morning. Fights quickly broke out, and occasionally escalated to mighty battles between clans. These fights, which, at first sight, seemed to have been caused by the competition for food, appeared in fact to concern the integrity of the harems. Intermingling of harems from different clans led to insecurity on the part of the harem owners. These tried to drive their females together, and began to attack other males. It is striking that the harem males from the same clan, who are usually related to each other, seem to 'recognize' each others' rights. This Kummer could prive by means of ingenious field experiments. The animals appear not to expect such respect on the part of males from other clans.

At the opposite end of the animal kingdom, amongst the social insects, circumstances have arisen which strongly favour the occurrence of intergroup aggression. That is the storage not only of large, concentrated supplies of food, but also of eggs and pupae, for instance, by ant populations. These form an attractive booty for all sorts of robbers. No wonder that some ant species have developed a soldier caste, with various kinds of weaponry. Other ant populations, whether of the same species or not, form an important threat too. These are not only after the food supplies, but also after the pupae, which are reared by the plunderers to become a cheaply gained labour force in their own colony (a functional analogy of slavery even though, of course, the mechanisms involved are completely different; see Brian, 1983, for other examples of evolutionary inventiveness in this kind of exploitation).

2.7 INVOLVEMENT IN INTERGROUP CONFLICTS

Conflicts between groups are on the whole fairly rare among animals. If it does come to a massive skirmish, it doesn't appear to be a purposefully coordinated undertaking. Rather, it appears to be the escalation of minor tussles between individual animals, with which members of the same subgroup interfere (Kummer, 1971). A number of years ago I was the chance witness to a massively escalated conflict between two groups of baboons during a visit to Shirley Strum's research site near Gilgil in Kenya. We saw two large groups, the 'Pumphouse' group and the 'Eburu Cliffs' group, each numbering 100 or more animals, engage in an impressive, at times deafening spectacle, which lasted for about half an hour. Animals of each group, but by no means all, stood facing each other along two fronts, more than 100 metres across. From there they threatened each other. Here and there one would make a sortie, while emphatically 'showlooking' alternately at its own neighbours and at the opponents, and letting out loud shrill barks and grunt barks. Suddenly, probably because one or more opponents shrunk back, a number of animals rushed into a simultaneous assault. Practically immediately the whole front surged forward and chased the other party about a 100 metres to their rear. This was accompanied by a massive swell of shrieking barks, in which all the members of the pursuing group took part with full voice, even those who were not at the front. In a short while the impetus had gone out of the chase, and the two sides were facing each other again. At intervals a similar wave of attack would repeat itself in the same or the opposite direction. The confrontation ended unspectacularly, not with a final decisive rush, but because participation died down, especially among the intruders. These withdrew to cliffs at the edge of their own territory.

Most striking was the small amount of physical contact in fights, which were moreover of short duration. Nor were there any serious wounds as far as we could tell. This did not seem to be the result of good natured restraint, but rather of the animals' fear of being caught up in the wave of aggressors. There was not a single indication that the obvious synchronization of the assault rushes could be ascribed to one or more coordinating leaders.

2.8 WHO PARTICIPATES IN BETWEEN-GROUP FIGHTS?

It was also striking that it was not primarily the males who stood in the front line. Contrary to what one would expect at first sight, the females were just as active in the skirmishes, if not more so. The same has also been observed in the defence of groups against predators. The classic picture of the adult male placing himself as a protective shield outside the group when it is threatened (Washburn and DeVore, 1961), does not seem to be a universal truth. Thus Rowell (1972) found that the long legs of the males in a threatened baboon group brought them soon to the head of the fleeing crowd. And Gouzoules, Fedigan and Fedigan (1975) discovered, in one of the rare cases in which humans have witnessed a predator attacking a member of a primate group, that animals of all ages and both sexes took part in the defence when a lynx grabbed a young Japanese macaque.

We meet an important problem here, namely: who is doing the fighting? If the fighting is for the weal and woe of the group at large, and if this involves energy as well as risks, then why doesn't each group member leave it to his fellows to do the fighting? The best strategy here would seem to be that one avoid as much of the costs and risks of the battle as possible but then however, reap the benefits. But who would fight at all then, if in fact those who do not take part have the greatest net benefits. If fighting for the sake of the community is not to die out, certain conditions must be met.

The simplest case of course, is when group members at large profit, even though an animal does not fight for some general good, but only, directly or indirectly, in furtherance of its own interests. A male could keep strangers away from its own favourite females only. Several males attacking strangers together is then nothing more than the sum of individual profit strategies

(Cheney, 1987). Where an individual seems to be acting in a self-sacrificing way, detrimental to its own direct survival and reproduction prospects, it may nevertheless be acting in its own interests. Such behaviour may therefore be promoted when there are compensations of an indirect nature; the 'altruist' furthers the survival and reproductive interests of group members with which it is related more than on average. Thus it furthers its reproductive interests, its 'inclusive fitness' in an indirect way (Hamilton, 1964). The problem is that as a rule relatives cannot be distinguished easily by external appearances. In most animal species, however, those animals with which one has been associated from early on are likely to be near relatives. Natural selection will consequently exert a positive selection pressure for altruistic inclinations towards nearest associates, because this will often result in kin selection.

Thus we can predict that especially in species in which groups have a relatively high level of internal consanguinity, there will be a greater willingness to take part in a fight which is at once dangerous and altruistic. And there are indeed examples to support such a view. Some species of birds, for instance, form breeding communes, where a number of animals help a breeding couple to gather food for their young, and to defend their territory. The helpers are mostly siblings of the same couple from a previous year. We can find a similar phenomenon in wolves, cape hunting dogs and hyenas. Another example is the naked mole rat, where groups show a high degree of inbreeding. The most spectacular examples, however, can be found in colonies of social insects (wasps, bees, ants, termites), where worker and soldier castes have lost their ability to reproduce in favour of mothers or sisters who can reproduce, the queens. Here the most specialized cases of massively coordinated intergroup aggression with blind sacrifice by individual group members can be found, for example in the form of 'ant wars' (see, for instance, Brian, 1983).

A second prediction is that those who will be the most prepared to take part in joint aggression, will be those with the greatest genetic interests in the group. Among the majority of primates (for instance macaques and baboons), males form the exogamous sex; they leave their group of birth around adolescence, and migrate to other groups. The adult males in a group are, therefore, as a rule, immigrants, and probably not strongly related to each other. And if they have not been with the group for long, they will certainly have no descendants there. It would be only too understandable if this faction were conspicuous by their absence on the front line. There is indeed evidence to suggest that the protective attitude of a male increases with the amount of time that he has spent with the group. One indication comes from recent research done in the Sumatran jungle on wild long-tailed macaques (van Noordwijk, 1985). Dominant immigrants start their career in a group by showing off near fertile females; they are little concerned with the rest of the group. After about a year they quieten down, they are often found in the centre of the group in the company of females. They increasingly take part in protective behaviour, such as contact- and warning-calls.

As we said earlier, in the majority of primate species intergroup conflicts are neither frequent nor serious and do not give the impression that they are coordinated in a planned fashion. This is not to say that they do not take place in primates at all. The shocking discovery by Goodall *et al.* (1979), that in a population of our closest relatives, the chimpanzee, the males went on raids into the territory of a neighbouring group, successively killing three adult males there, is by no means unprecedented. Nishida and Itani have observed aggressive fracas between groups during their long-term research. These were presumably the cause of the disappearance of a number of adult males. These encounters had the atmosphere of a 'skirmish in a war' (Itani, 1982). Both Goodall *et al.* and Itani draw attention to the strategic, deliberate, and co-operative patrolling behaviour that the raiders displayed on these occasions. They crossed the foreign territory in a marching band, which differed from normal hikes in its speed and the straightness of direction: 'they looked as if they were aiming for the best chance of encountering another group' (Itani, 1982, p. 366). The analogy to a raid in a primitive tribal war is striking (for example, Murphy, 1960).

Such organized forms of male intergroup aggression seem to be, at least for the time being, a remarkable exception among primates. This difference does of course raise the question as to under what circumstances certain forms of intergroup aggression develop, or rather, under what conditions individuals can join in such aggression and thereby obtain a reproductive advantage.

2.9 INTERGROUP AGGRESSION, ECOLOGICAL FACTORS AND ADAPTIVE SIGNIFICANCE

At first it may seem fairly simple to establish the adaptive significance of intergroup aggression. A similar answer is plausible as that given in the case of individual aggression. For, as groups compete with each other, and aggression against another group may improve or safeguard the existential needs of one's own group, then it seems to be obvious that between-group aggression is thus encouraged and that the development of the motivational mechanisms required is favoured as well.

We have seen already that the formation of groups amongst the majority of primates is not primarily promoted by the advantages which the group members might obtain in their competition with neighbouring groups, in that they can monopolize a better living environment and better resources. The effects of this upon the successful functioning of individual group members are far less easy to demonstrate than the effects which have to do with greater security in larger groups with regard to predators. What is also obvious is the high price which the animals have to pay for living in larger groups in the form of an increase in within-group competition, i.e. with other members of the own group (Janson, 1985).

Once groups have formed on the basis of a particular balance between security and competition, then a preponderance in the competition between groups will of course give advantages to the members of the strongest group. To answer the question, under what circumstances is this really the case, we have to examine the nature of the competition which can exist between individuals and between groups.

Competition can lead to two forms of contests: **scramble** competition and **exclusion** (or contest) competition, (Janson and van Schaik, 1988; van Schaik, 1989). Scramble competition occurs when the resources that are the object of the competition, such as water, nesting materials etc. but above all food, is found in small clump sizes which, moveover, are distributed in space in such a way that it is not worthwhile to invest time, energy and risks in defence of such a clump. Why make a fuss over a tuft of grass or a fresh leaf, if there are other tufts of grass or fresh leaves just a bit further away? Even so there is a greater competitive disadvantage in larger groups. More often an animal will find the grass is cut away before its feet. There is little it can do but to forage faster and more efficiently than its competitors. The ensuing scramble competition is a competition in terms of efficiency of exploitation.

If resources are clumped, it will depend on the size and distribution of these whether monopolization is feasible. Monopolizable clumps are, for instance, shrubs or trees with a concentration of fruit supply. Such circumstances may lead to contest or exclusion competition. In the same vein, it may be worth defending an area as a territory. For this the area must at the same time contain a steady but sufficient, exhaustible supply of necessities, and it must be small enough to be surveyable and controllable. The way in which the resources of a species are distributed can lead to either scramble competition or to exclusion competition. This again must have far-reaching consequences for the social organization of that species (van Schaik, 1989). If it is rewarding to engage in exclusion competition, then it will also be rewarding to invest in attaining a dominant position within the group. Then it is also advantageous for one's reproductive success if offspring and relatives occupy better positions than non-relatives. One can expect that under such circumstances there is a positive selective pressure acting in favour of emotional characteristics which increase willingness to support one's next of kin. The rise of 'matriarchal clan-systems' within groups, as found in many species of primates (for example, in baboons and macaques), can thus be easily understood. In these species the children attain a social position close to that of their mother (and grandmother etc.). A child will eventually become dominant over all those other members of the group against whom its mother and relatives can effectively support it, namely those over whom they are dominant themselves. Such clan-systems evolve around female primates, as opposed to males, because females are in general 'more certain' of precisely who their offspring are; kin-selection works more accurately along this line.

This in turn must have its effects on the migration system. In many species of animals, immediate relatives are not preferred as sexual partner. Strangers are more attractive than members of the same species with whom an animal has been in close contact from an early age. The adaptive consequence of such preferences is undoubtedly that it works against inbreeding. However, if females are strongly dependent on their family coalition system for their success in reproduction, they will almost always be worse off if they migrate to another group; as a solitary without family they will settle on the lowest rank position of the dominance hierarchy. And if the females are restrained from leaving their natal group, then the males (for whom their own strength is a more important determinant of fitness than family support) will be encouraged to migrate; they become the exogamous sex.

Comparative socio-ecological research to test this picture is badly needed. Nevertheless there is some evidence that primate species which primarily subsist on fruits, a resource that tends to be well monopolizable, also tend to live in matriarchal family clans with definite linear ranks in which relatives occupy neighbouring rank positions as a result of nepotistic social behaviour (for example, clear family coalitions), while adult males tend to be immigrants. On the other hand, in a number of primate species, such as the leaf-eating langurs, the gorilla and the chimpanzee, where competition is more of the scramble type, female relationships are more egalitarian, without outspoken hierarchies, and with individualistic relationships between the adults. The formation of female family clans is not very conspicuous, and females as well as males may emigrate to a more attractive group (van Schaik, 1989).

The degree to which animals take part in conflicts between groups will certainly be influenced by the factors mentioned; they explain why among primates which engage in exclusion competition, such as macaques and baboons, the females, as members of the resident sex, join in the intergroup conflicts just as ferociously if not more so.

Wrangham (1982) has already demonstrated that the fitness of female primates depends on different factors as compared to males. For females it depends, above all, on safety and access to food; these will, therefore, be the major determinants of their grouping patterns and social order. The maintenance of a good condition for a long period of time will be decisive in how many offspring a female will be able to rear consecutively. In contrast, males can reach a high reproductive success by fertilizing many females in a short timespan. Male animals are thus put before the evolutionary choice, whether to make long-term paternal investments in a few consecutively raised children (tending towards a monogamist strategy), or to invest in a maximization of the chance of fertilization (a polygamist strategy). This last entails a strong selective pressure favouring the competitive capabilities of the males, which enables them to monopolize access to as many females as possible, at least in so far as the distribution of those females makes it possible to prevent access by other males. If the females stay close together for reasons of safety, their groups can be guarded more easily, and the males can engage in exclusion competition for them and try to establish and maintain a one-male group. In the other case multi-male groups will arise. In a multi-male group two strategies may be successful. If the females tend not all to be fertile at the same time, then a male can attempt to build up an exclusive consort-relationship with the fertile female for the duration of her fertile period. Thus he may try to achieve that only his matings have any chance of success. If, on the other hand females tend to ovulate at the same time, then the balance can come down in favour of a sperm-competition strategy, where each male tries to fertilize more females than the other males by copulating as frequently as possible (a form of scramble competition). This is reflected in the interesting discovery that among primates who live in multi-male groups, the males have comparatively larger testicles than males from species with single-male groups (Harcourt *et al.*, 1981).

Cooperation between males to monopolize access to a female or a group of females, or to monopolize certain consorts within a group, is not often seen among primates. This is probably hampered in most species by the males being strangers to each other. There is then no easy solution to the question of how the profits of the cooperation are to be shared: 'who gets the consort, of which the coalition took possession?' A recent study of baboons, where such coalitions do nevertheless occur, shows that this can lead to complex negotiating situations (Packer, 1977: Noë, 1986, 1989). Coalitions for mates occur between males who have been in the same group for a long time, and have got to know each other so well that cooperation on a basis of mutual advantage becomes possible. A situation of male cooperation can evolve more easily when the males are closely related. Even if there is no fair distribution of the profit, such cooperation can bring gain to all those involved in the form of an inclusive fitness advantage by means of kin selection (see p. 25). This situation exists where brothers migrate together as they do in lions (Bertram, 1976) and turkeys (Watts and Stokes, 1971), or where males are the resident sex, and their family relations thus stay intact.

This last situation has been discovered in a few populations of chimpanzees (Pusey, 1979; Tutin, 1979). These fruit-eaters only form groups at times when the food supply is superabundant. In order to avoid competition for food, female chimpanzees usually travel alone or at most with a few relatives (offspring). On the whole the situation is characterized by scramble competition. Chimpanzee females are, therefore, hardly competitive amongst each other, and appear to migrate easily. Under these circumstances groups of related males can stay together and defend an area against other 'brotherhoods'. The larger and richer this area is, the more females will be able to live in it. The willingness of chimpanzee males to engage in between-group warfare now becomes understandable.

We know from studies carried out in the wild (for example, Goodall, 1986), and in captivity (de Waal, 1982), that dominant male chimpanzees can be tolerant, even if not always with heartfelt sincerity, of the sexual behaviour of their coalition partners. There is reason to assume that this tolerance of promiscuity makes it worthwhile for non-dominant males to keep lending their support to the coalition. Bearing this in mind, it is possible to comprehend the exceptional, purposeful cooperation which male chimpanzees give in intra-group conflicts. There is an amazing similarity to the situation among humans. The development of social structures, in which men join in discrete solidarity groups (fraternal interest groups), is regarded as a condition which favours the development of bellicose tendencies (see Ferguson, 1984a b, for further references).

There is a second fact, which sets the chimpanzee apart from most, if not all the other primates. And that is the fact that they are capable of coordinated cooperation on a hunt. From the Gombe study area, in Tanzania, came the first observations that chimpanzees regularly hunt baboons, colobus monkeys, and a number of smaller animals. Only the males did this. They surrounded their victim by keeping a close watch on each other, and by shutting off the prey's possible escape routes from the trees. (Teleki, 1973; McGrew, 1979). Over the last few years similar behaviour has been observed in other chimpanzee populations. In a West-African population, intensively studied by Boesch and Boesch (1989; pers. comm.), the males go hunting together every day, systematically looking for groups of colobus monkeys. Such hunting forays were almost always successful, after varying lengths of time. Afterwards, and this is remarkable, the booty is shared (!).

Baboons also hunt small mammals and birds occasionally. The development of a hunting tradition has been partially followed by observers of the Gilgil baboons in Kenya we introduced earlier. Gilgil is a mainly agrarian region where the African predators have been exterminated. Because of this, the baboons could safely occupy the vacated hunting niche. Gradually they developed an opportunist and unsystematic hunting style to capture antelope calves, rodents and the like, which a baboon might chance to discover in the long grass (Harding, 1973). Some years later, Strum (1975) described how this behaviour had developed into a more systematic routine; small groups of animals trekked across the savannah searching together. However, in baboons this behaviour has never taken on the proportions that it has among some chimpanzee populations, and it lacks the clearly coordinated methodicalness.

These observations show that a germ of the skills necessary for a lifestyle based on cooperative hunting is present in our closest relatives. In species like the chimpanzee these skills have already been developed to a high degree. This gives rise to an interesting thought. Chimpanzees may have developed skills and capacities which enable them to take account of each other's movements and positions, as a consequence of the social adaptation in which the males trek around together in closed combines under the pressure of intergroup competition. These capabilities can have formed the predispositions which could have been the basis for the development of cooperative hunting techniques. The question is, therefore, whether the development of cooperative hunting techniques has also promoted cooperation in intergroup conflicts, the tribal war, or whether, on the contrary, cooperation in tribal war has led to cooperation in hunting!

Indications about the order of development in chimpanzees may allow choices in speculations about the early evolution of man. The transition to a hunter-gatherer culture, in which the women acted mainly as gatherers, while the men took on cooperative hunting, is often supposed to have furnished the most important pressures for the development of specifically human characteristics (see for instance Washburn and Lancaster, 1968; Pfeiffer, 1972). This is supposed to have brought about traits which made possible social coordination, sex-linked task differentiation, sharing of food, marital care and male parental care, technical skills such as the manufacture and use of weapons and tools, and finally the evolution of a verbal communication system.

This scenario is partly plausible. It has often been claimed (among others by Peters and Mech, 1975), that in order to appreciate the conditions which determined the evolution of mankind, we could learn a great deal by looking at analogous developments in other social-cooperative hunters, namely the wolf (Mech, 1966, 1970; Zimen, 1978) and the cape hunting dog (Kühme, 1965; van Lawick and van Lawick-Goodall, 1971), two species which evolved independently. These animals do work together not only while hunting, often for game which is far larger than the hunters themselves, but also share in a far-reaching responsibility for the rearing of the young. The hunters do this, for instance, by bringing back food to the nurses and the pups. Cape hunting dogs, which have developed this communal aspect even further than wolves, will even take care of the sick and injured animals for long periods of time. Contrasting with this loyalty towards pack members is a bloody and merciless enmity toward non-pack members (Murie, 1944; Zimen, 1978).

2.10 PREDATORY CHARACTER AND MURDEROUSNESS

A sense of cooperation on a hunt, and, possibly stimulated by this, collaboration in bringing up offspring and in helping group members, togetherness required for this and the dedication to one's own group and its leader(s), these are all characteristics which we find above all among the cooperative carnivores. We find inclinations in this direction among our nearest 'apish' relatives. The suggestion is that by our transition to a more 'wolfish' lifestyle, these characteristics have been strongly enhanced.

A question that many have raised is whether this transition has had consequences for our willingness to participate in between-group conflicts and for the way we behave in these circumstances. This question can again be dealt with in two ways. On the one hand in the ultimate sense: has hunting as a lifestyle of evolving man created circumstances in which fighting other groups might have yielded selective advantage, and might it have furnished the motivational structures and skills needed? And, in the proximate sense: which motivational mechanisms and skills made it possible for intergroup conflicts to arise?

This way of arguing is based on the assumption that the development of hunting behaviour made the development of warlike behaviour possible rather than vice versa. The notion that bloody behaviour in war is a fruit of a carnivorous mentality which has developed among humans, has undoubtedly been brought about by the misleading idea of some authors, that the behaviour involved in the capture of prey is a form of aggressive behaviour; in both cases common motivational mechanisms are supposed to be at work. The mechanisms of satisfaction which reward the capturing and killing of a prey are supposed also to provide the rewarding effect to the warrior when a conspecific is slain. This view has been put forward by a number of writers, either carefully (for example, Washburn and Lancaster, 1968; Washburn and Hamburg, 1968; Hamburg, 1973) or more peremptorily (Ardrey, 1961; Pfeiffer, 1972; Koestler, 1978). In the last case man is pictured as a murderous 'killer-ape', in which a predatory instinct, evolutionarily gone rampant, has led to a bloodthirsty character, which is given expression in tribal warfare and cannibalism.

This view has met with hefty criticism, certainly inasmuch as it implies that there is a kind of murder instinct in man. Thus it appears that real predators are definitely no more aggressive towards conspecifics than are other species, nor do aggressive motivations play a part in the predatory behaviour when capturing prey and vice versa (Lorenz, 1963; Eibl-Eibesfeldt, 1967; Scott, 1974). A tiger is no more angry at a deer than a family father at the Christmas turkey on December 24th. In the same vein, there are no indications that human hunter societies are more aggressive towards each other than farmers or cattle-herders; on the contrary!

The recent observations on serious between-group conflicts among chimpanzees suggest that the development has gone the other way around. Ecological circumstances permitted male residency and the formation of coalitions between males. Together males could rule a territory and thus control access to the females who had settled there. Cooperation in tribal warfare subsequently set the scene for cooperative hunting.

Of course a subsequent interaction between both processes might have occurred which could have facilitated the further development of each of the behaviours in its own context. Thus, certain actions which are useful as an instrument in hunting might also prove useful in battle against members of the same species, and vice versa. Skills like enclosure tactics and the use of weapons can be applied in both contexts. A refinement acquired in one functional context can improve the way the same function is carried out in the other context. If such an improvement increases the efficiency of this behaviour pattern, then it might also shift the balance of costs and returns of this action. A group which has developed a method of attack which involves less personal risk, will come more easily to a decision to choose attack as a means of bringing a conflict to a favourable solution than another means (cf. Maynard Smith and Price, 1973; McEachron and Baer, 1982).

As we saw, intergroup clashes among chimpanzees can become very gory. There are several examples in the literature which reveal the enormous difference between the level of tolerance towards one's own group members on the one hand, and non-group members on the other hand. Conflicts between members of the same group occur regularly in many species. Yet, as Lorenz (1963) pointed out, relatively few of these lead to fatal aggression. However, this should not lead one to an idyllic concept of conflict resolution in animals. Conflicts between members of different groups can certainly be of a bloody and lethal nature (for example, lions, Bertram, 1976; hyenas, Kruuk, 1972). Xenophobia, a negative attitude towards strangers, seems to be widespread in the animal kingdom. This can easily flare up into a fierce, uncontrolled animosity with which, inexorably, all the group members join in (for example, Bernstein, Gordon and Rose, 1974; Southwick *et al.*, 1974; Wilson, 1975a). Apparently the group stranger cannot appeal as easily to the aggression-inhibiting mechanisms which protect the familiar fellow.

A surprising example of this has been observed in the way many primates, in particular the males, can treat infants of their species. Just as a male can be protective and tolerant towards children born of females with whom he has an attachment, he can be equally antagonistic in his relationship with strange infants. Among many primates, for instance the langurs, males who have fought their way into another group and have taken over the leadership, will kill all the nurslings at their mothers' breasts.

Initially this kind of behaviour was regarded as a pathological deviation due to stress of overpopulation (Dolhinow, 1977; Boggess, 1979, 1984). However, this kind of behaviour has since been observed among several species who live in normal conditions, amongst others the chimpanzees (Bygott, 1972; Goodall, 1977) and gorillas. (Fossey, 1984). There is growing support for the viewpoint that males display this kind of behaviour when meeting new females. Thus they may not only remove future unrelated group members which will compete with their own offspring, they may also shorten the period after which lactating females will be fertile again, and thus increase the chance of producing their own young. The selective advantage that an infanticidal man has in this case, can increase an infanticidal predisposition in a population (Hausfater and Hrdy, 1984; Struhsaker and Leland, 1987).

2.11 PRIMITIVE WAR; CAUSES, MOTIVES AND ADAPTIVE CONSEQUENCES

If one examines the phenomena of group aggression and war from a comparative perspective, then primitive war deserves special attention, that is war fought in human communities who still live under socio-ecological conditions not too different from those to which our species must have been subjected during its million year long evolutionary process. These conditions undoubtedly most resemble those of the present-day primitive hunter-gatherers.

In our species there is a great variety in the degrees of belligerence. Extremely bellicose, often cannibalistic tribes in the jungles of South America (Murphy, 1960; Chagnon, 1968) and the Indo-Australian archipelago (Gardner and Heider, 1968), offer a strong contrast with relatively peaceful tribes such as the Bushmen (Lee, 1968; Lee and DeVore, 1968; Lesser, 1967), and the Eskimo (Balikci, 1968, 1970). The word 'relatively' is used here because the view supported by Lee and Lesser that the Bushmen have no such concept as war is convincingly laid to rest by Eibl-Eibesfeldt (1974). Ever since Rousseau the idea has been alive that war and intergroup violence are pathological deviations which have been brought about by civilization, i.e. the development of organized states, and that war does not occur among peoples who live in harmony with nature. Eibl-Eibesfeldt has shown that this is a myth. This is confirmed by the extensive inventarization carried out by Divale (1973).

In any case, the view that war is a pathological deviation does not help us any further. It tells more about the negative appraisal which the phenomenon evokes than that it explains anything (Vayda, 1968).

An explanation for the phenomenon of war has to give an insight into the factors which determine the emergence of war. Here again, one can distinguish different levels of explanation. On the one hand one can ask which selective forces lead to the propagation of this phenomenon. What are the positive and negative consequences for those who take part in wars, and under what ecological and social circumstances do these issue their effects. On the other hand, one can ask what the mechanisms are that lead populations, and their members, into war. What are the causes, and what are the motives which are used to justify a war (these do not have to be the same!)? And what are the motivational mechanisms which bring those involved so far as to take part, that is, both the decision-makers and the actual combatants. We can recognize the ultimate and the proximate approach again here.

Van der Dennen (1984 a, b) has presented an extensive survey of the diverse explanations which are used in connection with war. I shall summarize a number of important points. In most explanations a sharp distinction between proximate and ultimate considerations is not made. It is often unclear that these levels of consideration must not be confused. Thus van der Dennen has ascertained, correctly, that the fashionable neo-functionalist school will never be able to explain the function of war out of the motives that those involved report. Function here refers to the consequences on psychological, ecological, economic and demographic levels; these are not necessarily identical to the effects participants are after.

Table 2.1 sums up a number of motives for war, and the main functions the war in question is supposed to serve (from van der Dennen, 1984 a/b, and

Table 2.1

Classification of war-motives (after van der Dennen, 1984a/b)

(A) PSYCHO-SOCIAL MOTIVES (function: the maintenance and strengthening of internal cohesion and integration of the group)

(a) Family feuds, hereditary enmity

–(a.1) as revenge for murder, manslaughter, etc.
–(a.2) as retribution for infringement of sexual codes: adultery, rape, stealing women
–(a.3) as punishment for trespasses
–(a.4) as a penalty for black magic, witchcraft, etc.

(b) Machismo, prestige, status, honour, fame.

–(b.1) raids to obtain status symbols and insignia of power: head-hunting and scalping
–(b.2) war as a part of rites of passage
–(b.3) war as sport and exciting adventure: an escape from boredom
–(b.4) war as a safety-valve for tensions originating inside the group

(c) Mystico-religious motives

–(c.1) for the satisfaction of bloodthirsty gods
–(c.2) ethnocentric-xenophobic aggression
–(c.3) tribalistic delusions of superiority
–(c.4) raids to obtain victims for ritual human sacrifices
–(c.5) as revenge for sacrilege, the rights of guests, etc

(B) ECONOMIC MOTIVES

(C) POLITICAL MOTIVES

based on the main categories distinguished by Wright, 1942). The extensive inventory research done by Wright reveals that in primitive peoples psycho-social motives play a major role. This may be taken to imply that the functional consequences of war are to be found in the integration and cohesion of one's own group.

However, a 'primitive' warrior may fight in order to restore injured honour, to make an offering to the gods, to avenge a family feud etc., without being aware that his actions will bring about changes at ecological, demographic and population-genetic levels. These changes may well influence his survival and his, and his family's, fecundity. These could be the adaptive consequences which gave a selective advantage to those, who kept to suitable cultural rules under circumstances in which these effects could arise, a selective advantage. At the same time the cultural system concerned would be successful. Even a primitive man, who cannot see through the maze of causal connections involved here, may nevertheless feel that things will be better for him and his kin, if he keeps himself to certain rules; therefore, he had better obey his gods.

The discrepancy between motive and adaptive effect will be greatest among the hunter-gatherers. As the socio-cultural revolution progresses, and man gains a better comprehension of the circumstances on which his existence depends, for example, farming, cattle-raising, storage of produce and equipment, the economic and political-strategic motives will start to coincide more and more with the adaptive consequences of war. But even in the most advanced cultures immaterial, psycho-social and ideological motives emerge, whether or not to disguise material motives.

Let's return to primitive cultures. Here one finds the greatest diversity of belligerence. And it is also here that immediate material profit from a war is least visible. That is why some find it useless to regard the development of belligerence as an adjustment to ecological circumstances, let alone as an aspect of phylogenetic adaptation. Den Hartog (1982), an avowed representative of this viewpoint, believes that such an approach is fruitless, and can only lead to improbable post-hoc explanations. He regards the great variety in belligerence as the result of developments with their own dynamics and causality. From the ecological perspective this diversity should be seen as purely accidental.

This view was promoted by a study by Murphy (1957, 1960) on the extremely belligerent South American Mundurucu head-hunters. Their bloody escapades seem to have no more tangible gain than a handful of warriors returning home honourably with their trophies, the heads of victims from nearby villages. No land is conquered, no loot is plundered. Murphy concluded, therefore, that war only served here as a safety valve for internal strains and to maintain the solidarity of the group. Vayda (1968, 1974) has pointed out that war can bring about several beneficial effects to the belligerent party, given, of course, that this party wins. He wondered whether the supposed function really reflects the effective advantage, for there are other means of ensuring social cohesion, which involve less cost and risk. Vayda supposed that the real favourable effect might have been a decrease of population pressure. If the hunters have to work harder to capture prey, and if this leads to other decreases in the standard of living, which in turn cause increased competition, then these factors may activate the proximate mechanisms leading to a head-hunting expedition. Those involved need not be directly aware of these factors; the proximate mechanisms may involve mystic motives.

Such considerations led to Durham reanalysing Murphy's material, and also comparing it to material from other peoples (Durham, 1976). The most important source of protein for the Munducuru is the peccary, a small kind of wild boar. This animal is the limiting factor of their population size. An increase in population pressure is reflected in an increased hunting pressure. It forces the hunters to roam much farther away from their villages. A head-hunting expedition can serve to decrease the amount of competition, both directly and because those defeated may leave the area. So the peccaries can increase in number again. According to Durham the warriors bringing home head-trophies get a greater share of the protein supply. The population are convinced that the trophies have a magical, favourable influence on the capture of peccaries. A greater share gives the warriors in turn a better chance of reproducing. Durham is convinced, therefore, that warlike behaviour in primitive tribes can definitely be explained as an adaptive response to ecological influences increasing the fitness of those involved.

In the same vein Harris (1984) argues against Chagnon's (1974) interpretation of warfare in the bellicose South-American Yanomamö. Whereas the latter maintains that Yanomamö warfare cannot be explained in terms of scarcity of resources, population density etc., Harris argues convincingly that the pattern of warfare is in accordance with ecological theoretical predictions.

It is exciting to trace how far the occurrence of war among primitive peoples can be understood as the result of a selection process influenced, on the one hand, by the nature and distribution of primary resources, their exhaustibility and the speed with which they regenerate, and on the other hand, the costs involved in keeping others away from those means of subsistence, and, as a consequence of this, the choice of means to safeguard the means of subsistence (war or trade systems: cf. Ferguson, 1984b). Sea-going Eskimo fishermen may not be belligerent because there is no way in which they can monopolize the schools of fish on which they depend. Neither the arrival nor the exhaustion of these fish supplies can be influenced in any way by the traditional fishing technology, completely different, of course, from the fleets of high-speed, high-tech trawlers which can trace every shoal in Icelandic waters; the term cod-war still has an awfully familiar ring to it!

Another interesting question is whether an analysis of the ecological dynamics, and the influence these have on hunting behaviour, can explain why the Pygmies in the Central African rain forests find it rewarding not to be belligerent.

2.12 THE MOTIVATIONS OF WAR-MAKERS AND FIGHTERS

An interesting aspect in the description of Munducuru raids is that the raiders go and hunt heads to use their magical powers for their own good. This is done by means of a surprise attack on a nearby unsuspecting village in the early hours of the morning. The attack turns into a mass-slaughter, during which as many heads as possible are collected, of men, women and children, without mercy or chivalry, as if they were hunting a completely different species. This calls forth associations with the cannibalism found among chimpanzees, and which is possibly facilitated by their habit of hunting young baboons.

This cool, purposive bloodshedding differs most strongly from the way meetings between enemy tribes take place, for example, in New Guinea. There the warriors meet their opponents in full martial adornment, and, in a whirl of both angry and fearful excitement threaten their counterpart in exaggerated and ritualized displays, occasionally coming to blows. The motivational processes involved may well be homologous to those of coalitional aggression found in many species of non-human primates. One would like to know if the restraint of violent behaviour, which appears to be present here, rests on the fact that the tribe members see their opponents as members of the same species (cf. Eibl-Eibesfeldt, 1984). Is the difference in behaviour in the two situations described above caused by differing perceptions of the opponent? Durham mentions several epithets given by the head-hunters to their opponents which seem to suggest 'dehumanization'.

Obviously, the causal background of a war between two modern, organized states is far more complex than that of a war between two neighbouring tribes who are still living in the hunter-gatherer stage. In the light of what we have discussed before, one could even defend the view that the distance between modern and primitive war (cf. Turney-High, 1971) is far greater than the distance between the latter and a raid undertaken by a group of male chimpanzees against a neighbouring group. For the purpose of warfare modern states maintain highly organized systems with logistically and tactically advanced programmes. The analysis of costs and returns, the development of strategies and the decision-making process can take place here with cool evaluation. Affective-aggressive motives appear not to play an important part (Stouffer *et al.*, 1949, cited in Sherif and Sherif, 1970; van der Dennen, 1986).

But nevertheless, we may recall that in the recent Falklands War a whole nation, noted for its rationality and stolidity, appeared to react collectively out of indignation and hurt pride. It gathered around its leader, let itself get carried away by communal emotions, and experienced the thrill of belonging together, to which sports club fans also aspire (Russell, 1981), and a war machine came into being which functioned very nicely. In short, the motives which work together to create a process of war can comprise both rational and irrational elements. These last, undoubtedly, share the same kind of foundations as those which we found among other higher species of mammals (Eibl-Eibesfeldt, 1984).

Those who take the decisions guide the motivations of those over whom they decide, but are in turn influenced by the motivations which rule the latter. The degree of motivational homogeneity between and within decision-makers and followers will undoubtedly vary (Pear, 1950). Different motivational dimensions guide those who are involved at different phases and levels. Kellett (1982) found such a diversity of dimensions in a study on the motivations of soldiers. Factors such as idealism, patriotism, moral indignation, hate and the like can take possession of the soldier going into battle. Kellett even suggests that these may be more important than we are given to think nowadays. For this, however, the soldier has to feel supported by the rearguard. The collective emotionality can strengthen the motivations of those involved. Kellett points out the vast difference between the moment the US joined in the Second World War, and the time they got involved in the Vietnam War. The army in action in Vietnam was far less effective (Balkind, 1978).

Once a soldier takes part in a war he may also be governed by different motives. He has chosen or is forced to do a job. Motives that have to do with his own personal interests, social, material, and career rewards, and costs in terms of investments and safety risks often prevail, and determine choices and attitudes. Beside this, his behaviour is governed to high degree by factors like discipline, example of leadership and above all by factors which have to do with the maintenance of his own social unit, that is the fighting group. Aggressive motivation in the restricted sense, i.e. as a desire to damage one's opponent, usually is of little or no importance. This can be instantly reversed when the social unit is directly hit by an identifiable enemy (for example, Shirom, 1976). Now, on the one hand, the soldier may be willing to risk his life for his unit. On the other hand, a generalized hate, supported by xenophobia, can give rise to the kind of bloody reprisals which we know from just about every war. In general, however, the destruction of lives and goods is of a mainly instrumental nature. This is strongly augmented in modern warfare by the increasing distance and instrumentalization: the enemy target is destroyed when the green spot on the monitor has been made to reach coordinate P.

CHAPTER THREE

Selfish cooperation in social roles: the vigilance game in continuous time

U. Motro and D. Cohen

3.1 INTRODUCTION

Social interactions in nature often involve conflict to a certain extent. Game theory can sometimes provide us with insight and tools to improve our understanding of how natural selection has resolved such conflicts. Undoubtedly, the major conceptual tool in this field is the concept of **evolutionarily stable strategies** (ESS), which was introduced by Maynard Smith and Price (1973). A strategy, in an evolutionary context, is one of a set of possible alternative behavioural programmes that an individual in a population can adopt. It is usually assumed that these are genetically determined. An ESS is a strategy which, when adopted by a large enough fraction of the population, cannot be invaded by any alternative rare ('mutant') strategy. A rare ESS strategy can invade a non-ESS population.

A certain class of evolutionary games consists of games having two pure strategies: 'defection', which yields only a personal benefit, and 'cooperation', which yields a common benefit to all the individuals in the group. Clearly, if the personal benefit from defection is greater than the personal benefit from cooperation, the evolutionarily stable strategy is defection. What happens, however, if the benefit from cooperation is the greater? In such cases it is more advantageous to cooperate, but even more advantageous to defect if other individuals in the group will nonetheless cooperate, thus enjoying both the personal benefit from one's own defection and the common benefit from cooperation performed by other individuals. Since this argument applies to all individuals in the group, it may seem that natural selection will always favour the selfish strategy of defection.

We shall briefly refer to three examples which will illustrate the ESS of these evolutionary games, namely the three brothers' problem, the conflict involved in dispersal, and the vigilance game.

3.2 THE THREE BROTHERS' PROBLEM

The theory of kin selection was introduced by Hamilton (1964) and has been developed further in many other papers; it is described in terms of help directed from one relative (the donor) towards another (the recipient). Let us consider the more complex situation in which an individual needs help, and this help can be provided (with some risk to the donor) by each of several relatives. In such a situation, even if Hamilton's condition for altruism between two relatives is satisfied, it is easy to see the advantage of standing by and waiting for another relative to take the risk and provide the necessary help. It is true, though, that if all are passive, Hamilton's original argument again holds and any potential helper can increase its own inclusive fitness by exclusively taking the risk and saving the relative in need. However, this entails an even greater increase in the inclusive fitness of the relatives which decided not to offer their help. It seems, therefore, that if there is any altruistic relative in the vicinity natural selection will always favour the other selfish relatives.

The analysis of situations involving more than one potential helper reveals that if Hamilton's condition for one-to-one altruism does not hold, the pure strategy of absolute selfishness is the only ESS, independent of the number of potential helpers. If, on the other hand, Hamilton's condition is satisfied, the ESS is a mixed strategy of altruism and selfishness represented by an evolutionarily stable probability of providing the needed help. This probability decreases to zero as group size increases. This is true both for cases where immediate help is needed (Eshel and Motro, 1988) and for cases of delayed help (Motro and Eshel, 1988). The former are situations in which each potential helper has to instantaneously decide whether or not to offer its help, without knowing what the other potential helpers are doing. In cases of delayed help, no immediate help is mandatory (yet any delay increases the risk to the individual in need) and, at any moment, each potential helper has full information on what the other potential helpers have done and on the situation of their distressed relative.

3.3 THE CONFLICT INVOLVED IN DISPERSAL

Upon dispersal of the parental site, a dispersing offspring leaves more room for its siblings (with which it shares, to a certain degree, the same genes), thus increasing their survival chances. Dispersal, on the other hand, is more risky for the dispersing individual than staying at home. Hence, whenever the decision whether to disperse or to stay in the parental site is made by the offspring itself, there is a conflict: the strategy of staying at home confers a personal benefit, whereas the strategy of dispersing yields a benefit to the other siblings (and, via the kinship component of the inclusive fitness, a smaller benefit also to the dispersing individual).

It turns out that if dispersal is too risky (i.e., if the survival chances of the dispersers are below a certain level), the only ESS is the pure strategy of staying at home. If, on the other hand, the survival chances of the dispersers are above that level, a mixed strategy, represented by a probability of leaving the parental site, is the only ESS. This probability increases as the risk involved in dispersal decreases (Hamilton and May, 1977; Motro, 1983).

3.4 THE VIGILANCE GAME

Vigilance for predators while feeding, also known as scanning behaviour, represents a similar evolutionary problem because a scanning individual gives up feeding. In choosing not to scan, an individual gains a personal benefit through increased feeding, whereas scanning yields a benefit to all individuals in the group. In other words, for an individual it is clearly more beneficial that others will do the scanning, and that it will spend all its time feeding. Again, since such an argument applies to all other group members as well, it seems that natural selection will always favour the pure strategy of not scanning at all.

The evolutionary aspects of vigilance behaviour are studied in several game theory models (Pulliam, Pyke and Caraco, 1982; Parker and Hammerstein, 1985; Motro and Cohen, 1989; refer to Hart and Lendrem, 1984 and Lima, 1987 for different approaches). Under the plausible assumption of diminishing returns with regard to feeding effort, a single evolutionarily stable vigilance strategy, represented by the probability of being vigilant at each time unit, is found to exist in each of the game theory models. This positive probability is a consequence of absolute selfish, short-term considerations, without any further assumptions concerning kinship, reciprocity, the gaining of prestige and so forth. Since it is assumed that even a single vigilant is sufficient to avoid predation, it is quite reasonable to find that in all the models the evolutionarily stable vigilance probabilities decrease as group size increases.

A common feature of these models is that at any time unit, the actual performance of each individual is independent of that of the other group members. Thus, at any moment, the number of simultaneous vigilants is a random variable which can be 0, 1, 2 or more. This model is an appropriate description in many natural cases (e.g., flocks of waders or winter groups of other small birds).

There are many cases, however, where each individual usually has full information on the current vigilance situation in the group (e.g., certain barbets and babblers (Wickler, 1985; Zahavi, personal communication), the dwarf mongoose (Rasa, 1986)). If a single sentinel is sufficient to avoid predators' success, we expect to find, in such cases, no more than one vigilant individual at a time.

In the next section we present and analyse such a model, that is, a continuous time, full-information vigilance game.

3.5 THE VIGILANCE GAME WITH FULL INFORMATION

Consider a group of n ($n \geqslant 2$) individuals, which expect an attack by a predator, but have no knowledge about its timing. If the attack has not yet occurred until time t, there is some probability, $\theta\,dt$, that the predator will appear during the time interval $(t, t+dt)$. Hence the time passed until the predator's appearance is exponentially distributed, with expectation $1/\theta$.

The predator is successful only if there is not a vigilant group member at the time of the attack. In that case, each group member has an equal probability of being captured. In any case, the game terminates after the appearance of the predator.

The vigilance strategy of an individual is represented by v, such that for any given time t, the probability of this individual starting a vigilance shift during the time interval $(t, t+dt)$ is $v\,dt$ (provided that the predator has not yet apeared until time t and no group member was vigilant at that time).

A vigilance shift has a fixed duration, and let r be the probability that the predator will appear during the shift. The vigilant individual bears a cost $c(c > 0)$, which reflects the deprivation from feeding while being on guard, and the extra risk of predation of the vigilant, especially while rejoining its moving, foraging group (Rasa, 1986).

In order to find the ESS, we assume that the prevailing strategy in the population is v, and consider the fitness $W(x, v)$ of a mutant having the strategy $x \neq v$. The fitness of the mutant is the weighted sum of its fitness in the three possible events:

1. the predator attacks before any of the group members go on guard;
2. one of the other $n-1$ group members goes on guard before the predator comes;
3. the mutant goes on guard before the predator comes.

As given in the Appendix, the fitness of the mutant is given by

$$W(x, v) = \frac{(1-1/n)\theta + (n-1)rv + (r-c)x}{\theta + (n-1)rv + rx} \tag{3.1}$$

The definition of the ESS, $v^\star$, requires that the fitness of a mutant $W(x, v^\star)$

is a maximum with respect to the mutant strategy x at $x = v^\star$. We find the maximum by solving the equation

$$\frac{\delta}{\delta x} W(x, v^\star)\Big|_{x=v^\star} = 0$$

We get that the ESS $v^\star$ is

$$v^\star = \frac{\theta(r/n - c)}{(n-1)rc} \qquad (3.2)$$

if $n < r/c$

$$v^\star = 0 \qquad \text{if } n \geqslant r/c$$

Not surprisingly, the ESS $v^\star$ is a decreasing function of the group size, n. Moreover, $nv^\star$ is a decreasing function of n (i.e., $1/nv^\star$), which is the expected time without vigilance between shifts, is an increasing function of n). Thus, a larger group increases the probability of a successful attack by the predator. These results are concordant with those of the non-information models.

Finally, the evolutionarily stable (ES) fitness $W(v^\star, v^\star)$ of any individual in an all $v^\star$ group is

$$W(v^\star, v^\star) = \begin{cases} 1 - c/r & \text{if } \quad n < r/c \\ 1 - 1/n & \text{if } \quad n \geqslant r/c \end{cases}$$

Consider now the more typical case in which it is conventional that an individual never performs two consecutive vigilance shifts. That is, after completing a shift, the individual does not take part in the next one. The analysis of this model (carried out in the Appendix) reveals that a single ESS ($u^\star$) exists also for this case, and that if $n < r/c$, $u^\star$ is larger than $v^\star$, and also the ES fitness $\tilde{W}(u^\star, u^\star)$ is larger than the ES fitness $W(v^\star, v^\star)$. (If $n \geqslant r/c$, both $u^\star = v^\star = 0$ and $\tilde{W}(u^\star, u^\star) = W(v^\star, v^\star) = 1 - 1/n$.)

Indeed, it is not very surprising to find that if the one to be last on guard never participates in the next vigilance shift, the evolutionarily stable vigilance strategy implies a greater tendency of the other individuals to go up on guard. Less self-evident, however, is the result that the conventional refraining of any last sentinel from participating in the next vigilance shift increases (at the ESS) the fitness of each group member. The predictions of the two models can be tested by field observations of vigilance behaviour in the two types of social organization.

Another interesting result of modelling vigilance behaviour is that the ESS in a group with a nonguarding individual is changed if the other group members are aware of the presence of such an individual (Motro and Cohen, 1989). In such a case, the evolutionary stable level of guarding increases. By adopting the new ESS, the guarding individuals suffer a smaller reduction in fitness (compared to the reduction in the case where the nonguarding individual cannot be detected). Over some ranges of the parameters, this increased level of guarding provides a sufficiently high benefit to the detectable nonguarding individual so that its fitness is larger than that of the guarding individuals in the population. Hence, for this range, both declaring the intention to abstain from guard duty and the detection of such an intention will be selected for. Since the advantage of the detectable defector is frequency dependent, the population will stabilize on a polymorphism in which both the guarding and the detectable nonguarding types coexist together.

APPENDIX
DERIVATION OF THE MUTANT'S FITNESS IN THE VIGILANCE GAME

We assume that the prevailing strategy in the population is v, and calculate the fitness $W(x, v)$ of a mutant having the strategy $x \neq v$.

If the predator attacks before any of the group members went on guard (the probability of this event is $\theta/[\theta+(n-1)v+x]$), the conditional fitness of our mutant is $1-1/n$.

If any of the $n-1$ other group members is the first to go on guard before the predator has appeared (the probability of this event is $(n-1)v/[\theta+(n-1)v+x]$), the mutant's fitness is $r+(1-r)W(x, v)$. (With probability r the predator appears during this shift, and the game is over, and with probability $1-r$ we are right back where we started).

Finally, if the mutant is the first to go on guard (the probability is $x/[\theta+(n-1)v+x]$), the fitness is $r+(1-r)W(x, v)-c$.

Hence the mutant's unconditional fitness $W(x, v)$ satisfies

$$W(x, v) = (1-1/n)\frac{\theta}{\theta+(n-1)v+x}$$

$$+(r+(1-r)W(x, v))\frac{(n-1)v}{\theta+(n-1)v+x}$$

$$+(r+(1-r)W(x, v)-c)\frac{x}{\theta+(n-1)v+x}$$

Rearranging, we get

$$W(x, v) = \frac{(1 - 1/n)\theta + (n-1)rv + (r-c)x}{\theta + (n-1)rv + rx}$$

Now consider the model in which no one individual performs two or more consecutive vigilance shifts. Using a similar (but somewhat more complicated) argument as before, we see that if the prevailing strategy in the population is u, the fitness $\tilde{W}(x, u)$ of a mutant having the strategy $x \neq u$ is

$$\tilde{W}(x, u) = \phi + \frac{ru}{\theta + (n-1)\,u + x}(1 - \phi)$$

where

$$\phi = \frac{Ax + B}{Cx + D}$$

and

$$A = r - c + (1 - r)\frac{(1 - 1/n)\theta + (n-1)ru}{\theta + (n-1)u}$$

$$B = (1 - 1/n)\theta + (n-2)ru$$

$$C = 1 - (1 - r)\frac{(n-1)(1-r)u}{\theta + (n-1)u} \qquad D = \theta + (n-2)ru.$$

The equation $\delta \tilde{W}(x, u)/\delta x|_{x=u} = 0$ has a positive solution ($u^\star$) if and only if $n < r/c$. In that case, this solution is unique, and is the ESS. Moreover, $u^\star > v^\star$ and also $\tilde{W}(u^\star, u^\star) > W(v^\star, v^\star)$. If $n \geqslant r/c$, the ESS is $u^\star = 0$ (i.e., the pure strategy of no vigilance).

CHAPTER FOUR

The biological instability of social equilibria

P.P. van der Molen

INTRODUCTION

4.1 OUTLINE

This chapter deals with a behavioural mechanism which thwarts any systematic attempts to prevent and put a permanent end to conflicts between social groups and organizations. Essential in this mechanism is a certain kind of social-role blindness, a peculiar unawareness of what we are doing on the level of social-role interactions, whereby attraction or repulsion are effectuated. As in Tiger's contribution (Chapter 5), special provisions in our behavioural system are discussed which prevent us from utilizing our intellectual and cognitive faculties for investigation of our innermost social tendencies. We shall return to these 'no entrance' signs built into our cognitive system below.

Other elements of this mechanism are involuntary incrowd–outcast selection reflexes and a 'trait dimension', which may be described as a 'readiness to comply with a submissive role'. This dimension is correlated with a great amount of social behaviour and a small amount of thing-oriented, individualistic and explorative behaviour. It is, by definition, of great importance for the distribution of social roles and for the social structure in a group; it determines, for example, the likelihood of drifting into an outcast position versus the likelihood of assuming or maintaining a compliant and socially accepted subordinate position. Knowledge of this personality trait dimension and of its effects in social groups and structures may increase our understanding of a wide range of intriguing and sometimes disquieting phenomena. These phenomena range from educational and organizational strategies to the often catastrophe-like collapses and turn-over phenomena in companies and other social structures, and from the way social roles and positions tend to be distributed to the resulting evolutionary consequences.

First I will explain why, from a purely biological point of view, differences between individuals are to be expected in any socially living mammalian species in the following situations: readiness to comply with a submissive role; sociability versus thing-orientedness; compliance versus self-will. It will be argued that the underlying biological organization must, from an evolutionary standpoint, be very old and elementary. We will investigate then the consequences of these behavioural differences on the level of social interaction. A life span theory of social structures and organizations will be introduced as one of the implications.

The first sections of this chapter comprise a concise outline of these mechanisms, omitting at this point experimental data and illustrations. The basic assumptions made will, however, be stated explicitly. In the following sections we will check these assumptions against experimental and empirical data from biological and psychological research. Finally, it will be pointed out why understanding of the way these interpersonal differences are behaviourally organized (and the way our awareness tends to be blocked in these respects) have such far reaching consequences; an increase in our understanding of the life cycles of social structures might be by far their most important result. Such understanding would enable us to map the processes underlying periodic catastrophe-like turn-over phenomena and to learn how to control their decreasing efficiency and violent backlashing on any level of organization.

4.2 SOME CONSEQUENCES OF LIVING SOCIALLY

Among socially living mammals each individual is by necessity saddled with a conspicuous bi-polarity in behavioural urges. First, being a socially living animal, drives for social contact and interaction are an important part of its behavioural–genetic endowment; but secondly, it has a set of perhaps even more basic drives to ensure the fulfilment of a range of non-social personal needs, e.g., water, food, cover, warmth, sex, territory, etc. As far as these latter needs are concerned, the amount of resources is often limited, thus causing competition and social conflict. For that reason a very basic functional conflict does exist in every social individual: between the urges to fulfil a great variety of personal basic (physical) non-social needs and the urge to maintain social contact and interaction. A socially living mammal inescapably has to shift between these two sets of urges much of the time.

Whenever some of the needed resources are scarce, the ensuing competition will put a strain on social relations. Under such conditions an individual frequently has to choose between continuation of peaceful social relations and receiving an appropriate share of the resources, eventually at the cost of social peace and harmony. Most of the time this dilemma boils down to the question of whether or not to submit to the initiative of other individuals at the cost of fulfilling personal urges and desires. In any socially living species this conflict of needs is inescapably present in each individual day after day, the average outcome determining how the individual will deal by and large with the social situation at hand. It is most desirable to have one's own way most of the time and still maintain close social contact and interaction, but that is more or less identical to what is generally understood by a dominant social role, and such roles are rather scarce.

It is, therefore, of theoretical interest to know what happens to the majority of individuals, the various types of subordinates (see Fig. 4.1), who are under regular pressure to comply and postpone or even abandon part of their individual desires and initiatives. For such non-dominant individuals, the balance between the strength of the desire for social contact and interaction, and the strength of the desires to fulfil other biological needs, determine the outcome of this continuous process of balancing one need against the other. Given a certain pressure to comply, it largely depends on this equilibrium of basic sensitivities within the subordinate individual as to what the behavioural outcome will be, either drifting gradually into an outcast position or assuming a compliant and socially accepted subordinate position. Such differences between subordinates have indeed frequently been observed in mammals (refer to section 4.7).

What is important for us to note here is that for any socially living mammalian species the competing sets of needs under discussion are very general and basic. We must therefore assume that the variance in the balance between those sets of basic needs has strong genetic roots. (After all, for many species, the behavioural organization is so simple that learning processes can only play a minor role in establishing behavioural variation. The equilibrium discussed above is therefore also an equilibrium between functionally

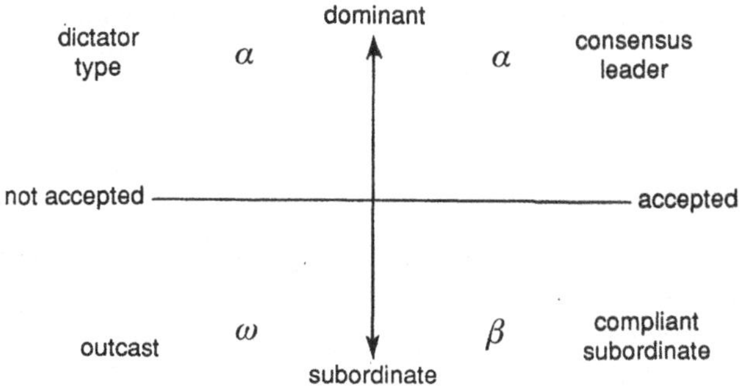

Fig. 4.1 Two dimensions of social-role behaviour.

competing parts of the genetic programme. As such we may consider this equilibrium, varying over individuals, as a trait in the classical sense. We could express this set of behavioural polarities as a set of (*inter alia* genetically based) trait differences which do have a clearly defined impact on the distribution of social roles.

Up to this point, three basic assumptions have been made about the behaviour of socially living mammals in general:

1. There is a strong functional link, on the level of behavioural orientation, between the frequency of social behaviour versus thing-oriented individualistic behaviour, and, on the level of distribution of social roles, between conformity and compliance with authority versus a self-willed attitude. These two polarities cannot be separated; they have the same behavioural basis. Therefore a range of personality characteristics have to be strongly intercorrelated, e.g., self-will, thing-orientedness, individualism and innovative creativity on the one pole, and compliance, person-orientedness, sociability, conformity, and adaptiveness to rules and traditions on the other pole.
2. Individuals differ from one another as far as the balance between these polarities is concerned.
3. This variation between individuals must have genetic components.

In the next part of this chapter we will check these assumptions against experimental data, but first we will investigate their logical consequences.

At this point one might justly retort: 'why so much ado about nothing? It seems self-evident that these polarities in behaviour are interconnected, and since for most broad behavioural characteristics it is likely that differences in behaviour are partly caused by genetic differences, in particular if they are of very old phylogenetic origin, which these behaviours apparently are, it is rather tautological to state that they have genetic roots.' The point is, first, that this notion of a biological basis of certain behaviours may be self-evident to behaviour biologists, but it is certainly not for large groups of sociologists and psychologists. Secondly, these three assumptions do have peculiar and important consequences if applied to the sociology of group structures — the incrowd–outcast dynamism and the concomitant behavioural reflexes in particular.

In order to discuss these consequences we have to add one more assumption, which is rather a definition, namely:

4. In what follows, social groups will mean groups over which individuals are distributed discretely. In other words, individuals can be recognized by one another as either belonging to the social group in question or not – and are treated accordingly.

HYPOTHESIS

4.3 LIFE CYCLES OF SOCIAL GROUPS AND STRUCTURES

If social groups are defined as above, the previous four assumptions imply that within such social groups there is exercised a more or less continuous selection pressure in favour of compliance and sociability. It is such because the most compliant – and thus most socially-oriented and rule-adaptive – individuals are most likely to establish long lasting accommodation within the group. Self-willed individualists on the other hand (also being innovative and thing-oriented according to assumption 1), are most likely to run into trouble and disagreement with the dominant individuals and/or habits and rules in the group. They are least prepared to pay the price of postponing or giving up personal urges and initiatives in order to keep up peace and social harmony. As a consequence, such individuals are the ones who are most likely to either fight hard for attaining a dominant position, or, if failing, to drift into marginal omega-like social positions and eventually become outcasts and leave the social structure. For any eventual influx of individuals into the social group or structure, the opposite holds. Individuals will be most readily accepted if they do not pose a threat to the individuals and/or habits ruling group life, which of course favours rule-adaptive compliants.

The effect of such a continuous selection pressure is that the average behavioural make-up of a group will shift gradually towards compliance and sociable rule-adaptiveness. Due to assumption 1, this also implies a shift towards less and less independent creativity and thing-oriented innovativeness; because of assumption 3, this shifting of group characteristics is (genetically) consolidated. What automatically happens then with every social group (structure) is a gradual loss of innovativeness and behavioural flexibility. In the end such a gradual ossification reduces the effectiveness of the group (structure), whether its function be the preservation of a territorial area with sufficient resources to keep a deme of mice alive, or, in man, the enhancement of some sport, the maintenance of political ideals, the aim to get a better share of the market, or the preservation of a political state. Such ossification especially matters whenever novel challenges turn up in the form of environmental changes or the emergence of competing groups. The disadvantages of a lowered flexibility and innovative creativity weigh most when, because of changing circumstances, innovations and a change of habits are urgently required. In such circumstances the advantages of the old social system in terms of experience, solidly established routines, compliance, malleability of all members, and sheer size, may easily be outdone by the innovativeness, flexibility and efficiency of a younger, and often much smaller, social group (structure) on which these selection pressures have not, as yet, been working for such a long period of time. At such a moment the old structure will yield to the younger structure in a relatively sudden way.

Therefore, provided the above mentioned assumptions are valid, social groups and structures only have a limited life span, and, as I shall try to show below, these assumptions indeed seem to fit most socially living mammal species with discrete group structures, including man. The life cycle of a social institution in human society then, can be indicated roughly as:

> foundation --> consolidation --> internal selection pressure --> increasing ossification and a reduction of flexibility of the social structure --> eventual attempts to compensate these effects by means of more striving for growth and power --> further increase of rigidity and ossification --> catastrophic collapse by sudden environmental changes or competition (Fig. 4.2).

Our model implies a departure from notions of mere gradual changes in societal structures. The probability of sudden catastrophic turn-over events increasing in time with cumulating selection effects can be graphically represented and mathematically described with help of the bi-stable models from the mathematical branch of catastrophe theory (Thom and Zeeman, 1974; Zeeman, 1976; Woodcock and Davis, 1978). Figure 4.3 shows a cusp catastrophe, visualizing the relation between the continuous and the discontinuous part of the cycle. After foundation of a social structure, the level of overt challenges tends to decrease and the stability of the structure tends to increase until the inefficiency begins to take its toll, after which the stability of the structure decreases again. During this process the average level of self-will decreases. An increase in the level of experienced challenges may then sooner or later lead to a catastrophic turn-over event. In the new structure the percentage of innovators (average level of self-will) starts again at a high level.

The selection rate determines the speed of ossification; the life expectancy of a social structure is, therefore, roughly inversely proportional to the

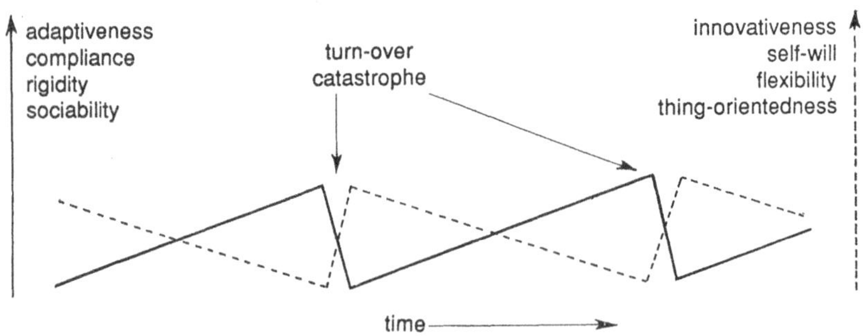

Fig. 4.2 Change in time of the average characteristics of the prevalent social group structures and their incrowd members.

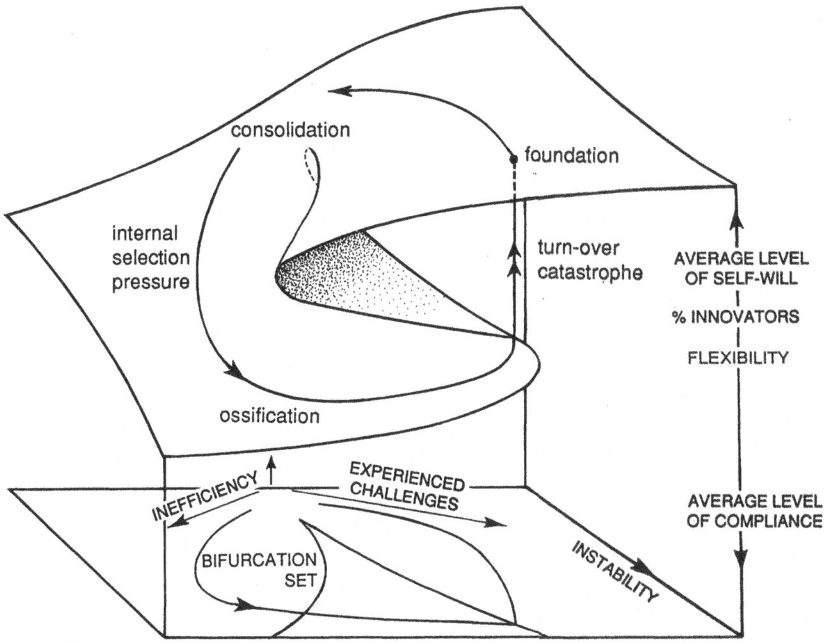

Fig. 4.3 Turn-over cycles in terms of personality characteristics and institutional functioning.

internal selection pressure. Such sudden turn-overs of social structures are therefore bound to happen at any level at which discrete social group structures are operating, as long as individuals can be recognized by one another as either belonging or not belonging to that group, and as long as there is some outflow or neutralization (and eventually a selected influx) of individuals. Depending on the level of organization, such a turn-over goes by the labels conquest, close-down, discontinuance, bankruptcy, revolution, subjugation or extermination.

Once the old, ossified social structure has been replaced by one or more younger competitor-structures, the individuals from the population as a whole have been reshuffled in favour of resourceful self-willed innovators who now occupy the 'incrowd' positions. The rule-adaptive compliants who formed the bulk of the establishment of the former social structure in power, have drifted into marginal positions and now run the worst risks. Thus the previous internal shift in genetic make-up has been undone, and a new selection cycle is started in these new structures.

The selective advantages for individuals are therefore different within and outside of social groups and structures, and are also different depending on the stage of the life cycle an institution is in. A compliant, adaptive and sociable temperament gives a selective advantage within a large, and especially older, social system, whereas a thing-oriented, innovative and self-willed temperament is selectively advantageous outside of the protective maze of established structures, or within small, young systems.

4.4 EVOLUTIONARY ADVANTAGES

Notwithstanding the above mentioned unpleasant aspects of the turn-over catastrophes themselves (in the case of man labelled bankruptcy, revolution, etc.), such a scheme of automatic and unavoidable cyclical changes in social-behavioural structures does also have evolutionary advantages. It is, for instance, clear that this mechanism keeps everything moving: structures, individuals and finally genes. After every turn-over event (or catastrophe) there is a thorough re-shuffling of individuals and when in the ensuing chaos new combinations of individuals reassemble in the newly emerging social group structures, novel combinations of gene sets are eventually formed also.

Apart from this advantage at the level of interpersonal social reorganization and consequently of ensuing recombination of gene sets, there is also an advantage at the level of migration, exploration and colonization of the environment (e.g., Lancaster, 1986). Most mammals are reluctant to go beyond the limits of familiar territory – their home range – and generally must be forced one way or the other to do so (Christian, 1970). Every time an old structure breaks down, a large number of individuals is forced to move and is therefore added to the extra-group surplus population. This will produce a sudden increase in interindividual competition outside of the group (structures) and will produce, therefore, a sudden increase in the pressure on other established (group) structures. This will catalyse the impending catastrophic collapse of more systems, thus locally adding to the already existing chaos. This spatial synchronization causes migratory and related pressures to occur spasmodically and strongly instead of continuously and rather weakly. This may be an advantage where, for example, geographical barriers need to be overcome in order to enable further migratory moves of the population or species as a whole. Many authors have commented on the importance of surplus individuals in producing strong pressures for dispersal (Darlington, 1957) and from the model discussed above it may be clear that social hierarchies constitute by themselves a major force for dispersion. This is also stressed by Christian (1970) in a review of population dynamics research in mammals. He concludes that it is in general primarily the low-ranking individuals that are forced to emigrate from their birthplace (often maturing young animals), and whereas they have an extremely high rate of mortality, it also follows that by their expulsion increasingly more marginal and submarginal habitats should become occupied as density and migration pressures increase. Moreover, this process facilitates speciation:

> . . . once in a great while a dispersing individual may, one would suppose, harbour a mutation or genetic change that increases its ability to adapt to the new surroundings and improves its chance of survival. It is such individuals that should be the basis for evolutionary changes. A sub-optimum area could be invaded repeatedly by countless numbers of individuals before a genetic change permitting adaptation occurred. Thus, the dispersal of large numbers of socially *subordinate* individuals into new environments may provide the wherewithal for natural selection, in contrast to the relative conservatism of dominant individuals in an optimum habitat. (Christian, 1970, p. 86)

The implication is that the Darwinian 'struggle for life' is in fact a process with much irony and relativity, since those individuals with, apparently, maximum reproductive success (the dominants) create by the very violence of their success the outcasts that carry on the process which we call evolution (Hoffschulte, 1986). Likewise, the ethologist and social psychologist Calhoun (1974), p. 302–3) comments on our own origins:

> The strong remain where conditions are most salubrious to preserving the old life-style. The weak must emigrate – bodily, behaviourally or intellectually. Our more distant ancestors swung from trees. Slightly less distant ones lost that race and won another. Population pressure forced them out of forest islands to wander across the African plains in search of another patch of forest where they could renew the old ways. Successive losses and successive demands for adjusting culminated in upright walking creatures much like ourselves. So it has been through all of evolution; the weak [eventually] survive, changed, to open new routes into the future. The meek do inherit the earth.

The evolutionary advantages described here are of course most important in species inhabiting niches of a temporary character, which require the regular invasion and colonization of new environments. Christian (1970) reports that the above mentioned strong fluctuations in population density and migration pressure are indeed most conspicuous in species living in habitats which are ecologically transitory and thus of a strongly temporary and changing character. A species dependent on that type of habitat depends more on regular migratory moves for survival than species living in extensive and stable habitats. In this light it is noteworthy that the human species, by colonizing the most extreme sorts of habitats, has, in its recent evolutionary history, managed to inhabit virtually all of the earth's surface. The mechanism of social selection pressure and expulsion of outcasts should, therefore, have been of great importance in man's evolution.

This is also implied by Coser's (e.g., 1956, 1978) and Girard's (e.g., 1982) comprehensive works on scapegoating in man. Girard describes how throughout human history the distribution of social role positions has been brought about by means of violent acts of social repression. Not only is the dramatic shifting of non-average, deviant subordinate persons into outcast positions just as common as in lower mammals, but, according to Girard, the very development of our culture even depended on it. Only through acts of violence and the collective commemoration of the victim-outcast or scapegoat do human groups find the social-cognitive norms and unanimity from which culture can develop. Culture in our species is not to be considered, therefore, as an immaculate attainment with which we have overcome primitive forms of violence. On the contrary, it is precisely through the violent social collisions themselves that human culture emerged from the animal background. The threatening circle around victims who are found guilty of social disorders is, so to say, the daily bread of social cultural order (Hoffschulte, 1986, on Girard).

In summary, the mechanism of population- and group-cycles, as postulated above, would facilitate speciation through genetic adaptation to marginal habitats, would help to overcome migratory bottlenecks and would, in the case of man, serve to motor the evolution of culture. The actual turn-over catastrophes themselves may not be pleasant for the participants at all, but that is irrelevant from an evolutionary perspective. On this grand scale it is not the feeling and suffering of the individual involved that counts, but the long-term behavioural and behavioural–genetic output that does.

Having outlined these intriguing and also somewhat disquieting consequences of the four assumptions made, I will now present some data from ethological and psychological research that may help us assess the validity of those assumptions.

DATA

4.5 EXPERIMENTS WITH BEHAVIOURAL DIFFERENCES IN HOUSE MICE

Some 33 groups, each containing 4 male and 2 female housemice of the same age, having grown up together from the same litter, were each placed in large observation cages in order to investigate interindividual differences in behaviour and the way these differences come about (van der Molen, 1981, 1988). The study investigated:

1. how social role differences within such groups could be manipulated;
2. which part of the behavioural differences had to be ascribed to those role differences;
3. which part of the behavioural differences was due to innate trait-factors.

Dominance appeared to determine the behaviour of an individual to a great extent, thus being an indispensable tool for ethological descriptions of interindividual differences. It could also be shown experimentally that becoming dominant or subordinate was mainly dependent on coincidence and contingencies, and only to a limited extent on individual characteristics such as body-weight, social- and fighting-experience, self-will, ferocity, etc.

Within the categories of dominants and subordinates there appeared large differences in tolerance for other individuals. Some dominant mice behaved far more aggressively towards their subordinates than did others and these differences determined to a large extent the number of subordinates eventually holding out with such a dominant.

Another role-difference which could rather easily be manipulated experimentally was the Incrowd/Outcast difference, or rather, the difference between beta and omega-subordinates (the usual terms in mouse research).

Detailed ethological data on the behavioural characteristics of 36 individuals were factor-analysed, using factor rotation with the experimentally found social-role differences as anchoring points. The remaining factors of (within-role) differences in behaviour were interpreted as *active versus non-active* and as *self-willed versus compliant*. As far as the latter dimension is concerned, self-willed conflict-proneness was found to be strongly correlated with a high frequency of exploratory and thing-oriented behaviour, whereas compliance was found to be strongly correlated with a high frequency of social and partner-oriented behaviour.

Every time a group of four males and two females was placed in a large observation cage for the first time, there were at first no clear alpha-, beta-, or omega-roles. In the course of the following days (or weeks) an alpha male would emerge and the differences in behaviour between the subordinate males would still be rather vague. Subsequently, differences would gradually evolve between the behaviour patterns of the subordinates. These differences occurred in the amount of resistance to the initiatives and the manipulations of the alpha, the number of fights they had with the alpha, and the amount of patrolling and exploration by themselves through the territory. Some subordinates sat still and allowed the alpha to groom them whenever he chose to do so and in return groomed the alpha if he offered himself by 'crawling under', which is the mouse way of saying something like 'please scratch my back'. Other subordinates tended to walk away more often when the alpha started to groom or to crawl under. The latter type of subordinate eventually appeared to be attacked by the alpha more regularly and subsequently showed more 'fleeing'. Such individuals then remained for increasingly longer periods of time in their hiding places, especially when the alpha was walking around, and eventually they ended up as inhabitants of an uncomfortable and, for the alpha, rather inaccessible hiding place.

The subordinate mice who adapted to the initiatives of the alpha behaved submissively more regularly and underwent the maipulations of the alpha more often. They were however less often disturbed by aggressive attacks from the alpha, and did not much care whether the alpha was awake or asleep. The subordinates who put up more resistance towards the alpha showed on the other hand a conversely adjusted type of activity pattern; they kept silent for as long as they sensed that the alpha was active, and walked around when he was asleep. These gradually developing behavioural differences between subordinates can be described as differences in staying (beta types) and fleeing (omega types), since the latter type showed a tendency to flee the territory if possible. It should be noted here that the emergence of extreme omega behaviour patterns was an artifact of the experimental setting, owing to the fact that the mice were unable to escape. In natural settings they would probably have disappeared from the territory before showing such clear omega type reaction patterns. Indeed, in experimental situations in which opportunities for fleeing are provided, a large proportion of the (young) subordinate males do indeed flee the territory (Van Zegeren, personal communication). This is similar in many other rodent species (for example refer to Healey, 1967; Ewer, 1971; Wilson, 1975a, p. 278; Barash, 1977).*

In the process of a subordinate gradually becoming an omega, the behaviour of the alpha gradually changes towards treating the omega ever more as a stranger. What is important to note here however, is that the behavioural differences between betas and omegas seemed to develop *before* the alpha began to treat the subordinates in a different way. This suggests that these beta/omega differences are caused by differences between the individual subordinates themselves. It could, in principle, also be explained by assuming that an alpha male initiates these differences by having a dislike for one of the subordinates, and that this subordinate thereupon avoids the alpha more than the other subordinates do. These differences in treatment by the alpha might initially be of such a subtle nature that even though the subordinate in question reacts promptly with increased avoidance behaviour, these differences have escaped our attention. It could however be shown in a cross-breeding experiment that the differences between omegas and betas originate primarily from the subordinates themselves (see van der Molen, 1987; 1989). (Of course these two hypotheses are not mutually exclusive; they may both be valid, supporting each other's effect.)

*In human societies there are also many occasions when a fleeing pattern is as difficult to achieve as it is with the artificially restricted mice in these experiments. Ghettos must consist of groups unwilling or unable to integrate fully and unable or unwilling to disappear. And whereas enforced ghettos are an extreme case, it exemplifies the general thresholds existing in any social structure, were it alone for overcoming the psychological bonds of habituation and attachment to the old situation and the extra risks and feelings of insecurity concomitant with breaking out.

4.6 EXPERIMENTS WITH BETA- AND OMEGA-ROLES

In 30 populations (or groups of 4 males and 2 females in a large observation cage, as described above) observations were undertaken to determine whether or not the subordinate males did indeed develop into 2 distinct classes of betas and omegas. We used groups from the inbred C-57-black strain and the inbred CPB's-bagg albino strain and also from their F-1 and F-2 hybrids. These two inbred strains were chosen because of the conspicuous differences in the patterns of their aggressive behaviour and in their level of inter-individual tolerance. Two similar populations of wild mice were also incorporated in the experiment. Ethologically verifiable and clearly recognizable differences between betas and omegas developed in:

1 out of 9 CPB's populations
1 out of 8 C-57 populations
4 out of 8 F-1 hybrid populations
3 out of 3 F-2 hybrid populations
2 out of 2 wild populations

In the CPB's populations subordinate males tended to take up an (outcast-like) omega position, whereas in the C-57 populations the subordinates tended to take up a (compliant) beta position. The development of distinct classes of subordinates occurred quite clearly in half of the F-1 populations, and seemed to be normal in the F-2 as in the wild mice.

The hypothesis that subordinates from F-1 populations showed less individual differences in this respect than subordinates from F-2 and wild populations was tested by means of Fisher's exact test for independence. The statistic in question, having a discrete, hypergeometric distribution when the zero-hypothesis is true, rendered a significant value for stat.alpha = 0.10. This is in fact the most significant result that can be obtained with these numbers of populations.

These results suggest* a segregation and recombination in the F-2 generation of the genetic factors that determine the likelihood of subordinates becoming omega versus the likelihood of becoming a beta. An explanation of these effects in terms of differences in behaviour of the alpha mice does not make sense because in these data, subordinates were distinguished in behaviour *only* in relation to the alpha. Apart from this, an increase in behavioural variance of alpha males in the F-2 generation would imply more populations of the F-2 lacking either omegas or betas. This is contrary to what was found; thus the differences between omegas and betas stem, at least for a

*See, for example, East and Nilsson-Ehle in Srb, Owen and Edgar, 1965, pp. 450–74; or any other handbook on the basics of population genetics.

greater part, from genetic differences between the subordinate individuals. We label these differences accordingly as self-will, intolerance, tendency to have one's own way, or for that matter, tendency to dominate.

In these experiments it was found furthermore, that tolerant, compliant males, apt to take up a beta role instead of an omega role when in a subordinate position, were tolerant of the subordinates when performing an alpha role, contrary to males with a high level of self-will or tendency to dominate.

4.7 OTHER ETHOLOGICAL RESEARCH DATA

In many species differences between individuals have been found which resemble the above mentioned differences in male mice. From ethological field research it appears to be a general characteristic of social mammals that some individuals exert a lot of aggressive dominance, bullying their subordinates much of the time, whereas other dominants act as a sort of controller, governing the social relations in the group by social skill, sustained by the appreciation from companions rather than by aggressive intimidation. These differences are, for instance, reported from ethological research on mountain gorillas by Fossey (1972), on chimpanzees by Reynolds and Luscombe (1969), on a number of species including man by Chance and Jolly (1970) and Wilson (1977a, pp. 311–13) and exclusively on man by, for example, Lippit and White (1958), Krech, Crutchfield and Ballachey (1962, Chapter 12), Gibb (1969), Strayer and Strayer (1976), Hold (1976), and Sluckin and Smith (1977). Wilson (1975a) comments on these differences (p. 294):

> It is not wholly imprecise to speak of much of the residual variance in dominance behaviour as being due to personality. The dominance system of e.g., the Nilgiri langur (*Presbytis johnii*) is weakly developed and highly variable from troop to troop. Alliances are present or absent, there is a single adult male or else several animals coexist uneasily, and the patterns of interaction differ from one troop to another. Much of this variation depends on idiosyncratic behavioural traits of individuals, especially of the dominant males (Poirier, 1970).

Itani *et al.* (1963) and Yamada (1966) describe the behaviour of extreme beta-type males in Japanese monkeys (*Macaca fuscata*) and indicate that a compliant temperament seems to be conditional for assuming such a role. Yamada further points out that, when such individuals eventually achieve a dominant position, a tendency for independence sometimes seems to exclude a tolerant attitude towards subordinates.

Differences of this sort between dominant males have also been described in stumptail macaques (*Macaca speciosa*) by Bertrand (1969), who describes both 'bullies' and 'fair alpha males' and stresses that aggressiveness is not always a necessary factor for dominance. She states that stumptail macaques differ considerably in the amount of intolerance and aggression displayed, and that in certain cases the sustained aggressiveness of some individuals, who were followed up for several years, seemed a personality trait that appeared early in childhood. Furthermore she concluded that the amount of investigative behaviour shown by an individual also depended upon the predisposition of each monkey, apart from social rank, age and conditions of captivity. Some individuals were far more adventurous than others. This personality dependence of investigative behaviour overruled age and rank dependent behaviour in particular when the stimuli were frightening or ambivalent.

In animal and social psychological research alike, variation in tolerance and acceptedness is reported between individual subordinate role styles. In general, it appears that individuals who do not manage to attain a dominant role (α-position in Fig. 4.1) may either stay in a subordinate position while (incrowd-)subordinates are often observed to gradually grow into a semi-(incrowd-) subordinates are often observed to gradually grow into a semi-outcast or outcast position. Such outcast-like subordinates are then the potential migrators, running all the risks implied, whereas the better accepted incrowd-type subordinates, who show a better adaptation to existing hierarchical pressures, may eventually succeed the present dominant(s) if the latter should become incapacitated or even die. Such differences in social-role types have been observed frequently in relation to dispersal mechanisms operating through young individuals in particular (Wilson, 1977a; Barash, 1977).* Bertrand (1969) reports the occurrence of scapegoats in stumptail macaques and de Waal (1975) in Java monkeys. The latter reports that high-ranking individuals often formed alliances against the lowest ranking adults or adolescents, notwithstanding the fact that each of the high-ranking monkeys clearly dominated the scapegoat in question also without any help of others.

4.8 HUMAN BEHAVIOUR

The significance of the self-will versus compliance or individualistic, thing-oriented versus social dimension in the domain of temperament traits is not only corroborated by a substantial amount of ethological research data on animals, but also by ethological as well as personality–psychological literature on *human* behaviour.

Gibb (1969), Strayer and Strayer (1976), Hold (1976) and Sluckin and Smith (1977) report differences in dominance-styles of children, and of

*Similar descriptions have been given for e.g., deermice (Healey, 1967), free-living populations of black rats (Ewer, 1971, pp. 135–137), free-living lions (Bertram, 1975), rhesus monkeys (Vandenbergh, 1966), free-living Japanese monkeys (Itani *et al.*, 1963; Yamada, 1966) and by Eisenberg *et al.* (1972) for a number of primate species.

adolescents (Savin-Williams, 1977a,b, 1979, 1980); they are similar to those described above for mammalian behaviour. Gibb (1969) calls the two antagonist styles leadership and domination. With leadership, authority is spontaneously accorded by fellow group members whereas with domination there is little or no shared feeling or joint action and authority derives from some extra-group power.

Turning from dominance styles to more general differences in behavioural style, Abrams and Neubauer (1976) report that human infants differ considerably in the way they divide their attention between persons and objects. This trait dimension, which they called thing- versus human orientedness, was manifest as early as in the second month of life. They found that the more thing-oriented child shows a greater freedom in exploration. Therefore we might label this dimension of thing- versus human orientedness (or sociability) also as explorative versus social, parallel to the vocabulary in Bertrand's (1969) longitudinal research on macaques. Abrams and Neubauer (1976) furthermore report that learning processes are shaped in a way which is different for each type of child:

> Training issues are characterized essentially as 'tasks' for the more thing-oriented child; for the human-disposed infant, they are characterized as acts in the spectrum of approval or disapproval . . . If earlier impressions were that the more thing-oriented children are more outer-directed, by the third year of life they appeared more inclined to be motivated by inner determinants and resources, a distinction which seems to persist thereafter . . . The dispositions of infants are re-inforced in the milieu, as implements in evolving strategies are cycled back into the psychological system and thus inevitably emerge as traits of character.

From her long range ethological research, Hold (1976) reports that children who rank high in the attention structure tend to set initiatives instead of complying to the initiatives of other children and that they

> . . . prefer to play alone when the leading role was already taken by another high-ranking child. It seems that these children do not like to be commanded by other children.

This runs essentially parallel to what has been said in the introduction in that self-willed, thing-oriented individuals are more prone to become either dominants or loners than to become beta-type compliant subordinates. A similar trait contrast is employed by Edwin McClain (1978, 1979) in his detailed longitudinal study on the behaviour of adult women. He distinguishes between women who are dominated by a need for independence and women who are dominated by a need for affiliation. McClain, like Ausubel (1952), points out that two basically opposing patterns of maturation already occur in the parent–child relationship during a youngster's early years. He terms the resulting personality types as satellizers who tend to adapt to existing rules, versus nonsatellizers who tend to behave more individualistically.

> The satellizing child establishes her life orbit about her parents, whom she perceives as the benign source of all that is good in her life. In contrast, the nonsatellizing child rejects this kind of dependency because she *believes* that her welfare lies in her freedom to choose her own course. (McClain, 1978, p. 436)

The material of McClain's study was derived from the behaviour of women. Kirton (1976, 1978b, 1987a) investigated a somewhat related dimension, namely the balance between adaptiveness and innovativeness in adults in general. The K.A.I. (Kirton Adaption–Innovation Inventory) was developed as a psychometric instrument for these investigations. Kirton based his instrument on the notion that a person confronted with a problem has a choice: he/she can do things 'better' or 'more' to solve the problem (adapt; the social-oriented approach) or he/she can do things 'differently' (innovate; the thing-oriented approach). Doing things better implies the acceptance of the old framework, while doing things differently means breaking accepted patterns. As Kirton says:

> The Adaptor is right at home in bureaucracies, which tend to become more adaptor-oriented as time goes on . . . whereas . . . the natural position of high Innovators seems to be out on a limb.

Kirton's work is of special significance for the performance of leaders (Kirton, 1961, 1977, 1987a; Thomson, 1980; De Ciantis, 1987). He shows that innovators tend to become initiating and directing task-leaders whereas adaptors tend to become consideration-oriented maintenance-specialists of social relations. This is in line with the differences between leader types as described by, for example, Bales (1953), Halpin and Winer (1957), Thibaut and Kelly (1959), Krech *et al.* (1962) and Reddin (1970, 1987). From a conceptual point of view, innovativeness may be considered, furthermore, as a positively appreciated creative variant of non-conformism and disobedience.

Conformity as defined by Krech *et al.* (1962) in their research on the dimensions of social interactive behaviour, is also related to the trait dimension thing-oriented and self-willed versus social and compliant. They found that some people are more resistant to group pressures and demands (the hard-core independents and the deviants) than are others (the easy conformists). Their research offers strong support for the proposition that conformity tendencies are significantly related to enduring personality factors in the individual. The relevance for our model becomes especially clear where they define conformity as a 'trait of the person' as opposed to conformity as a 'trait of the situation' (or social role dimension in our words).

> . . . conformity might be thought of as a 'trait of the situation'. [but] There are also marked individual differences in general readiness to conform, over a wide variety of situations. These differences . . ., reflect conformity as a 'trait of the person'. This distinction between conformity as reflecting the conformity-inducing properties of a situation and as reflecting the conforming propensity of a person should be kept well in mind. Much of the controversy and misunderstanding about the facts and theories of conformity stems from a confusion of these two aspects of conformity. (Krech *et al.*, 1962)

Of particular interest is the existence of a similar dimension in Factoranalytic Personality Trait Research. Feij (1978) compares the trait models of Heymans (1932), Eysenck (1953), Zuckerman (1974), Strelau (1974a, b), Buss, Plomin and Willerman (1973), and Buss and Plomin (1975), amongst others. Although these authors often use different classes of subjects and prefer different final rotations of their resulting factorial models, some of their dimensions appear closely related to our dimension self-willed and individualistic and thing-oriented and explorative versus compliant and social (Fig. 4.4).

For instance, a high score on Zuckerman's (1974) and Feij's (1978; Feij *et al.*, 1979, 1981) trait dimension of sensation seeking indicates a strong need for change, exploration and new experiences, a tendency towards independence of other people and an anti-authoritarian attitude, while low sensation seeking implies a tendency to comply with conventional values and rules. Feij (1978) stresses that extreme sensation seekers may on the one hand be antisocial, drop-out delinquents, but may on the other hand be unconventional but fully accepted creative innovators. This is in agreement with what was

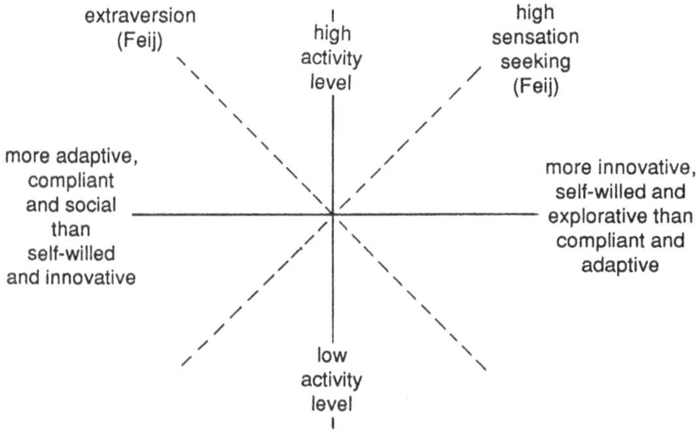

Fig. 4.4 Feij's dimensions 'extraversion' and 'sensation seeking', the balance between them – our dimension adaptive – and their relation with 'general activity level'.

postulated above, namely that highly self-willed individuals tend to become either drop-outs (omega-role) or accepted innovators in the focus of attention (alpha-role), and that individuals with a low self-will tend to assume beta-roles compliantly.

Buss and Plomin's (1975) trait dimension sociability, indicates a strong need to be together with others, a high responsiveness toward others and a predilection for social interaction above non-social reinforcers (Feij, 1978).

In Cattell's sixteen-personality-factor set, the dimension labelled as 'liberalism' (Q_1) is supposed to measure an underlying tendency toward nonconformity and independence versus a need for affiliation (Cattell, Eber and Tatsuoko, 1970; Karson and O'Dell, 1976; McClain, 1978). At least three other dimensions from his 16PF battery also relate to concepts discussed here, namely Cattell's higher order factor IV, indicating subduedness versus independence, the factor assertiveness (E), indicating cautious humbleness versus abrasive assertiveness, and the factor superego (G), indicating conscientiousness versus expedience (Kirton and de Ciantis, 1986; Kirton, 1987b).

In most other factoranalytic classification systems one or more dimensions may be discerned which are related to our concept of self-willed and thing-oriented versus compliant and social. Moreover, the empirical work of *inter alia* Goldsmith (1984, 1986, 1989; see also Kirton, 1987b, 1989) shows that the concepts emerging in all these factoranalytic dimensions from the various authors on personality are indeed statistically correlated, while forming a coherent web of conceptually intertwined behavioural characteristics.

4.9 GENETICS

The above mentioned data from factoranalytic personality research are the more relevant because various writers point out that a genetic basis of these dimensions has repeatedly been firmly established (Eysenck, 1967; Vandenberg, 1967; Buss *et al.*, 1973; Buss and Plomin, 1975; Feij, 1978; Claridge, Canter and Hume, 1973; Eaves and Eysenck, 1975; Wilson, 1975a; Plomin and Rowe, 1977, 1979).

The empirical findings of Kirton (1976, 1978a, 1987c) and Ettlie and O'Keefe (1982) are also in line with the notion of a biological basis. They report that differences in innovativeness versus adaptiveness are not significantly related to IQ, to level of education, or to previous experiences, but are apparently of a more basic (personality-trait) nature (Kirton, 1978a, 1987b, 1989; Kirton and De Ciantis, 1986). In this respect innovativeness, indicating the type of creativity differs from instruments which measure the level of creativity (Kirton, 1978a; Torrance and Horng, 1980).

In section 4.2 it was pointed out that in socially living mammals at least two sets of basic urges have to be postulated, which, independently from one another, vary over individuals, thus producing *inter alia* the adaptor/innovator differences. The first set contains drives for social contact and interaction, leading to gregarious types of behaviour; the second set contains the drives for thing-oriented behaviour.

From recent neuroanatomical and endocrinological research it appears that there is probably a strong link between these two distinct sets of drives on the one hand, and specific neuro-endocrine systems on the other. Cloninger (1986, 1987) presented a biosocial theory of personality, based on a synthesis of information from family studies, studies of personality structure, as well as neuropharmacologic and neuroanatomical studies of behavioural conditioning and learning in man and other animals. He describes three dimensions of personality that are genetically independent, two of which, the novelty seeking dimension and the reward dependence dimension, relate to the two distinct sets of basic drives mentioned above.

One of his dimensions of personality trait differences is principally ruled by the monoamine neuromodulator dopamine. This system determines the heritable tendency towards intense exhilaration and excitement, leading to frequent exploratory activity (novelty seeking) and avoidance of monotony. Individuals high on this dimension are generally also characterized as impulsive, quick-tempered and disorderly. They tend to neglect details and are quickly distracted or bored. They are also easily provoked to prepare for fight or flight. The other dimension is principally ruled by the monoamine neuromodulator norepinephrine. This system determines the heritable tendency to respond intensely to signals of social reward and approval, sentiment and succour. Individuals high on this dimension are generally characterized as eager to help and please others, persistent, industrious, warmly sympathetic, sentimental, and sensitive to social cues, praise and personal succour, but also able to delay gratification with the expectation of eventually being – socially – rewarded.

According to Cloninger, a person high on novelty seeking (the dopamine system) and low on reward dependence (the norepinephrine system) is characterized as: seeking thrilling adventures and exploration; disorderly and unpredictable; intolerant of structure and monotony, regardless of the consequences; frequently trying to break rules and to introduce change; quick tempered and strongly engaged with new ideas and activities; socially detached; independent nonconformist; content to be alone; minimal ambition and motivation to please others; insensitive to social cues and pressures. Conversely, a person low on novelty seeking (dopamine) and high on reward dependence (norepinephrine) is characterized as: dependent on emotional support and intimacy with others; sensitive to social cues and responsive to social pressure; sentimental; crying easily; rigid; orderly and well organized; trying to impose stable structure and consistent routine; rarely becoming angry or excited; an analytic decision maker who always requires detailed analysis of complete information; slow to form and change interests and social attachments. The striking similarity of this polarity with descriptions of Kirton's innovator vs. adaptor dimension is obvious.

In summary, the available data, including data from neuro-endocrinological research, support the view that a biologically based trait dimension thing-oriented, explorative versus social or, in different terms, self-willed versus compliant is indeed conspicuously present, and *does* have genetic roots.

4.10 SELECTION WITHIN HUMAN SOCIAL STRUCTURES

The first three assumptions made at the beginning of this chapter apparently find ample support in ethological and psychological literature. Therefore, in any class of social (group) systems in which there are clear differences between members and non-members (prerequisite 4), cyclic changes should occur in the sense that each separate social group or structure only has a limited life-span, which is inversely proportional to the effectiveness of the selection pressure within the (group) structure in favour of compliance. The life cycles are then separated by turn-over catastrophes which go by various names, depending on the level of organization: territorial conquest; close-down; discontinuance; bankruptcy; revolution; subjugation; extermination; etc.

In the literature on animal ecology and population dynamics, the research data on population explosions and emigration waves, at more or less regular time intervals, are renowned (for example, Christian, 1970, on various species of lemmings, mice and voles). Whereas Christian points to the importance for evolution of these periodic changes in density and migration activity, the proximal causation of these conspicuous phenomena has up to this moment not been explained satisfactorily. It shall be clear that the present model constitutes, among other things, an attempt to fill this gap.

That selection forces do indeed operate within social groups against non-compliant, non-adapted individuals and other deviants, has experimentally been shown in various social mammals and birds (Neumann, 1981), in non-human primates (Kling and Steklis, 1976) and also in man (Schachter, 1951; Scherer, Abeles and Fischer, 1975; see van der Dennen, 1987, pp. 28 ff. and Flohr, 1987, pp. 200–2, for a discussion).

In the psychological literature we can also find many comments referring to the relevance of the discussed selection processes for the way our human society is run,* including data on the personality dimensions these selection processes operate upon.

White and Lippitt (1960) and Scheflen and Scheflen (1972) give detailed

*(See, apart from the authors quoted here, also e.g., Snow, 1961; Etzioni, 1964; Weick, 1969; and Tiger, Chapter 5).

behavioural descriptions of the process of creating chronic scapegoats as a fundamental process in the functioning of human social groups. They describe the physical as well as the cognitive and communicative aspects of the processes that lead either to getting stuck in a superdependent immobilized scapegoat-role or to becoming outcast (ω-type). In their opinion there is a conspicuous contrast between, on the one hand, chronic superdependent immobilized persons who tend to neuroticism by accepting guilt and assuming the scapegoat role and thus getting stuck in cumulating 'double-binds' (Laing, 1967, 1970; Watzlawick and Fish, 1973), and on the other hand anti-social types who tend to deny guilt, generally refuse to be immobilized in a scapegoat role and tend to stay socially mobile, although in peripheral social roles. This is indeed what would be expected from our theory.

Parallel to what de Waal (1975) suggests in the case of Java monkeys, Scheflen and Scheflen (1972) explain how in their opinion every human social group or society generates automatically its own neurotic scapegoats, deviates and outcasts as a necessary by-product of continuous consolidation and reaffirmation of internal (cognitive) values and social order. Such marginal social roles serve for the society in question as a necessary external frame against which the internal social values and role criteria may be projected and by which the 'shoulds' and 'should nots' for all its members are continuously exemplified (Erikson, 1966).

Milgram's (1974) famous experiments in which he asked subjects to administer heavy and supposedly life endangering electric shocks to stooges 'for the sake of scientific progress', are also enlightening in this respect. According to Milgram, this general readiness to obey and even to torture fellow men, if urged and backed up by the authority of common opinion, . . .

> . . . is the psychological mechanism that links individual action to political purpose. It is the dispositional cement that binds men to systems of authority. Facts of recent history and observation in daily life suggest that for many people obedience may be a deeply ingrained behaviour tendency, indeed, a prepotent impulse overriding training in ethics, sympathy, and moral conduct.

This dependence of strongly repressive systems on a strong and dependable compliance of its employees and agents, explains what is often considered a paradox in the literature on holocausts. What is, for example, surprising is that the people who in 'das Dritte Reich' were in charge of the extermination machinery, quite generally appeared to be extremely docile, middle-class, adapted, morally rigid and reliable house-fathers and exemplary husbands, with an aversion to adventure and violence, and who more often than not were friendly and kind to children and pets in their daily social interactions, with an overtone of sentimentality. As shall be clear from the present theory, this is indeed the only type of person – the highly compliant, non-innovative, non-self-willed adaptor – that can be relied upon to carry through orders ('Befehl ist Befehl!') in situations where obedience strongly conflicts with morals and ethics. Under such extreme circumstances the selection pressure on personality characteristics, therefore, is extreme also, the not-so-compliant individuals trying to avoid such ghastly agentic responsibilities. As Koestler (1967, in van der Dennen, 1987) eloquently stated:

> It is not the murderers, the criminals, the delinquents and the wildly nonconformists who have embarked on the really significant rampages of killing, torture and mayhem. Rather it is the conformist, virtuous citizens, acting in the name of righteous causes and intensely held beliefs who throughout history have perpetrated the fiery holocausts of war, the religious persecutions, the sacks of cities, the wholesale rape of women, the dismemberment of the old and the young and the other unspeakable horrors . . . The crimes of violence committed for selfish, personal motives are historically insignificant compared to those committed *ad majorem gloriam Dei*, out of a self-sacrificing devotion to flag, a leader, a religious faith, or a political conviction.

Milgram (1974) labels this compliant, subordinate style of functioning the agentic mode, which expresses that somebody in that mode functions as the agent of some (personal or impersonal) authority. He points out that individuals tend to function in any one situation in either this mode or in its opposite, the autonomous mode. Milgram explains that the readiness to shift from the agentic mode into the autonomous mode in certain conflict situations differs considerably between adults, that people differ in the amount of time they spend in either mode, and that there is a complex personality basis to obedience and disobedience.

These differences between individuals in their tendencies either to comply with social standards most of the time, or to act autonomously and independently most of the time, are also of crucial importance for the way in which bureaucratic structures and other social institutions are run (Kirton, 1978b, 1987a):

> . . . the 'adaptor' personality . . . who can be relied upon to carry out a thorough, disciplined search for ways to eliminate problems by 'doing things better' with a minimum of risk and a maximum of continuity and stability . . . [whereas] . . . innovative change . . . leads to increased risk and less conformity to rules and accepted work patterns (Bright, 1964), and for this reason it rarely occurs in institutions on a large scale . . . (Kirton, 1978b, p. 611)

It is said that organisations in general (Whyte, 1957; Bakke, 1965; Weber, 1948 (published in 1970); Mulkay, 1972) and especially organisations which are large in size and budget (Veblen, 1928; Swatez, 1970) have a tendency to encourage bureaucracy and adaptation in order to minimise risk. Weber (1948), Merton (1957) and Parsons (1951) wrote that the aims of a bureaucratic structure are precision, reliability, and efficiency. The bureaucratic structure in its nature exerts constant pressure on officials to be methodical, prudent, and disciplined, resulting in an unusual degree of individual conformity in that situation. (Kirton, 1987a)

Therefore institutions tend to become more adaptor-oriented as time goes on because of selection, training and promoting policies which are in line with those aims (Drucker, 1969; Schumacher, 1975, p. 243). A negative selection pressure is continuously exerted against innovators.* Even when an innovator finds badly needed novel solutions for pressing problems, it will often fail to render him social approval, *inter alia* because of inherent (sometimes insurmountable) communication problems with his more adaption-oriented colleagues. Instead of winning social approval when coming up with such solutions, the innovator finds that tolerance for his innovative style of approach is at its lowest ebb when his adaptor-type colleagues feel under pressure from the need for quick and radical change (Kirton, 1987b). Even when the novel solutions in question *are* accepted, it does not generally lead to a suspension of the above discussed selective forces. In an empirical study to investigate the ways by which ideas leading to radical changes in some companies were developed and implemented, Kirton (1987a) found that:

> There was a marked tendency for the majority of ideas which encountered opposition and delays to have been put forward by managers who were themselves on the fringe, or were even unacceptable to the 'establishment' group. This negativism occurred not only before, but after the ideas had not only become accepted, but had even been rated as highly successful. At the same time other managers putting forward the more palatable (i.e., conventional) ideas were themselves not only initially acceptable, but remained so even if their ideas were later rejected or failed.

It can thus be seen how, much unlike the fate of innovators, failure of ideas is less damaging to the adaptor, since any erroneous assumptions upon which the ideas were based were also shared with colleagues and other influential people (Kirton, 1984).

As a consequence of these differences in selective pressure, ageing institutions suffer from the disadvantages of not having innovator type creative output available in times of change when policy and methods are required to change as well. Such necessary changes, therefore, are often

*A similar process of selective isolation was seen by Rogers (1959), and reported in his account of the 'creative loner'.

brought about only when a precipitating event, or a crisis, occurs when at last the adaptor needs, and so collaborates with, the innovator (Kirton, 1961).

4.11 OSSIFICATION

Science is an outstanding example of a branch where these considerations about systematic intra-group selection pressures seem relevant. The very goal of scientific research is to find even better conceptual and instrumental frameworks, but, as Kuhn (1970) points out, changing the paradigms which are hitherto accepted without question by an entire scientific community requires a breakdown of previously accepted rules. Such breakdowns are the very process of scientific revolution and this revolutionary process is fundamental to scientific advance; thus, as a social institution, science stands out as an extraordinary oddity (Tiger, 1985). The consolidation and preservation of group cohesion and established rules are not its primary goal, but the creative expansion of conceptual boundaries. In scientific institutions, the innovator type input is not only needed at the rare times of inevitable change, but very regularly, since precipitating events or conceptual crises are the very thing that scientific efforts are supposed to be aimed at.

When ageing institutions become too adaptor- and compliance-oriented by the resulting unconscious bias in selection and in promotion policies, it may not be initially very disastrous in the case of factories or bureaucratic units. For as long as no drastic external challenges turn up, they can just go on producing their output as they formerly used to do with excellent results. But in research units it is eventually disastrous if the cognitive climate becomes more and more adaptor-oriented. The innovator type creative output, consisting of (often disquieting) conceptual challenges and explorations of the unknown and unthought, will in that case gradually be replaced by compliant adaptor type output consisting of puzzle-solving and quasi-discoveries without any conceptual threats. This means that scientific units finally reach an efficiency close to zero when becoming more adaptor-oriented with increasing age. After an initially fruitful phase of consolidation, the prevailing paradigms will become rigid and dogmatic. It is clear moreover, that in a government-protected scientific community competition does not operate freely. This may postpone organization–structural turn-over catastrophes considerably, and thus the timely rejuvenation science continually needs. As a result, . . . 'The Church of Reason [science] like all [ageing] institutions, is based not on individual strength, but on individual weakness. What's really demanded is inability. Then you are considered teachable: a truly able person is always a threat' (Pirsig, 1974).

This description of the process of ossification also fits perfectly for most of the established and institutional religions, as has been recognized by many philosophers and other scientists. Kierkegaard for instance, heavily criticized the organized churches, pointing out that in our society authentic existential religiosity nowadays has two great enemies: philosophers indulging in mere abstract speculations of a strict and limited rationality on the one hand, and church-going fundamentalists and uncritical believers on the other. Kierkegaard believed that the Church has degenerated into a bunch of unthinking fanatics, or even worse, a flock of passionless and anemic herd-mentalities, who dutifully walk into church for no other reason than that was the direction most others were walking. He resented that the church does not have the decency to recognize that whatever its teaching of watered-down, polite moral humanism has become, it isn't Christianity any more (Wilber, 1983).

Christianity may have been founded for the enlightenment of mankind, as an attempt to raise people to a higher level of autonomy and socio-psychic health, and for overcoming the frequent social tendency towards hateful and revengeful cultivation of deviants, scapegoats and other presumed enemies by institutionalized practices of denunciation and mobbing (Coser, 1956, 1978; Girard, 1982; Hoffschulte, 1986). But, like any other social institution, the Church has gradually deteriorated into a system, preoccupied with its own propagation as a system, and thus – contrary to its original goals – with power and the binding-in-dependency of its members in uncritical docility (Toynbee, 1972). The Church does not invite, any more, to mysticism or to experiencing the 'void', instead it imposes 'belief' in God and promotes conformity and respect for 'respectability' (Laing, 1967). Similar considerations hold for other traditional religions and belief systems.

4.12 SOCIAL-ROLE BLINDNESS

Apart from these specific ossification phenomena, many more areas in human society can be found where such effects of the selection mechanisms, as discussed in section 4.10, are manifest. Since these selection mechanisms are operative in lower social mammals as well as in man, they must be anchored quite solidly in the behavioural system. This is not surprising because this mechanism does have considerable evolutionary advantages, not only in animals, but, at least up to recent times, also in the case of man. As mentioned in section 4.4, it facilitates speciation through genetic adaptation to marginal habitats, helps in many species to overcome migratory bottlenecks, and even serves to motor the evolution of culture in the case of man.

It seems plausible therefore, that *if* in the case of man, our superior capacity for learning plays a modifying role in these matters, then the organization of our intellectual capacities will have evolved in such a way, as to enhance the occurrence of selection cycles, rather than to thwart them. The mechanism of selection cycles and periodic turn-over catastrophes is basically powered by the involuntary forces of attraction and repulsion between individuals within social groups and structures. It must have been evolutionarily advantageous, therefore, for behavioural and cognitive 'masterprogrammes' to develop, serving to prevent the newly evolved intellectual capacities from interfering with the (phylogenetically very old) involuntary biases in social interactions.

This is indeed what can be found. Human beings appear to be peculiarly unable to assess objectively the quality of their own social-role behaviour and the behaviour of other people they are dealing with in the social group. There is a sort of 'social-role blindness', of specific blind spots in our cognitive capacities, safeguarding primitive, elementary tendencies of being either attracted or repulsed by other people, depending on the own and on the other's social role and position. As in the experiments with mice described in sections 4.5 and 4.6, the omega-like subordinates, the peripheral non-conforming types, are also in man most likely to be disliked by the established leaders as well as by the conforming and compliant subordinates. This is, in fact, a tautological statement, since drifting into a marginal or an outcast position (ω-type in Fig. 4.1) is just another way of saying that one is less acceptable to the in-group.

As a tool for this mechanism, a considerable part of human communication consists not of transferring pure information, but of more or less involuntary emotional expressions of praise, admiration, criticism, ridicule and insults, as is shown for instance in the ethological work of Weisfeld (1980) on social-role behaviour in adolescent boys, or in the sociological investigations by Segerstråle (1986; Chapter 14) into the Wilson–Lewontin scientific debate as part of the sociobiological controversy. To a large extent the use of language serves to support or to camouflage non-verbal actions, actions for manipulating other people and for staking out and sustaining social roles (Scheflen and Scheflen, 1972; Mehrabian, 1972; Argyle, 1976a,b).

In factoranalytic research on the social interactions between people, the first and by far the most conspicuous principal component is the so called evaluation of positive–negative (Good–Bad) dimension, describing to what extent one appreciates or disappreciates the rated other. In questionnaire research where elucidation of the actual social behavioural attitudes and social-role distributions is the primary goal, the raw data are, in general, firstly corrected for the positive–negative evaluation or social desirability dimension by partialling out its influence (Benjamin, 1974, p. 419). Those who rate other persons or questionnaires in general colour their judgements with appreciation or disapproval to indicate, explain and consolidate the social relations between themselves and the rated person.*

The importance of this negative or positive bias in the way we think about

*In the case of dominating individuals this cognitive distortion of the own and the other person's qualities is labelled by Kipnis 'the Metamorphic Effects of Power' (1976, ch. 9).

our companions is also expressed by the fact that most behavioural attitudes and personality characteristics can be expressed in positive as well as in negative terms. We virtually have, therefore, a double set of conceptual labels for other people's actions and behavioural attitudes; a positive set and a negative set. This cumbersome and at first sight inefficient cognitive organization, in which the pure assessment of other people's behavioural qualities is blurred to a great extent by the strong involuntary bias of appreciation or disappreciation, can only be evolutionarily advantageous if it serves an essential purpose. From the above, it may be concluded that this purpose may be found in protecting the involuntary attraction and repulsion reflexes, which direct our social behaviour, against our intelligent faculties. Indeed, human individuals are hardly aware that the way they assess the other person's qualities is to a large extent coloured by their positive or negative feelings towards that other person, resulting from the involuntary forces of social attraction and repulsion in operation. People do not realize that their, say negative, labelling of (the behaviour of) an important other can easily be changed into its positive counterpart by simply regarding the same behaviour from the point of view of a supporter, and vice versa. They tend instead to attach a sense of permanence and absoluteness to their (categorically negative or positive) judgement, and in particular they are not aware of the relativity of the judgement in terms of its dependence on the mutual social positions of the rater and the ratee.

In summary, the postulated blind spots and no entry signs in our intellectual faculties apparently do indeed exist. Despite our vaunted intellects and our protestations of rational and scientific know-how, we humans show a disturbing tendency to reserve our intellectual powers strictly for certain specific tasks. In other specific areas of functioning, like the mechanisms of social attraction and repulsion mentioned above, we tend to rely on involuntary biases while allowing the intellectual faculties to be effectively blocked. Therefore, ironically and paradoxically, this specific stupidness, this social-role blindness in us humans, should probably be regarded as a special adaptation to our great cleverness.

In daily life, the result of these cognitive biases is that, in many instances, we cannot help but foster, involuntarily, a lower esteem for other persons if they happen to be less 'in-crowd' than ourselves. In more extreme cases, we cannot help tending to join others in mobbing or in scapegoating. We tend to justify the actions taken through our (biased) evaluations of the outcast's or scapegoat's qualities, attitudes and behaviour (unless we incidentally happen to be one of the outcast's supporters). Being in the agentic or systemic (Milgram, 1974) or compliant (Apter and Smith, 1976, 1985) motivational mode while dealing with a victim, we involuntarily tend to see the person in question to a larger or lesser extent as inferior, or even repulsive, detestable or evil. This cognitive distortion can in fact be considered the behavioural basis of torture. Without this psychological effect, the role of torturer would be impossible (Amnesty International, 1973), various built-in inhibitions on aggression would then in most cases take precedence. What happened in the extermination camps of 'das Dritte Reich' is an extreme – though unfortunately not very rare – type of event, exemplifying what this human faculty for selective blindness may facilitate.

The Nazis called their victims 'Untermenschen', but likewise, we ourselves have in turn a strong tendency to label the people who were in charge of the Nazi extermination projects as incorporations of evil, as devils incarnate. As was stressed in section 4.10 however, they were, if they can be characterized at all, rather the opposite, or they would not have been fit for a task where ethics and personal norms would most likely conflict with obedience. Like the spectator-subjects in Milgram's experiments, we cannot imagine ourselves doing the same in similar circumstances, but the facts are that most of those Nazis were not beasts, but very quiet middle-class, social-oriented adaptors, and we are no saints, but ordinary people, who in similar circumstances are rather likely to do similar sorts of things. The difficulty we have with acknowledging that those Nazi employees were not so very different from ourselves, and the other way around, exemplifies the all overruling strength of this type of social-role blindness within ourselves.

All in all, it seems that we cannot help but hate our (self-created) enemies and that we cannot help but love primarily just those individuals which the described selection-cycle mechanism urges us to appreciate. The effects of this same blinding mechanism can also be recognized in less extreme contexts, like for example, the social interactions between scientists or between managers. The contribution by Segerstråle gives us a very illustrative and piquant example of how this mechanism works out in the case of scientific colleagues with strikingly different cognitive styles. In her account of Wilson's and Lewontin's respective contributions to the sociobiology debate, Lewontin plays the part of the thorough and careful adaptor whereas Wilson plays the part of the creative, speculative and daring innovator, and the subsequent mutual denunciation of each others scientific qualities is prototypical.

Thus, the turn-over catastrophes keep happening unhampered, in the case of man just disguised in cognitive ornaments which we, erringly, take for true. Whenever a turn-over catastrophe is at hand, there is still another type of social-role blindness in operation which ties in with the cognitive biases discussed above. It is discussed in more detail in the contribution by Tiger (see also e.g., Janis' book *Group-think*, 1982; Tiger, 1985). According to Tiger the evolution of the human intellect must of necessity have been accompanied by a simultaneous development of a set of awareness blocks, safeguarding groupthink tendencies and safeguarding the unhampered compliance with social habits and prejudices. He argues that the human intellect has primarily evolved as a tool for enhancing coordinated social action, not for independence and for critically observing the social processes one is involved in. Reason was designed to improve consent with the overriding purposes of kinship, not to challenge them. The effect of these particular blocks is that eventual intelligent attempts by any non-conforming individuals, trying to stop the precipitating catastrophe from happening, are likely to be futile. In most persons involved, this particular limitation to the use of intellectual faculties will overrule any capacity to rationally assess the personal risks and general consequences of the turn-over catastrophe at hand, and that will impel them to join compliantly in concerted mobbing or warring actions, not unlike lower social mammals.

It is clear from the aforesaid that the former type of social-role blindness ties in here seamlessly. The concerted actions of animosity towards scapegoats, outcasts or enemies are of course greatly facilitated by the involuntary and uncritical denouncement of the supposed opponent's qualities. And at higher levels of organization, for example political states, the degrees of blindness of the collective are even more disquieting, not only with respect to the systematic and collective denunciation of supposed enemies, but also with respect to what are desirable and effective political courses of action (Janis, 1982), in particular as soon as the ideals and goals chosen become fixed and rigid (Talmon, 1980). As Popper says: 'The attempt to make Heaven on earth invariably produces hell'. The most extensive, quixotic and disgusting violence is justified with the invocation of an Utopian ideology, a paradisaic myth, a superiority doctrine, an eschatological or millenarian ideal state, or other highly abstract political/ethical categories, metaphysical values, and quasi-metaphysical mental monstrosities: national security; raison d'Etat; freedom; democracy; God; Volk und Heimat; Blut und Boden; peace; progress; empire; historical imperative; sacred order; natural necessity; divine will; and so on and so forth (van der Dennen, 1987). In view of their tasks, the stream of information governments take in is even more biased and unbalanced, and their tools are even less effective, than they are in the case of individuals (Deutsch and Senghaas, 1971). This unbalance has become particularly precarious under the present late-20th century conditions.

DISCUSSION

4.13 NATURE AND NURTURE

It can be argued that, at least in the case of man, the same social structure cycles with their turn-over catastrophes might occur merely because of mechanisms on the cognitive psychological level. In that case one would not need to postulate a genetic background for these gradual shifts in social structures to occur.

Indeed, as we saw above, social-role blindness and related cognitive biases are a very powerful influence in man. Moreover, we can also find a wealth of empirical and experimental data on the various constraints on learning in man, on habit forming, traditions and the transfer of cultural information, on perceptual biases like the cognitive dissonance theory, etc., all showing that our behaviour is organized in such a way that a great inertia of ideas, concepts and habits is safeguarded in spite of our capacity to keep learning. These data would suggest that enough mechanisms, at a purely cognitive and cultural level, can be traced as to make social-structural cycles likely to occur. Admittedly, the basic requirement for the postulated selection cycles to occur is not so much that there is a genetic basis to it, but rather that individuals, once their phenotypes in these areas of functioning have established themselves, can not be reshaped into their opposites. As we saw above, this inflexibility aspect, irrespective of its causes, has been firmly established by psychological research. However, the evidence for genetic influence cannot be neglected. These mechanisms therefore are most probably implemented on the genetic as well as on the learning level. It goes without saying that in man, the learning animal *par excellence*, the influence of learning will be relatively important in that case.

Another, related, critique is the argument that where a multi-gene basis of these differences should be expected, a strong enough selection pressure and a quick selection response are difficult to imagine. However, no high mortality, low fecundity or whatever on the part of the declining morph needs to be assumed at all. Basic to the model rather is the existence of differences between incrowd- and outcast-individuals. No physical elimination whatsoever needs to be assumed to let the cycles run. The only thing which needs to be postulated is that the in-group/out-group and the incrowd/outcast distribution of social roles and positions is subject to reshuffling; it depends on the level of organization we are talking about whether the postulate of a genetic effect needs to be included in a description of the cycles or not. In man this will, in my view, probably only be indispensable in the case of very long-term cycles on a very large scale. On most levels of human social structures the individuals selected against just need to be shifted into outgroup or outcast positions, relative to the unit(s) of organization in question, for the selection process to proceed. The very presence of the removed individuals in the organizational periphery then increases the likelihood of a turn-over catastrophe.

4.14 PERSPECTIVE

As was pointed out above, the duration of social-structural cycles is predicted to be roughly inversely proportional to speed and intensity of selection for the trait under discussion. In an industrial company the intensity of selection and the take-on/dismissal percentages are much higher than for example, the selection intensity and the immigration/emigration percentages in the much larger units of political states. Therefore the average cycle periods are likely to vary from a few decades in companies (for illustrative material refer to e.g., Schumpeter, 1939; Kirton, 1961, 1976) or in political parties (e.g., Ostrogorski, 1982), to a few centuries in political states (e.g., Olson, 1982), or even to one or two millennia in whole civilizations (refer to e.g., Spengler, 1918; Toynbee, 1972;* Darlington, 1969; Davis, 1974). The small-scale turn-over cycles with a relatively stronger and quicker selection effect are superimposed, therefore, on the larger-scale turn-over cycles with a longer life span. Thus, individuals may be outcasts in terms of some small-scale social structure while at the same time being totally accepted incrowd members in terms of some larger-scale social structure. The small-scale cycles may be seen as the ripples on the surface of the long range waves of the large-scale cycles. What happens with a person at the social-role level of a sports club is not necessarily parallel to what happens to him at home or at the level of the village community, and what happens to a person at the level of a company does not at all need to be parallel to what happens to him on the level of the political state. In fact, being an incrowd group member on some small-scale level of organization may be vital for a person to keep functioning properly in case of struggling with an outcast position on a larger-scale level of organization.

If it were possible to manipulate these – hitherto involuntary – selection mechanisms, it would be possible to stop or to speed up population cycles at will. This might for instance be relevant for personnel management in industrial companies or for measures at the level of political nations. The latter might be of particular significance in our nuclear age, since population cycles at this broad level tend to be worked out and consolidated by means of war and other economic strangling techniques. Mankind as a whole, up until now, has been able to afford this luxury of genocidal praxis, but war and economic asphyxia, nowadays, threatens to come close to total nuclear destruction. It might be worthwhile, therefore, to take the pressure off the dynamic population cycles kettle and to search for a way to replace or short-circuit nature's hitherto applied selection tricks with which it powered our

*Toynbee disagrees with what he calls Spengler's 'determinism'. Though he (Toynbee) gives abundant material to illustrate the point made here, he emphasizes that one cannot convincingly speak of some or other predetermined and fixed life span of societies. The present theory would support Spengler's view. But it would also give room for something that Toynbee stresses, namely that, as far as their life span and their spin-off in terms of disseminative effect towards other societies is concerned, societies differ greatly from one another. According to the present theory, it very much depends on incidental environmental factors how effective selective emigration can be, and how strong the differential propagation within the structures themselves. And it depends on the actual presence of competing structures how quickly the effects of the internal selection processes will precipitate an eventual turn-over catastrophe.

evolution and our spatial spreading. It seems about time to substitute alternative and less dangerous mechanisms for it.

It will be clear however from section 4.12 that such is easier said than done. The mechanisms in question apparently are anchored in our behavioural system quite solidly. Many authors are however of the opinion that this is no reason to sit down in utter despair. Girard (1982) for instance, points out that in some respects social-role blindness is gradually losing its grip on our behaviour. He calls attention to the exemplary function of Christianity. On the one hand, Christianity has its roots in ancient Jewish traditions, suffused with admonitions towards and justifications of revenge and genocide (see e.g., Deuteronomy, 20, 17, 7, 12; Joshua, 1–3, 6, 8, 10; Kings, 3, 22, 23; Isaiah, 61). The Old Testament can in fact serve as a school example of militant-ethnocentric delusions of racial superiority. On the other hand, a novel phenomenon has emerged from the Jewish tradition, and even more so in Christianity, which is the attempt to replace organized spite, hate and revenge by love and compassion. This scheme may not have been completely successful as yet – were it only for the systematically organized violent blindness in and through the Christian religious organizations themselves – but it certainly has had some effect in overcoming the all-out violent tendencies towards deviants and scapegoats. Christ's example and admonitions like 'love your enemies', and the attempts to break the old tradition of revenge and the resulting vicious spirals of violence and counter-violence, counteract the ordinary selective forces within social groups we have discussed here. They put the primordial tradition upside down by denying the guilt of the victims and scapegoats and by putting the blame on the persecuting society. They de-sanctionize social violence; but what is most important, this tradition, though not reversing our behaviour instantaneously, has opened our awareness to what is actually going on at the social-cognitive level. It has opened our awareness for the fact that we do not like to give up our scapegoats, that we are attached to them and find it utterly difficult to refrain from denouncing and persecuting them (Girard, 1982). It constitutes therefore a massive attack on certain blind spots, on aspects of social-role blindness that, since aeons, have been the cornerstone of the cyclic selection processes themselves.

That is surely not the only glimmer of hope that may be discerned. The involuntary selection forces discussed are under attack from more sides. On the level of personnel management Kirton's work – as described in the previous paragraphs – may also be interpreted as an attempt to extend our awareness beyond its age-old confines into the realm of the dynamics of social attraction and repulsion, and what is more, his approach provides practical scientific tools to undercut the involuntary selection effects, tools that are likely to be utilized more and more because of their profitable effects on the output of the social structures (industrial companies) involved.

It is my hope that this chapter may also add to our understanding of the mechanisms underlying periodic turn-over catastrophes. Admittedly, the present theory, in part, is still tentative, but its relevance for our very existence might urge us to search for further experimental evidence against or in favour of its basic assumptions.

PART TWO

Sociobiology and Enmity

CHAPTER FIVE

The cerebral bridge from family to foe

L. Tiger

Endless controversy has accompanied the broad questions surrounding the issue of conflict and its possible relationship to the evolution of human characteristics, particularly and most recently those involved with cognition and analysis. Therefore it must be asserted instantly that nothing written here should be taken to mean that if there is a biological element in the motivation and sustenance of conflict, that such conflict is inevitable or indeed likely (though it is, on the face of it, difficult to avoid perceiving some biological quality to the action of any living creature). To analyse a situation is not, perforce, to recommend the results of the analysis. Put another way, diagnosing a pathological disease is presumably only a step towards its possible cure, not an admission of defeat or an assertion of the value of the disease because it is found in nature.

Nevertheless, one must deal with the endless controversy which has surrounded efforts to examine human biology. These appear to become particularly colourful where violence and aggression are concerned. As long ago as 1971, in *The Imperial Animal*, Robin Fox and I called those of our colleagues who refused to consider the possibility that in some degree aggression and violence were natural 'the Christian Scientists of sociology' (Tiger and Fox, 1971, 1989). The vigorous riposte to this was harsh even from advocates of human pacificity such as Ashley Montagu and Alexander Alland. The aggressiveness and even headiness of the dialogue has not abated much since. For example, as recently as November 1986, with stunning and preposterous naïvete, the American Anthropological Association, virtually unaminously at its annual meeting, adopted a resolution endorsing the 'Seville Statement on Violence' (Anthropology Newsletter, 1987). The propositions of this statement may be familiar to many but I will cite them briefly nonetheless since they illustrate the manner in which well-meaning half-truths can become relatively coercive in restricting scientific investigation and speculation. At the very least, their utterance may produce a sense of self-righteous comfort with the world which neither its complexity nor tragedy will permit on cold inspection.

These are the propositions:

1. It is scientifically incorrect to say that we have inherited a tendency to make war from our animal ancestors.
2. It is scientifically incorrect to say that war or any other violent behaviour is genetically programmed into our human nature.
3. It is scientifically incorrect to say that in the course of human evolution there has been a selection for aggressive behaviour more than for other kinds of behaviour.
4. It is scientifically incorrect to say that humans have a 'violent brain'.
5. It is scientifically incorrect to say that war is caused by 'instinct' or any other single motivation.

Now, the problem with these propositions is that they are as true as they are untrue. If they are the starting point of a programme of investigation, they will doom it immediately; and if they are the endpoint, then they foreclose the possibility of new findings or new theories with which to interpret old findings. The consequence of this style of manifesto, having decided that the cup is half full, is that anybody who concludes the cup is half-empty is, by definition, some form of scientific rogue, irresponsible for sure, possibly in the pay of armaments dealers, possibly an active apologist for bellicose regimes, in all cases dangerous to the body politic because they support or at least legitimate the crudest and most dangerous enterprises of destructive people who cling to power against the broad interests of humanity. In a letter commenting on this Statement, Fox underscores its classic if unintended nature by noting 'It is ironically appropriate that this document should have originated in the sordid center of the Inquisition, Seville' (Fox, 1987).

Unquestionably there have been people who employ science to justify political and military ends. Scientists have often been involved in profoundly anti-social activities and used the status and mystery of their craft to abet disastrous human actions. For example, Robert Jay Lifton in his recent and authoritative study of the Nazi doctors who conducted horrible experiments on human beings shows how their justification of what they did was that it would lead inevitably to various forms of improved understanding of human medicine and nature. Of course, most importantly, this would benefit Nazi Germany, but it would also provide a lift to the rest of the dignified world. Anti-social behaviour became pro-social within a context of fantasy medical justification (Lifton, 1986).

There is also no question that notions of racial superiority were linked to general Victorian conceptions of evolution, progress, levels of sophistication and the role of competition in the lives of animals and people. The relationship between biologism and racism was certainly apparent for a time; but it could have scarcely been causal, if only because the Darwinian biology, usually castigated in this context, did not exist for virtually all of the world's bloody history. Its lessons barely penetrate still the political councils of warring powers, for example, Iran and Iraq, who could hardly be charged with applying bioscientific principles to their course of action.

As a matter of fact it is precisely contemporary synthetic biology, with its punctilious demonstration of the community of human genetics and physiology, and the irrelevance of notions of superiority, which is the main intellectual opponent of racist assertions. Again, human warfare far antedates its scientific exploration. Belvoir Castle in Israel, built by the Crusaders in 1215 was not constructed because of the existence of ethology or sociobiology or the study of evolution. Nor is it likely that the PLO members whose quarters it was for a while were avid students of comparative biology, and it is surely true that its next (Israeli) occupants did not depend on theories of biology in order to sustain their claim to the structure and its environs. Thus, as a working principle, we can minimize the problem of scientists *causing* violence simply *because* they study it, even if on grounds of proven theory they may discern in violent encounters elements of behaviour also associated with actions producing successful reproduction and hence evolution in the long run. Indeed, if any scientists are responsible for assisting warriors, it is most likely those who practise technologic, not biologic. The social and biological scientists are the least important of all, but more about that later (Tiger, 1987).

I have dwelt on the issue of the relationship of science to the causation of violence and aggression because it permits me to introduce the assertion at the core of my chapter: it is precisely what are widely thought to be the most unusually highly evolved biological characteristics of *Homo sapiens*, our cognitive and symbolic skills, which offer the readiest facilitation to violence and aggression. The same zest for analytical skill and strong commitment to group norms, which is the essence of science, is at the root of the successful construction of the social and ideological boundaries which are the effective prerequisite to large-scale persistent aggressive interaction. In effect this statement denies that the decisive stimulus for aggression/violence is at the lower-end of the evolutionary scale, for example at the gonadal level. Rather, it locates it at the higher end, for example at the cortical. This is not unreasonable because it means that the evolutionary adaptations which distinguish us most prominently as a species are also those which were and are most centrally involved in the basic processes of our survival.

Few scholars informed about the pervasive and subtle forms of bioprocess would be so naïve or foolish as to deny that factors such as gonadal secretions play a discernible basic role – for example in structuring differences between males and females in the extent and force of aggressive encounter (Konner, 1982; Steklis, in press).

The focus of attention here is not individual acts of violence and aggression, such as domestic fights or those between friends or even those mediated by substances such as alcohol or amphetamines. Such events are certainly explicable in clear if partial ways by studies of physiology, fatigue, stress, and the like. A bounteous bevy of books describes, in a wealth of detail, the theory of how behaviour is affected by the intervening variable of the body. There is emerging in the US a broad consensus about the close relationship between drug use and criminality, and violence and substance abuse. There is also strong indication that violent killers suffer from brain damage with disproportionate frequency.

The earliest robust empirical work on these behaviours was the basis for an extensive discussion of body/behaviour linkage. It also stimulated considerable discussion of the instinctive nature of aggression, that is, that aggression was associated with the lower-order systems to which I have referred. Here such important findings as David Hamburg's early discovery of the enormous increase in testosterone during the adolescence of male chimpanzees – up to 200 times increase for males, but a doubling for females from a lower base to begin with – began to offer a sense of the internal secretions which affected external behaviour; we could appreciate the biochemical platform upon which the behavioural structure rested (Hamburg, 1970). Moreover, the dramatic experiments of the Harlows on maternally deprived monkeys illustrated a comparable impact of social behaviour or its absence on the physical calm of an animal and its related social demeanour (Harlow, 1962). It was understandable, therefore, that a relatively clear model of aggressive motivation and process could emerge. Even the broad scope of a theory such as Lorenz's could be contained within a framework of close, almost hydraulic, interaction between an aggressive drive bubbling up to the surface of an inhibiting community, rather along the lines of the Freudian Id lurking in the shadowy depths of the psyche and awaiting an opportunity to overthrow the coercive precision of civilized society – as cryptically described in Freud's own *Civilization and its Discontents* (Lorenz, 1965, 1966). To this day, judges sentencing violent criminals may admonish them by saying 'You acted like an animal'.

A view of aggression, again from the bottom up, but now let us turn to a view from the top down. First of all there is the issue of State Violence or State Aggression or State Terrorism to which I will return later because this category of behaviour is in a statistical sense far more important in perturbing the planet than private domestic or criminal endeavour. Also, this group-based activity presumably depends more firmly on higher order cortical processes for its maintenance than on internal secretions such as that of testosterone, serotonin, or adrenalin.

A few propositions of my own: the first has to do with what is a rather widely-held notion that the brain evolved to think, particularly about technical matters such as hunting and gathering, tool-making, painting on cave walls, and the like. This is likely to be wrong. It evolved to act, not to think; not only to act, but to interact; not only to interact, but to interact sexually, in the imperatively pleasurable and ramified action which is the immediate engine of change in species.

Compare thinking with acting. Recall the legendary experiments of Wolfgang Köhler with primates whom he discovered had the technical capacity to secure bananas from their cages using sticks, boxes and ingenuity. Their ingenuity seemed to be the point of the experiment. His findings made Köhler famous. It considerably accelerated the industry of trying to determine just how ingenious and cunning primates could be; but does anyone recall serious discussion of how the animals interacted with their bosses? Do we know how skilfully or otherwise these creatures assessed the demands placed on them by their keepers and how they determined their response to them? To this day endless arguments persist about whether chimps have humanlike kinds of language (Sebeok, 1987). Psychologist Herbert Terrace has concluded about his consort, Nim Chimpsky, that he was principally interacting with his keepers – a version of the Clever Hans phenomenon, the circus horse who could count because he perceived his master's highly subtle gestures (Terrace, 1984, 1985). What is surprising is that anyone should have been surprised that highly sociable animals should use a technique of social communication for – social communication. Of course technical communication could also occur; but first things first.

So the initial proposition is that cortical tissue developed to facilitate action. The most basic action to which it had to attend was courtship and reproduction. The next proposition is that while the most obvious outcome of effective aggressive behaviour is on the outside environment – either people or things or places may be violated and coerced – it is possible that a more parsimonious process is at work: to solidify internal solidarity rather than to conduct external adventure. Thus, while there might be obvious advantages to securing captives, booty, territory, for hunters and gatherers, these would necessarily be limited because of the nature of that kind of economy. Throughout virtually all of our evolutionary formation, imperial conquest would have been more trouble than worthwhile, captured property more difficult to transport than to enjoy, and captives likely to be normal mouths to feed at the same time as they were likely to be workers with subnormal enthusiasm. Here there could be a real sex difference since female captives could become integrated into the group and produce offspring, a pattern described in basic terms for rain forest chimpanzee society in which cooperatively territorial and murderous males of one group were observed to kill the adult males of a smaller group and then absorb their reproductive females (Ghiglieri, 1988).

Not until pastoralism or agriculture would it have made economic sense to conduct aggressive enterprises as a way of securing resources; but by then we humans had already developed our wonderful brain! Different processes than that would have led to rapid selection of cortical tissue. One such process could well have been, in this model, a set of social skills supporting co-operation, even relatively bellicose cooperation, which would in turn translate into reproductive access and success. It is folklorically significant that in the first episode of the saga the belligerent American hero Rambo concludes his epic destruction of the enemy alone. His mate is dead. He physically faces away from his own tribe. He moves into another territorial area. Though he is a success as a killer, he is a failure as a citizen and as a male. The real war is conducted by less colourful and flagrant souls, by family men. The point is not more complicated than the truism that successful aggression demands excellent cooperation. But the next question is not so truistic. Which was more useful for cortical evolution? Being able to cooperate skilfully with members of one's own group? Or being able to create destructiveness outside the group?

Obviously a straight either/or answer is impossible but on balance the simple logistics and biologic of reproduction would favour an inward-looking strategy. I imply that the real advantage to ancestral aggressors and violents was internal not external. If we remain similar in our enthusiasms and skills today, a similar calculus should apply. Thus we should be prepared always to ask first not what is the realistic function for national defence of a particular defence policy, but what impact does the policy have on domestic politics and economics? These are after all real and chronic. Threats and wars are episodic and are often about issues which the conduct of aggression is itself unlikely to solve. The city of Jerusalem is intense testimony to the extraordinary power of ideas of mythic grandeur in articulating the political and military behaviour of Muslims, Jews and Christians.

Few politicians or even religious leaders can be seen to be reluctant to mobilize aggressive forces for the protection of the perfect most moral community. Even though most people are scared to death of real wars, it is at the same time political death for an American or French or Israeli or Syrian or another politician to somehow imply that the protection of the central glory, which is our group (whatever it is), is not worth any sacrifice, however demanding, even if it is the ultimate one.

Of course there are discernible differences in the civility, sanity, equity, and practicality of defence policies of different countries, but the net result of the nature of bellicose interaction is that a kind of military Gresham's Law applies: bad policies in some communities drive out good ones in others. Good and peaceful politicians in Country A must without question respond practically and with appropriate military firmness to the bad politicians in Country B. It is as simple as that. Even if it is not simple, that is what usually happens.

Now let's ask why it should be the case that politicians will find it politically acceptable to buy armaments and the time and bodies of citizens in order to play a potentially warlike role in the community of nations. Is it because this is a direct way of appealing to patriotic and moral sentiments which is so beguiling that the world's remarkable expenditure on armaments betrays nothing more than that all human beings are frightened and unwilling to be scorned by foreigners? Hence they will invest heavily in armament which is in many cases the form of government expenditure least likely to provide discernible benefit to the majority of voters? In a real sense it could be held that the world's expenditure on armament is profoundly illogical. No doubt in many ways it is; that is not the issue. Perhaps the behaviour is biological.

This leads me to my final point: the brain evolved in the context of a powerful, ramified, overwhelming, concern with kinship, with the reproductive lives of family members – this was more important even than their productive lives. It is only in the present era that everyone, male and female, has become independent contrators to the economic system; they must first solve their productive problems before turning to their reproductive options (Tiger, 1987). The reverse used to be true. There were no independent quasi-legal contracts which were not mediated by kinship realities, and since we lived in relatively small communities, all articulated public discourse was mediated by familiarity, intimacy, and a sense of the nameability of all known forces in a system (Terrace, 1985).

In this setting the brain evolved, and this is the setting in which it currently seeks to make itself useful, under very different circumstances indeed. What does it do? It creates ideologies, religions, brilliant certainties. What has been seen as the great organ of dispassionate ratiocination is put in the service of one ideology or another, to demonstrate the nature and value of various opinions about the received truth, to describe and produce exegesis about a particular glorious certainty (Tiger, 1985). This was what all the universities of our great tradition were all about for most of their history. They were the first religious foundations (and many still are); great investigators of problems already solved. Malcontents could be burned or tortured or banished.

Real science, as a flourishing activity, is still extraordinarily difficult to sustain and genuine scientific communities are robustly established in relatively few places in the World. It is useful to recall that Lysenko held scientific sway in the USSR two years before Lorenz's *On Aggression* and eleven years after Tinbergen's *The Study of Instinct.* In 1988 the United States Supreme Court, after many years of legal battle, finally ruled that it does not violate the religious freedom of creationists when evolution is taught in schools. It had been argued that the Bible was literally true and that therefore evolutionary biology was sacrilege. While this case is settled, the creationist enthusiasm remains effective in many ways. In more general terms, George Miller has described differences between scientific and lay thinking which imply the exoticism of the scientific variety (Miller, 1986).

The difficulty of sustaining academic freedom is another indication of the fact that the brain is designed for coordinated social action, not for independence. It is an instrument of gregariousness not individualism, far less likely to have evolved in a tribe of independent thinkers than one in which coordinated social action was inextricably linked to coordinated thought. In fact, from a broad biological perspective, the animating function of the brain and the culture it supported was to permit a species with considerable capacity for variation to reduce this variation in local circumstances. The brain/culture system permits and indeed drives people to learn the central tendencies, values, beliefs, and limits of their tribe. Again, the function of the great gift of learning encoded in the brain is to make members of groups more the same, rather than to stimulate them to cultivate differentiation.

This brings us in a full circle to conflict and violence, because these depend on the assured commitment of members of groups to their own indigenous certainties and on their willingess to commit themselves strenuously and even fatally for the defence of their groups and their dignity. The requirement and its mechanism are revealed most fully in military training; but why do smart animals – animals with the capacity to analyse fully the intrinsic madness of self-sacrifice for, say, a nation of 50 million or of 5, or a revolution of 50 thousand – do this? The reason the analysis stops is because reason was designed to improve consent with the overriding purposes of kinship and its variants, not to challenge them.

To this end, emotions of public sentiment, patriotism (or matriotism (the use of family terminology in seemingly inappropriate contexts is forever fascinating)) can be successfully mobilized for conflict on behalf of large groups. However, what is brought into use is the emotional and analytical equipment developed for the defence of groups of kin for which one's personal actions, one's personal sacrifice, could make a perceptible difference. We know from the biology of altruism how this can have a direct impact on the genetic nature of social nature itself.

The result is that the confounding, terrifying fact of human bellicosity exploits some formative prosocial processes at the heart of evolution. The problem this creates is made almost surrealistically difficult to control because the impersonality of modern weapons systems appears to allow them to be used as instruments of pure reason, or what is thought to be pure reason by the mathematically skilful strategists of defence. What is somewhat reassuring about the human opera, however, is that often unexpectedly a wholly primitive primatological event occurs which appears to change many of the values of many of the equations which had seemed so durably unassailable for so long. These primatological events may even be bodily functions, such as that Sadat and Begin met in Jerusalem and Gorbachev and Reagan in Washington.

CHAPTER SIX

The evolutionary foundations of revolution

J. Lopreato and F.P.A. Green

6.1 INTRODUCTION

No one who understands Darwinian theory now denies that organisms, including human beings, tend to behave so as to optimize their inclusive fitness; the genes of those who behave otherwise, or merely fail in that quest, are deleted by natural selection. This logic proposes a variety of corollaries. For example, culture is adaptive or else, contrary to known fact, it would act negatively on population growth. Again, human beings tend to behave according to self-interest, for that has a positive influence on reproductive success.

This chapter defends the hypothesis that the quest for fitness is the ultimate cause of revolutionary activity, and thus of the sociopolitical revolution; a recurring human phenomenon that on the surface appears located at the outer edges of phenotypical extremities. The argument will unfold in three stages. First, we shall briefly examine some explanations of revolution prevalent among sociocultural scientists, with a view to singling out evolutionarily relevant variables. Secondly, we shall endeavour an ultimate explanation of revolution and, whenever feasible, underscore the biocultural, or co-evolutionary, nature of the phenomenon. Finally, we shall consider the relevance of the sociopolitical revolution to inter-societal, or group, selection.

6.2 SOCIOCULTURAL SCIENCE ON REVOLUTION

The fountainhead of much theorizing on revolution is the work of Karl Marx and Friedrich Engels. It is marked by at least three stultifying, often tacit, assumptions:

1. revolutions take place in industrializing societies;
2. revolutions are characterized by radical transformations of both form and substance that have epoch-making consequences, at least for the revolutionized society;
3. revolutions are class or collective phenomena executed for the good of a whole society or at least of a whole social class.

A meticulous examination of the most celebrated revolutionaries of the 19th century would reveal various conceptions and nuances of the sociopolitical revolution. Fundamental, however, is the postulation of a 'contradiction' or historical tension between 'the material forces of production' and 'the existing relations of production' (Marx, 1859). The contradiction may be grasped from various perspectives. From an evolutionary viewpoint, the following makes the most sense:

> The rapid growth of productive capital brings about an equally rapid growth of . . . wants . . . The social satisfaction [of the workers, though greater than in the past] has fallen in comparison with the increased enjoyment of the capitalist . . . Because [our desires] are of a social nature, they are of a relative nature. (Marx and Engels, 1847)

That is, the industrial workers were thought to become rebellious in direct proportion to the growing disparity between their ability to fulfill their needs and the corresponding capacity of the bourgeois owners.

In keeping with Marx and Engels' (1845–6) early sensitivity to the salience of the biological substratum in human behaviour, this hypothesis of relative deprivation was based on the logic of the first historical act: the production of material life itself along with the fact that 'as soon as a need is satisfied . . . new needs are made'.

Useful variations of the relative deprivation hypothesis have been offered by various scholars (e.g., Edwards, 1927; Davies, 1962; Brinton, 1965). These students alert to the fact that revolutions tend to occur when advances in socioeconomic conditions stimulate expectations for additional gains that are subsequently thwarted or not realized quickly enough. They also point to a variety of sociocultural variables that can block rising expectations. These include a rigid class structure, economic downturns and an inefficient governing class, *inter alia.* Moreover, Brinton's analysis suggests that the ensuing discontent tends to be stronger in higher-ranking members of the subject class than in the more humble ones.

A complementary tradition, drawing largely on the work of Vilfredo Pareto (1916), emphasizes the inevitability of 'class circulation' along with the tendency by the governing class to perpetuate its rule, even as it no longer possesses popularly desirable characteristics, and thereby to interfere with other people's interests, especially those of the nongoverning elite, namely politically adept members of the subject class who have governing aspirations (see, e.g., Edwards, 1927; Brinton, 1965). As Pareto (1916) puts it:

> Revolutions come about through accumulation in the higher strata of society . . . of decadent elements no longer possessing the residues suitable for keeping them in power [the traits, biological and cultural, that once earned them the support of the masses], and shrinking from the use of force; while meantime in the lower strata of society [i.e., the governed] elements of superior quality are coming to the fore, possessing residues suitable for exercising the functions of government and willing enough to use force.

Another scholar focuses on inter-societal competition and writes that 'revolutionary crises' develop because strains between the State and the landed upper classes prevent an effective response to challenges posed by more economically developed societies. In its efforts to respond to such confrontations, the State interferes with the interests of the dominant classes, thereby causing the latter to withdraw their support from it. The old regimes are then either dissolved through defeat in war, or deposed by the powerful and emboldened landed upper classes at home (Skocpol, 1979).

Ironically, it would seem, revolutions take place in times of relative prosperity and reduced repression; prosperity and mild repression accentuate the interests of the nongoverning elite and their inclination to challenge the rulers. The latter typically respond by restricting upward mobility and co-opting subjects who bring into the governing class the decadent traits that it already possesses in excessive quantity: for example, the tendency to eschew the use of force in favour of the cunning and the ruses that eventually fall prey to force.

The probability of the revolution increases when high-ranking individuals, including the intellectuals, withdraw their support from the incumbent regime. During periods of stability, intellectuals use a variety of ideological formulas that support existing relations of dominance and subordination. In revolutionary times they propagate social myths that envisage a social order more beneficial to the society as a whole.

Under siege, the governing class feels increasingly isolated and self-doubt sets in. Thus, it responds to the challenge from below with sporadic and ineffective shows of force, but a class that can neither defend itself nor maintain internal order encourages the opposition and invites the populace to shift its allegiance to new masters.

6.3 REVOLUTION IN EVOLUTIONARY PERSPECTIVE

6.3.1 Theoretical preliminaries

Without exception, all such explanations stress proximate causes and the collective or group nature of revolution. We shall disregard neither focus in this chapter; however our own emphasis will be on the ultimate and individual nature of revolution.

We define revolution as a forcible action, taking place within a dominance order, that is precipitated by the desire of individuals excluded from first access to the politically controlled resources in order to replace those who do have such prerogative. The action may succeed or fail. The ultimate cause of the desire, and hence of the revolution, is the quest for fitness optimization. The proximate causes are many and, with one exception, vary in time and place. The exception refers to what has been termed the 'struggle for satisfaction' (Ruyle, 1973; Langton, 1979) and the quest for 'creature comforts' (Lopreato, 1984, pp. 256–7). We shall touch on a number of proximate causes in due course, but in the meanwhile, we must justify the intended emphasis on ultimate causation.

The most trustworthy proof would show that the leaders of a successful revolution have, as a result of their revolutionary behaviour, a higher fitness than they would have acquired had they maintained a subordinate position. Such demonstration is entirely out of the question. Alternatively, it may be adequate merely to compare the fitness of the successful revolutionaries, or indeed of ruling individuals in general, to that of subordinate individuals in their own society.

The latter type of data is not entirely lacking, though it is not all of high quality. It is, however, strongly suggestive. Moreover, as we shall presently see, we intend to adduce other evidence and arguments in favour of our thesis. At any rate, the data in question are of two basic types. The first, compiled by ethnographers as well as historians, literateurs, journalists, and such other sources of soft data, reveals that ruling powers, including successful revolutionaries and their entourage, typically have sexual access to a disproportionate number of mates and therefore contribute an extraordinary number of offspring, legitimate or otherwise, to the population pool (see, e.g., Bullock, 1964; Speer, 1969; Cronin, 1972; Brodie, 1974; Betzig, 1986, especially pp. 32–8).

The second demonstrates that people in general, and those rich in resources in particular, tend to practise numerous forms of nepotistic altruism that favour kin selection. Upon succeeding to what was the English throne, in all but the name, Oliver Cromwell, for example, put his sons and sons-in-law in high positions of government, and shortly before his death he named his son Richard as successor (Fraser, 1973). Such nepotism also grants relatives the influence and power that permit them to engage in sexual liaisons of their own and to contribute extra-marital offspring to the population.

Such facts, as noted, are strongly suggestive, if not entirely reliable, in favour of the hypothesized link between revolutionary activities and the quest for fitness enhancement. We now wish to argue, therefore, that an actual causal connection between (revolutionary) dominance and enhanced fitness is not necessary in order to sustain the hypothesis that the ultimate cause of revolutionary behaviour is the thrust to optimize fitness. This statement, seemingly absurd, is of crucial importance for sociobiological studies of cultural behaviour.

For the basic logic of our demonstration we turn to the study of mechanics and to the distinction made in that science between real movements and virtual movements. A mechanical system – for example, the solar system – is determined by its force(s) (e.g., gravity) and its conditions (e.g., mass, distance). If we assume that the force(s) and conditions are determined, movements in the system are likewise determined. They are real. If then, for theoretical purposes, we further assume as checked or controlled some condition, the system will show movements different from the real. These are virtual movements.

Now consider a social system in which DNA replication is the force (implicitly expressed in the optimization principle), and marriage, hierarchy, polygyny, etc. are the conditions that individuals therein are subject to. The general principle of sociobiology predicts that in such a system highly ranked individuals are more likely to practise polygyny than low-ranked individuals. But what happens if polygyny is suppressed? The suppression is a legal fact in many present-day societies. As a result, without data that might show the effects of *de facto* polygyny in the absence of *de jure* polygyny (i.e., data on the paternity of illegitimate children), the system reveals a different (a virtual) movement, and the prediction based on real movements, i.e., the sociobiological prediction, does not hold true. But is it false? Not necessarily. The following hypotheses are compelling: 1. if polygyny had not been suppressed, the prediction would hold; 2. if we had data on the effects of the condition that may have replaced *de jure* polygyny (i.e., extra-marital affairs), we would find that the prediction held after all. Strictly speaking, facts on leaders' illegitimate reproductive behaviour and nepotism lend no help on the first hypothesis; but they are more promising on the second.

That is not all. Our general hypothesis (on revolution and fitness) receives support from a related line of reasoning. We must be mindful that the optimization principle does not state that organisms seek to have offspring or other blood relatives. It states rather that they tend to behave as if they had such goals, by virtue of the fact that the resultant of their behaviour is congruous with such (assumed) ends. In short, although matters may vary between one species and another, what we observe is that human beings compete for resources – including power, wealth, and sexual services – and are more or less successful in this competition. Moreover, we do not accomplish reproductive activities mechanically. Sex is pleasurable. The same may be said of power, wealth, influence, and so forth. It follows that to be dominant is to be successful and to enjoy the tokens of that success (resources). In short, success in resource acquisition and the optimization of fitness are, analyticaly, separate facts. If matters stood differently, neither sex without reproduction nor the acquisition of resources beyond sustenance needs would make very much sense at all.

Such being the case – namely if *stricto sensu*, the competition is for resources – it follows, further, that the quest for resources (for the pleasure of success) can in the course of evolution have become an end in itself. We have thereby conceptualized the in-principle autonomization of behaviour from the strict drive to reproduce. Autonomization, however, is a convenient but imperfect term for what we wish to maintain. We do not think that autonomization is complete; the leash is not broken. The origin of the behaviour is still likely to be genetic-reproductive in nature. Thus, to return explicitly to our general hypothesis, human beings, like other primates, rebel, as our comparative data will suggest, because they have evolved to rebel. But human revolutionaries may be said to behave according to the logic of fitness optimization even in their failure to optimize their fitness.

6.3.2 The comparative ethology of revolution

A note on the ruling oligarchy

In its basics, human revolution bears a striking resemblance to the phenomenon of forcible power displacement in the dominance hierarchy of other primates. The basic causal cluster consists of, on the one hand, a recurring challenge to the constituted power on the part of one or more aspirants to the prerogative of power and, on the other, the diminution of leadership and self-defence capabilities on the part of the reigning power. When so viewed, contrary to typical sociological understanding, human revolutions are quite common. Indeed, one historian points to approximately 150 cases in the course of the last four centuries alone, even when only the most glaring examples are singled out (Blackey, 1976).

The starting point for an understanding of revolution is necessarily an established dominance order – or stratification system, as social scientists say in view of their organizationally more complex units of analysis. The system features the following basic components:

1. a dominant individual typically heading a more or less precarious coalition or governing class;

2. a mass of low-ranking individuals who normally submit to the will of the ruler(s); and
3. between these two segments, a small number of individuals who have, or are about to have, political (i.e., dominance) competences at least equal to those of the ruler(s), plus the ambition to promote such qualifications to a ruling status.

In view of the shifting nature of coalitions, this third segment often overlaps with the subdominant membership of the coalition. Whatever the form of the ruling power, however, it is always a minority that governs. This oligarchy, in keeping with the optimization principle, behaves so as to maximize the probability of maintaining its position of privilege (see, e.g., Michels, 1914; Pareto, 1916; Mills, 1956).

In studies of non-human primates, the oligarchy is known as the central hierarchy. The concept seems to have originated out of Hall and DeVore's observations of baboon troops in Kenya. One troop had six adult males of approximately equal size, but differing age and physical strength. Three (Dano, Pua, and Kovu) supported one another in aggressive encounters and typically acted in concert. Thus they were able to control access to incentives, determine group movement, and so on. Whenever 'Dano, Pua, and Kovu or Dano and Pua combined, they were 100% successful against the three other males who rarely combined' (Hall and DeVore, 1965, p. 61). Similar observations are common in primatology (e.g., Simonds, 1965; Goodall, 1971, 1986; Bernstein, 1976; Riss and Goodall, 1977; Bauer, 1980; Bernstein and Gordon, 1980; de Waal, 1982; Nishida, 1983; Trivers, 1985).

Whatever name we give to oligarchs, the fundamental properties of their tenure and eventual fall are clearly interspecific. The first priority of newly risen rulers is to ensure their position. They bestow therefore, privileges upon those whom they depend upon for support. In human oligarchies, supporters are placed in key administrative positions and granted various privileges (e.g., Lefebvre, 1947; Djilas, 1957; Fraser, 1973; Volgyes, 1978; Chou, 1980; Hinnebusch, 1982; Voslensky, 1984). Other primate oligarchs defend allies in case of attack and share oestrous females and other resources. For instance, in the Arnhem colony of chimpanzees studied by de Waal (1982, 1986) a coalition between Yeroen and Nikkie enabled the latter to displace Luit from alpha status. To secure Yeroen's continued support, the new leader then defended his ally against attacks by Luit and granted him mating privileges.

The clash

The coalition in principle outranks any one individual (e.g., Hall and DeVore, 1965; Trivers, 1985), but as long as it supports the oligarch's position the latter is safe. When coalition members withdraw their support, the leader's position becomes tenuous. Critical to the success of the French, Soviet, and Iranian revolutions, for example, was the withdrawal of support by the upper class, i.e., the coalition (e.g., Skocpol, 1979; Stempel, 1981). By way of analogy, when Nikkie no longer granted Yeroen the sexual privileges to which he had grown accustomed, the latter withdrew his support, thereby enabling Luit to assume the alpha position (de Waal, 1982).

A capital obstacle to a successful revolution is a sense of orderliness and peacefulness in the populace. Such a condition is, in turn, enhanced by the allegiance of 'the intellectuals' to the reigning oligarch. Among humans these are the major ideologues, such as pundits, party theoreticians, public media experts and scholars, whose livelihood and/or research funds often depend on the good will and the coffers of the oligarchs. Their counterparts among other primates are the top oligarch's allies who hover around him* and direct the attention of the lower-ranking individuals toward him (e.g., Tiger and Fox, 1971, p. 29). This almost constant focusing upon the activities of the leader is, in turn, reflected in such human activities as erecting monuments, the diffusion of effigies, periodic celebrations, and, more recently, radio and television appearances.

Intellectuals, however, are rarely effective if their messages are contrary to the followers' interests, which include domestic peace, common defence against the outside and a distribution of resources that benefit them, though of course in varying degrees (Lopreato and Horton, 1987). For example, the failure of the Shah of Iran to end the rioting and secure the personal property of his loyal supporters 'was the conclusive disappointment that demolished public respect for the monarch' (Stempel, 1981, p. 135). Military defeats, foreign invasions, or both accompanied the revolutions that occurred in England (1640), France (1740–8, 1756–63), Russia (1905, 1915–17), China (1905–6, 1930–7), Hungary (1918), Germany (1918), Turkey (1918), Egypt (1949) and Cambodia (1973), among other places. 'The old leadership is discredited by defeat, and the appeal for radical social change and national reassertion thus falls on fertile ground' (Laqueur, 1968, p. 502). Moreover, when 'economic distress arises, contemporaries always blame the government' (Labrousse, 1943, p. 71). Intellectuals can advertise the good qualities of the reigning oligarch. By the same token, they render a service to the revolutionary movement by 'concentrating the general irritation' on him (Edwards, 1927, p. 46).

Among other primates, the masses comprise infants, juveniles, peripheral males, and especially the females who are the most coveted reward of leadership. Frans de Waal (1982) shows, for example, that Yeroen's dominant

*Our focus is on male revolutionaries, but this limitation is not necessary, as the case of the current leader in the Philippines shows for the human case. For other primates, see, e.g., Ehardt and Bernstein (1986).

position in the Arnhem Zoo colony of chimpanzees was safe as long as he had the support of the females. He received their support by virtue of his ability to defend them in case of attacks and through techniques reminiscent of the human politician's predilection for baby-kissing and analogous activities. He indulged, supervised and played with the youngsters. Luit, a challenger, attacked and eventually vanquished Yeroen by undermining, with Nikkie's aid, his female support. An 'alpha male who fails to protect the females and children cannot expect help in repulsing potential rivals' (de Waal, 1982, p. 125). By the same token, Luit's strategy in challenging Yeroen included his willingness to groom the females and play with their children (see also Bernstein, 1976; Cheney, 1977; Estrada and Estrada, 1978; Watanabe, 1979; and Nishida, 1983, among others.)

It is this necessity to touch the heart of the mothers, especially the higher-ranking ones, that sometimes results in what is essentially the primate counterpart of human political nepotism and thus of relative class closure. The offspring of dominant rhesus females, for example, have a better than average chance of growing into dominant adults (Sade, 1967; see also Kawai, 1965; Kawamura, 1965). There may be a genetic component in this result (e.g., Marsden, 1968), but to an extent the success follows also from the extraordinary amount of time that the children of dominant females spend in the favourable company of dominant males.

Facts show that the need for alpha position tends to increase in direct proportion to one's proximity to the reigning power. Thus, in human stratification systems, contrary to received wisdom (e.g., Marx and Engels, 1845–6; Dahrendorf, 1959), the highest rebelliousness and the least acquiescence to the existing authority structure are found not among those devoid of authority but among those situated near the top of it (e.g., Lopreato, 1968; Lopreato and Hazelrigg, 1972). In Europe, for example, the 'eighteenth century saw a growing attempt by the various sections of the nobility to challenge the royal Government' (Hampson, 1963). The same message was conveyed by the secular struggle of the bourgeoisie against the *ancienne noblesse.* Likewise, the challenge to the reigning oligarch among our cousins comes typically from the beta level. Such was the case with the four power displacements observed by Jane Goodall (1986) among the Gombe chimpanzees in the period 1961–84, and with de Waal's (1982) observations in the Arnhem Zoo colony. Departures from this scenario are usually only apparent upon close examination. With few if any exceptions, they refer to gamma or even lower-ranking individuals who are reaching full adult maturity and possess the cunning as well as the physical prowess to work their way to the top. Thus, Nikkie, an ambitious young male in the de Waal study, has a low standing in the troop when his climbing manœvre begins. He engages in an open coalition (de Waal, 1982) with Luit, the beta challenger. For a while, such arrangement describes fairly parallel interests: both parties profit from their reciprocal altruism. Eventually, however, Nikkie not only displaces Luit, who with Nikkie's help had displaced Yeroen but his excellence in cunning also undermines Luit's attempt to strike an alliance with his old enemy Yeroen by practising 'divide-and-rule' strategies against them (de Waal, 1982, p. 146; see also Nishida, 1983 for an analogous case).

The parallels among human beings are numberless (e.g., Szyliowicz, 1975; Zartman, 1975; Stempel, 1981). The following case from the French Revolution of 1789 is typical. The bourgeoisie entered into an open coalition with the nobility in its quest for more civil liberties and representative assemblies. The aristocracy, seeking to weaken the Crown, sought at the same time to reestablish the parlements and the provincial Estates. The meeting of the Estates General, however, brought to a head the conflict existing between the allies. The Third Estate rejected the secondary position that was offered to it by the nobility. In response to the latter's demands that public office be restricted to men of noble birth, 'the only recourse was to suppress the privilege of birth, to "make way for merit" ' (Lefebvre, 1947, p. 48).

The evolutionary significance of the open coalition cannot be overestimated. Its message is clear: 'I shall be your ally as long as it is beneficial to me,' the more cynical 'me' being the more cunning and the more adept at visualizing a singularly beneficial future situation. Equally revealing, in view of fitness optimization theory, is the fact that challenges to reigning oligarchs take place despite the inertia that tends to attach to existing coalitions. The central hierarchy does not preclude change in political systems, but the coinage of the concept was inspired by the observation of considerable stability in the dominance order of baboon troops in Kenya (Hall and DeVore, 1965). When Robert Trivers, who joined DeVore in Kenya in 1972, observed that Arthur, a powerful adult, challenged the middle-aged, reigning Carl, Trivers thought that very soon the troop would have a new leader. But DeVore, wiser on these matters, proceeded to disabuse him of such a notion: 'No dramatic change is going to take place. The system is stronger than you think' (reported in Trivers, 1985; see also Poirier, 1970; Bauer, 1980; Nishida, 1983; Samuels, Silk and Rodman, 1984; Goodall, 1986).

Indeed, if it were not so, incumbents in political systems would have no advantage over challengers. To challenge, especially in the forceful manner that we have defined as revolutionary, is to engage in rash behaviour. Many revolutionists fail and often lose their liberty (e.g., normal status in the group) if not their life as a consequence. In fact, even when they are successful and the prize has been won, the assisting coalition proceeds to disintegrate; then 'heads start rolling.' The French, the Soviet, the Cuban, the Iranian, all human revolutions are positive proof of the great danger of aiming audaciously for the top. The internecine war in George Orwell's celebrated allegory, *Animal Farm*, is a faithful expression of a general political fact. The guillotine falls not only upon the likes of Louis XVI; there is 'political murder' among our primate cousins too (de Waal, 1986). The climbing manœvre, a mechanism of the drive to optimize fitness (Lopreato, 1984, pp. 110–20), must be therefore very intense among certain individuals.

The ultimate cause of revolution, to reiterate, is cogently expressed by the optimization principle. But as we argued earlier, the quest for fitness is necessarily filtered through the search for resources. In non-human primates, the resource that most directly explicates the concept of fitness, namely, sexual access to females in oestrus, is not obfuscated by a plethora of ancillary resources. There is little doubt that alpha males mate more reproductively than lower-ranking individuals (e.g., Hall and DeVore, 1965; de Waal, 1982; Trivers, 1985; but cf. Bercovitch, 1986). Little wonder then, that revolution in our cousin species typically marks the sexual and physical maturation of ambitious and courageous males. Mike's rebellion as well as those that followed in Goodall's (1971, 1986) observations and Nikkie's revolt in the de Waal (1982) study are cases in point. The experience of exclusion from first access to oestrous females among such individuals is analogous to the sense of cramp felt by prosperous and ambitious segments of the human subject class (Pareto, 1916; Edwards, 1927; Brinton, 1965).

Human ambition, on the other hand, is bombarded by numerous and rich classes of resources: untold wealth, influence frequently far beyond one's society, the opportunity to shape institutions if not entire historical periods, a life of extraordinary comfort and luxury, and of course attractive members of the other sex, among many other privileges and prerogatives. It is hardly surprising then, that so many human revolutions have taken place not when socioeconomic conditions were poor but when they were relatively good (e.g., Brinton, 1965). The English, French, Soviet, Libyan, Iranian and other revolutions prove the point (Brinton, 1965; Stempel, 1981; Hinnebusch, 1982). As conditions improve, a society generates an abundance of individuals whose mettle has been tested by competition and whose ambition grows *pari passu* with success. They may be politicians, economic magnates, and in some instances great intellectuals and/or ideologues; having enjoyed the proximity to great power, they, politicians especially, now apply pressures that the ruling power may be unable to cope with.

The same considerations also help to explain the oft-cited fact that some of the best-known revolutions have taken place during the radical institutional transformations that accompany major technological breakthroughs, for example, the Industrial Revolution. This has been a period of intense innovation and expansion, of exceptional opportunities for self-enhancement, of radical redistributions of wealth; the need for self-enhancement is a fundamental cause of such transformations. Its further intensification is, in turn, a result of them. As animal studies show (e.g., Barash, 1977, p. 212), aggression rises as the supply of resources varies from the pre-established equilibrium.

The breakdown of a social order is not merely a phenomenon of power; it breeds normlessness, or 'anomie' (Durkheim, 1897). Given a crisis of authority, individuals do not know how to behave. Little states then grow up within the state. A profound sense of insecurity sets in among the masses. The challenger is thus emboldened at the very time that discontent, disorientation, insecurity in the populace, for example, females and juveniles among chimpanzees, are becoming rampant. The success of the challenger is to a large extent a direct function of such malaise.

Pretenders to the 'throne' use various means to intimidate the ruling oligarch and raise the level of disaffection in the masses. Unlawful demonstrations, refusals to pay taxes, destruction of property, armed assaults against the government are typical among humans. Displaying, refusing to greet and actual fights are favourites among some of our primate cousins (see, e.g., Simonds, 1965; Goodall, 1971, 1986; Riss and Goodall, 1977; de Waal, 1982; Nishida, 1983). At the same time, the challengers make promises of a collective utility: none is more generous or more cynical than the French *Liberté, Egalité, Fraternité*. Aspiring chimps are especially assiduous in grooming females and playing with their children. Humility, or charity with a hook, abounds. A baboon (for example, Arthur), promises friendship to another (for example, Rad), whose alliance he self-servingly covets, with a gesture that speaks more eloquently than any logorrheic promise of human favour: he presents his rear end (Trivers, 1985, p. 370).

Arthur's message is complex: in presenting he also shows his canines to indicate the alternative to cooperation. The extraordinary efficacy of such body language suggests that the importance of human language, while prodigious, is probably exaggerated by social scientists. Thus, the emphasis placed on the intellectuals' transfer of allegiance as a factor in the downfall of human oligarchs is correspondingly excessive. As diffusers of information, intellectuals play a crucial role in advertising the ineptness of the governing class, but their importance lies much more in this fact than in the implied message in social science that they are decisive in the revolutionary process as holders of values and opinions. As myth makers, they necessarily specialize in the construction of highly complex and abstract disquisitions. The masses are far more concrete and practical, especially in periods of crisis, when life and livelihood are at stake. In short, intellectuals have a role to play in human revolutions because the space and population size over which information must travel are exceedingly greater than the corresponding factors among other primates. In the small primate troops, there is no need of them. If there is any semblance of a counterpart, it is represented by such individuals as disgruntled members of the reigning coalition, rising subdominants and females high in the dominance order. When the ruling oligarch no longer serves their needs, they withdraw their support and the probability of a successful revolution increases greatly.

Revolutions are rarely, if ever, the sudden events that the pages of history typically portray. The Chinese Revolution, for example, spanned at least a period of 26 years. In the Arnhem Zoo colony, it took Luit at least 10 weeks to displace Yeroen. The challenge is slight at first; it intensifies in time. There are various defenses available to ruling powers, and many are used. But in the last analysis none work, precisely because of the logic inherent in the optimization principle. The more competent will strive; the less competent will fail. A stopgap measure, making room at the top, runs into the inexorable implications of kin selection and reciprocal altruism. We seek and favour those who are like us. As a consequence, threatened oligarchs practise cooptive mobility at best: they prefer potentially dangerous individuals who consent to become defenders of the political citadel (e.g., Woolrych, 1973; Stempel, 1981).

This strategy amounts to heaping ruse on stratagem when strength is in fact the deficiency. It adds to the decomposition. It also accentuates the sense of distributive injustice among those who feel competent but are excluded – especially those whose ambition was stimulated by socioeconomic expansion and intellectuals whose craft comprises the prerogative of conveying information faithfully or otherwise. Moreover, the top is by definition small, and if some must go up others must come down. Sacrifice, however, is costly and painful; many of those who suffer it are bound to join the intellectuals to promote the opposition, just as Karl Marx (1867; Marx and Engels, 1845–6) argued in relation to the alleged grievance of the petty bourgeoisie against the bourgeoisie.

As the challenge increases in intensity, and the coalition shows signs of disintegration, the self-confidence of the oligarch under siege falters. His strength lay not only in his intrinsic qualities but also in the support he received from the group. As the group wavers in its support, the sense of superiority, based on the intrinsic qualities, starts to fade; the right to hold office is no longer a certainty in the 'officer' (Michels, 1914; Pareto, 1916; Edwards, 1927; Brinton, 1965).

With respect to the contention between the chimps Yeroen and Luit, de Waal (1982, p. 111) puts it succinctly as follows: 'The prevailing social climate affected the self-confidence of the rivals. It was as if their [fighting] effectiveness depended upon the attitude of the group . . . [Luit's] success in the final fight was more than just a demonstration of brute strength: he made it quite clear to Yeroen that the attitude of the group had changed radically.'

The decline in self-confidence manifests itself in various ways. Basic, and common to many if not all primates, is indecisiveness: a fitful use of force that, associated with an excessive and ineffective use of conciliation, really amounts to the sort of capricious violence that fuels the revolutionary spirit. The Shah of Iran provides a classic example. He vacillated continuously between conciliation and the sort of violence best represented by SAVAK torture (Stempel, 1981). Social order breaks down to a critical point, and the new leader steps in to fill the vacuum.

6.4 CONCLUSION

With some partial exceptions (e.g., Pareto, 1916 (especially sections 1220–28); Edwards, 1927), social scientists have explained revolution mostly in terms of group costs and benefits. This tendency flows from the old Hobbesian stress on the basic functions of governing powers: (1) the organization and maintenance of internal social order and (2) the organization of defence against hostile powers. Accordingly, ruling oligarchies are fundamentally alleged to fall on the failure of one or both such functions; and revolutions are implicitly viewed as actions 'directed against principles and institutions, not individuals', which take place when 'legitimate means for effecting changes break down or function poorly' (Blackey, 1976). This tradition also accords with the Marxist view of revolution as both class action and action for the good of a class so large that it is nearly synonymous with society itself.

It follows that blocked mobility for example, harms not only the individuals directly concerned but society as a whole. Likewise, the failure to successfully defend the existing social order against external predators represents a harm perpetrated against society (e.g., Skocpol, 1979). Social orders, therefore, are viewed as moving through history in rhythms that mark first and foremost the shifting quality of relations between the rulers and the ruled as a whole.

Such conceptualizations are at best of indefinite meaning; at worst they overlook the fact that societies are constituted by individuals who are involved in what the great American sociologist, William Graham Sumner (1906) termed 'antagonistic cooperation'.

Deficiencies aside however, such conceptions of revolution do underscore the usefulness of introducing system-states (or group selection) arguments into sociobiological explanation. Provided that we do not equate the idea of group selection with the idea of the good of the group, as sometimes happens (e.g., Trivers, 1985), it can be argued that group selection is a fact to be stressed if bridges between neo-Darwinism and sociocultural science are to be built durably. Human history must be apprehended ultimately at the level of the natural selection of genetic units of heredity; but there are also levels of analysis of major proportion grafted on the irreducible fact of natural selection. We refer to the selection of genes at the level of social organizations of organisms held internally together by interlocking sub-systems of reciprocal altruism. It is this phenomenon that we here call 'group selection,' and we reject, with few exceptions (e.g., Lopreato, 1981, 1984), notions of individual behaviour for collective good that does not in principle, and typically in practise, redound to the benefit of the individual genotype. But self-serving behaviour at this level compels theoretical attention because indirectly it contributes to the differential reproductive success of whole populations.

As a consequence, revolutions must be understood not only as effects of competition for fitness optimization between individuals. They are also adaptive mechanisms that tend to maintain sociocultural systems within certain bounds of organization, of relations of dominance and subordination, and of social order in general that condition the individual quest for fitness optimization. Life conditions in some societies are more adaptive than in others.

As the palaeontologist George G. Simpson (1971) showed, inter-societal competition is one of the most common initiators of societal extinction in the entire biotic world. The observation is predictable from Lotka's (1925) principle, according to which natural selection favours those populations that, on the proviso of the availability of energy, are most adept at converting and controlling energy sources and forms.

History and ethnography show that neighbouring populations are very attracted to each other's energy sources. A governing oligarchy that cannot maintain internal order and organize the collective defence cannot thereby boast that degree of conformity that is required for the persistence of sociality. 'When conformity becomes too weak, groups become extinct' (Wilson, 1975a), and group extinction often subsumes the extinction of genotypes as well. Revolution is a group as well as an individual adaptation that tends to re-establish conformity through the rise of new cheerleaders or cohesion builders to the prerogative of command.

From another perspective, intra-societal aggressiveness appears to be not just a mechanism of reproductive competition between dominant and aspirant individuals. The adaptiveness of such a mechanism also radiates to the society as a whole by insuring a certain level of preparedness within the context of inter-societal competition, thereby compounding the adaptiveness of individual competitive behaviour.

From a biocultural perspective, therefore, human revolution appears to be an extension of a phenomenon rather widespread in primate society whereby social aggregates experience a periodical deterioration of their executive functions and a concomitant amelioration through the rise to power of highly competitive individuals. This conjunction coincides, in turn, at a more purely Darwinian level, with a recurring competition between group members for the resources whose control maximizes the probability of fitness. It also sensitizes us to the competition for reproductive success that goes on between members of different groups.

Finally, revolutions convey a view of human history as a series of pulsations corresponding to periodic increases in the expenditure of organizational energy. Thus they give an image of natural selection at the level of groups that is consistent with the more recent understanding of the evolutionary timetable (e.g., Stanley, 1981). The evolution of populations appears to reveal extended periods of relatively inactive selection that are periodically punctuated by relatively rapid transformations. From the perspective of the sociopolitical revolution, sociocultural systems seem to follow an analogous timetable.

CHAPTER SEVEN

Loyalty and aggression in human groups

Y. Peres and M. Hopp

Few ideas generate more resistance on cross-disciplinary grounds than the proposition that chauvinism and xenophobia in humans are selected for by group selection. The majority of sociobiologists are opposed to group selection on theoretical grounds (Lack, 1966; Trivers, 1985) while social scientists tend to be suspicious of natural selection as an explanation of human social behaviour (Van den Berghe, 1978). The normative implications of a notion which holds that the loyalty people feel for their own cultures and the aggression they exhibit towards outsiders might be heritable, further raises resistance to such a postulate.

We shall defer discussion of the social and moral implications of the theory and only address in this chapter the theoretical arguments against group selection and propose that a special case for group selection in humans can be furthered to explain group oriented behaviours. Our arguments are heavily based on the interaction of culture and biology and therefore we make no claim to extend the proposed concept to other species.

Much debate has raged in the past century over the concept of group selection. It has been used to reinforce early arguments for cultural supremacies by the Eugenesists (Trivers, 1985), has been formulated as a special case of kin selection (Haldane, 1932) and finally posited in the evolutionary-ecological model by Wynne-Edwards (1962) as a reproductive self-denial of individuals at a time of plenty.

We shall avoid the pitfalls of Wynne-Edwards' data-dependent arguments, and define group selection, theoretically, as: a system of natural selection in which the inclusive fitness of an individual is positively related not only to the frequency of his/her own genes in the population but also to the genes of unrelated group members. The distinction between selection at the individual level and selection at the population level is, in this sense, analogous to the distinction, made by Dawkins (1976), between the individual gene and the whole organism. For each individual gene the other genes in the body constitute a vital part of the environment in which the gene exists and on which it depends. Survival of a single gene is no less dependent on the survivability of the other genes than on its own adaptiveness.

We propose that the survival of a group-living individual depends as much on its own adaptiveness as on the survival skills of the group as a whole. The specific contribution of group survival to the survival of an individual can be termed group selection. The circumstances in which such a selection should have the most impact on individual survival are:

1. When individuals are incapable of surviving without constant interaction with other group members; or alternatively when the fitness of an individual is seriously curtailed if the size of the group diminishes below a certain number (Rasa, 1987).
2. When individuals cannot freely move from one group to another.
3. When the individual life expectancy to group existence ratio is high compared to other species. (Refer to group extinction rates in Maynard Smith, 1976, 1978.)
4. When competition between groups is high while competition within groups is limited.
5. When individuals can be personally rewarded for their contribution to the group, and thus altruism to the group can be reinforced.

In the following pages we attempt to examine to what extent these conditions hold for humans and how they can be applied to explain some of the facets of human social attitudes and behaviours.

Cut throat competition and violent aggression are familiar and well documented features among human groups: tribes, nations or cultures. Success and survival under these conditions depend, among other things, on group cohesiveness, for example on the proportion of individuals in every group who are loyal to other individuals and to the group as a whole. We posit that the importance of group cohesiveness is inversely related to the discrepancy between competing groups in other qualities. That is, the greater the similarity between groups in size and technology, the more critical are the differences in group fusion and patriotism. In a conflict among neighbouring tribes, therefore, a group would win or fail on the degree to which it can operate as a well coordinated unit; while in a clash between nation-states, for example, the result will more likely be determined by differences in equipment and technological sophistication. Group loyalty, a term we use to denote conformity, patriotism, social altruism, ethnocentricity etc., must therefore have been even more central in simple pre-technological conflicts.

These postulates could be incorporated into the well known functional model in sociology and anthropology. Functionalists tend to minimize any discussion of the origins of social systems under the assumption that social beginnings are not directly observable. However, the functionalist approach would accept as an explanation for social evolution a proposition that different social systems can rise out of one original population (one gene pool). It would further admit a scenario by which emergent systems with a higher level of group loyalty have an advantage over systems with lower levels of group loyalty. Unfit groups would fragment and disappear; fit groups would grow by recruiting and attracting members of splinter groups. In this way, the original gene pool changes its distribution in space but does not significantly alter its composition. (For a concise discussion of various forms of social selection see Stinchcomb, 1968.) Such a convenient middle-of-the-road solution by which a concept borrowed from biology is used to explain an evolutionary process, while avoiding the tightrope pitfalls of associating social behaviour with direct genetic selection, has found favour in sociobiological circles and has been termed cultural selection (Durham, 1979).

We are motivated to proceed beyond cultural selection by two main shortcomings of that approach: first, it holds that individuals can move freely among groups in their quest for the fittest social system, while in reality most societies pose severe restrictions on individuals seeking to join them; secondly, the theory expects social clash to result in no major change in the general population gene pool. However, in reality, conflict among human groups often results in an extreme reduction in the overall fitness of the losers. These two divergences from the central concept of cultural selection and their far reaching implications for the theory itself, are discussed below.

Modern technology largely transcends intergroup boundaries by affording individuals an opportunity to move freely between regions or countries. Such movements are well documented in the annals of history; however, equally well recorded are the obstacles placed in the way of migrants by the indigenous residents of desirable regions. The resistance to accept strangers can be explained as a form of human territoriality, but the concept requires a clarification of the screening mechanisms by which members of the group – insiders – are differentiated from non-members – outsiders. Today formal criteria of citizenship, education, etc. are applied for the selection of immigrants; but the actual and informal acceptance of a newcomer into a social circle is still based on the more subtle identifying markers of external looks, behaviour and vocalization that are less amenable to copy by aspiring forgers.

If group-hopping were part of the human strategy, evolution would have favoured an adaptation for integration; i.e. individuals with a greater capacity for mimicry would have fared better than individuals with a low integration quotient. Instead, a strong selection for behavioural rigidity was selected for; this rigidity, termed social imprinting, badging or marking in the literature, is a process by which basic patterns of behaviour and speech, form and set during ontogenesis. Once a critical stage of maturation is passed the behaviour pattern is no longer pliable to change or extinction. It forms a marker by which the individual will be recognized and labelled for life as a member of a certain group. Social imprinting closely parallels classical imprinting theory (Eibl-Eibesfeldt, 1975) and is considered to be largely complete by the time sexual maturation is reached (Smith, Udry and Morris, 1985). The timing thus creates a structural rift between children who can be moved from culture to culture with relative ease, i.e. have a high integration quotient, and adults who cannot move freely from culture to culture and therefore have a low integration quotient.

Though recent history suggests that the losers of a war may, sometimes, end up better off than the winners, documented history (as well as indirect evidence from pre-history) suggests that the inclusive fitness of vanquished groups is usually lowered by the victors. Acts of fitness reduction comprise a wide spectrum of primary and secondary behaviours perpetrated on losers. These include wholesale or selective extermination (for example, the Indians of North and South America, the Australian Aborigines, the New Zealand Maoris); prevention from breeding of captive populations (for example, forced separation of slave families or castration of male captives); gene dilution of the losers by systematically abducting and raping their women (a strategy which reduces the fitness of the loser while enhancing that of the victor). Not of least importance is progressive banishment to marginal ecological niches, etc. The spoils of battle can be measured in many ways and often include a direct increase in the progenitive resources of the winners. The infanticidal rage of a *coup d'état* in langurs (Hrdy, 1977) is mirrored repeatedly in the Bible. In the book of Numbers (ch. 31) after the Israelites had '. . . killed every male among them' (the Midianites) and 'made prisoners of their women and children', Moses raged at the officers: 'You had let all the women live? . . . Go, kill every male child and every woman who has known a man. Keep alive only virgin girls'. Earlier, in the book of Exodus (ch. 1), the Egyptians designed a similar strategy against their Hebrew slaves. Pharaoh ordered his people to: 'throw every son born to the Hebrews into the Nile but to save every girl alive'. Sometimes non-virgins were also saved but precautions were taken to ensure that the victors will not mistakenly invest resources in raising the vanquished enemy offspring: 'When you set out to war against your enemy . . . and God hands them over to you . . . if you see among the prisoners a beautiful woman whom you desire . . . let her bewail her mother and her father for one whole month, then you may have intercourse with her'. A similar strategy is seen in the rape and abduction of the Sabine women – an annihilation of the males of another group and an abduction of the females to be used as a breeding resource of one's own gene propagation.

Against the background of ruthless conflict between groups, all known human cultures temper conflict and rivalry within groups. This strategy, xenophobia coupled with a structured internal cooperation, is typical, almost mandatory, of group-living species under conditions of scarce resources (Trivers, 1971; Axelrod and Hamilton, 1981). We must here examine how it affects and what demands it makes of individual members of any group.

First, any individual selected for group living would have a poor capacity to survive alone and therefore little inclination to do so. Such an individual has two options: to remain with the group of origin or to try and join another group. Individuals with stronger group markings, i.e., adults, stand the poorest chance of integrating successfully into any other group and thus can be expected to be most loyal to their own group and most reluctant to out-migrate. Conformity with group norms (cultivated by socialization and social sanctions) enhances, therefore, the ability of individuals to survive intact within their group and thus improves their overall fitness. By default conformity also increases the overall survival chances of the group itself, insofar as a group of committed members is more likely to last than a group made up of egotistical opportunists. Secondly, the need to conform with group norms, while emanating from the fear of expulsion, conflicts with the equally strong need to compete within the group for dominance and resources. Competition between bickering members has a divisive potential and can threaten the very existence of a group. Thus its continuance is predicated on the ability of members to strike a fine balance between their own personal aspirations and their commitment to the very existence of the group in which most of their interests are invested. The balance is affected by the group's security; individuals operate and strive primarily within the group environment – any real or imagined threat to that entity preempts personal goals and loads the scale, temporarily, towards patriotic dedication. Altruism in this sense is merely egoism once removed. As the danger to human groups always comes from rival human groups, the appropriate response will be the reaffirmation of inward loyalty and outward animosity – the two alternate faces of patriotism.

A major indication of external threat to a group is the appearance of a rival group. The emergence of outward hatred and internal loyalty at such times are thus two sides of the same adaptational coin. At this stage we must differentiate between our approach and kin preference. Our premise is that inward loyalty and outward aggression both contribute to the inclusive fitness of individuals. Irrespective of the process by which such a socio-psychological mechanism may have evolved, it can be sustained without any reference to kin selection. Patriotic behaviour under certain conditions can be justified from the individual's point of view even in a group composed of genetic strangers. This analysis may be even more useful in explaining the emotional aspects of behaviour in a collective crisis. In describing such events, a sense of spiritual elevation, unity and relief from guilt are often mentioned. People at such times tend to refer to such crises as 'the best time of my life'.

One of the criticisms of group selection is that it cannot be maintained since it invites exploitation by social parasites and would ultimately lead to a takeover of any group by egoists and therefore back to selection of individuals (Lack, 1966; Johnson, 1986). To establish group selection successfully, a group has to develop some defence mechanisms by which the collective can defend itself against internal social enemies. A simple logical postulate can be furthered to show that such sensitivity can only be achieved through kin selection, thus producing a circular argument in which group selection is reduced to an ability to recognize and eliminate outsiders.

The concept of social parasitism in humans is far too complex to be answered by a simple evolutionary model. Humans have developed cognition, memory and abstract reasoning that affords individuals infinite means by which to turn on the group from within; but on the other hand, it has also sensitized them to seek out, identify and penalize unsocial behaviour. Sometimes group protective action against free-loading and treason can take extreme forms and can itself be used by social manipulators. Characteristically groups are more prone to uncorroborated witch hunts in times of external threat. Actions which are merely frowned upon in times of peace are likely to be regarded as high treason at times of war and warrant summary execution. Socially unacceptable behaviour in this sense can constitute any act which is interpreted as harming the group; for example, a tax evader, in times of an economic crisis, is likely to be branded and treated as an enemy of the people. The very fact that humans are so sensitive to the unpatriotic behaviour of fellow group members, suggests that the propensity to parasitize on the group is always present and that means exist to partially control it. This in itself weakens one of the stronger arguments against group selection.

7.1 CONCLUSION

The previous discussion has attempted to explain the following phenomena:

1. Social behaviour in humans involves various acts of altruism which directly benefit the group while reducing the relative fitness of the actors in comparison with others.
2. Group loyalty is not constant over time but varies as a function of the threat to the group. The appearance of a hostile group serves as a trigger for a display of loyalty broadly described as patriotism.
3. Loyalty is elicited by and directed towards abstract social symbols (for example, flags, anthems, constitutions etc.) and not only towards other group members.
4. In all cultures there are mechanisms which produce extreme reductions in operation of natural selection within groups.

How can natural selection explain these phenomena? It has been suggested that humans were selected for group living in conjunction with the development of mental capacities and a non-specialist physiognomic adaptation. The adaptive superiority of groups in humans exists even if members are genetically unrelated. In the historical competition for scarce resources among human groups, some groups have fared better than other groups and have often systematically reduced the fitness of members of rival groups. The inclusive fitness of the individual human is thus affected not only by belonging to a group but also by the particular attributes of this group. People with inferior qualities stand a better reproductive chance than superior conspecifics if they belong to a better group.

We must now look for adaptive group traits. Three types will be mentioned in brief, they are:

1. **Cohesiveness** – a group composed of cooperative and conformist individuals is likely – all other things being equal – to prevail over a group composed of non-cooperating members.
2. **Optimal size** – the functioning size of a group is determined by various technological and economic constraints. All other things being equal, it is expected that larger groups will have an edge over smaller ones.
3. **Symbolic culture** – an affinity to common symbols such as language, rituals and values enables a collective cohesion beyond personal acquaintance. Kin preference also benefits by the development of a symbolic culture insofar as it promotes the codification of a recognizable genealogy. The consolidative social function of such a genealogy is irrespective of its factual correctness. Symbolic culture allows social systems to expand over time and space and thus to overcome smaller groups.

An analysis of patriotic rhetoric will reveal precisely the above mentioned themes as sacred goals and values. Cohesiveness will take the form of 'one for all and all for one'. Concern with size leads patriots to espouse pronatal policies even when inconsistent with economic considerations; and finally, culture becomes a yardstick by which all other groups are measured. In this framework, neighbouring, often quite similar, cultural practices often seem barbaric and inhuman while ones own cultural heritage (language, music, art, laws, landscapes, etc.) assumes an ultimate quality with strong sacred overtones.

7.2 THE BALANCE BETWEEN INDIVIDUAL COMPETITION AND COOPERATION

We can now further refine the argument on the effects of group cohesiveness on overall fitness. Though a strategy which limits intragroup competition has advantages for the group and therefore for its individual members, total cohesiveness and altruism are not necessarily optimal at all times. A certain degree of competition among members allows for selection and hence a direct improvement of group members' qualities and, indirectly, an improvement of the group as a whole. The optimal balance between competition and cooperation is probably different in different social domains (one may safely assume that competition plays a greater role in the economy than in religion) and in different social situations. A propensity for both competition and cooperation are included in the behavioural repertoire of every individual and can be released by manipulation of cultural symbols. In times of external threat, when an internal conflict would be most damaging, the symbolic universe turns to highlight unifying themes, and focuses in particular on the familial motifs. All members become brothers, the territory takes on the entity of a motherland, the leader a father figure, etc. We do not maintain therefore that individuals are selected for consistent patriotism or that groups are selected for everlasting cohesiveness. Rather we propose that both have survived the severe test of natural and cultural selection because they exhibited a flexible balance between individualistic and collectivistic tendencies.

CHAPTER EIGHT

Territoriality and threat perceptions in urban humans

M. Hopp and O.A.E. Rasa

8.1 INTRODUCTION

Most investigators intuitively view territory and territoriality as two different sides of the same conceptual coin. Territory is seen as implying some measurable spatial entity, i.e., area, distance or volume; while territoriality is used to refer to a behavioural propensity of individuals or groups to monopolize a territory by excluding others from it (Altum, 1868; Howard, 1920; Mayr, 1935; Nobel, 1939; also see review in Malmberg, 1983). Thus defined, territoriality is a form of aggression (Wilson, 1975a) which, by default, has to be acted out or at least advertised for a territory to exist. An area that is not defended is not, strictly speaking, a territory and an animal which defends no area cannot, according to this view, be called territorial.

This circular definition can be taken one step further and used as an indicator in the study of behaviour. Presence or absence of territorial behaviour can be used to establish the existence and the size of a territory; and, alternatively, the dedication animals exhibit in defending their territory can be used to derive a species specific measure of territoriality.

However, the relationship between the two concepts cannot transcend the limitations of their mechanistic juxtaposition. Territorial behaviour merely indicates a behavioural trait; it does not in itself explain the behaviour's function, its evolutionary significance or its adaptiveness. Nor can it provide an insight into the complexities of territorial behaviour in humans; whether humans perceive the space itself as a territory and therefore defensible or whether the contents of this space, considered here as resources, are of primary importance. Indeed, the very symmetry of the argument loses ground when faced with the cognitive ability of humans to define hierarchies of multiple abstract territorial concepts and defend distant, metaphysical territories in lieu of close, real ones.

In this chapter we discuss some of the concepts which underlie human territorial behaviour, in particular man's unique abilities to perceive implied types of territorial encroachment and to assess and respond to an imaginary danger. It is this ability to conceive and respond to imaginary threat situations that provides an overall picture of what urban humans consider as threats to themselves, something which could not be achieved by direct observation alone where only one situation at a time can be dealt with. The data presented here are the results of a quasi-experimental study on how threat situations are perceived and illustrate gender-specific differences in the responses of adult urban Israelis to imaginary anxiety provoking situations.

8.2 TERRITORIALITY IN ANIMALS

Territoriality appears to be an old evolutionary attribute. Although it was first described in birds (see Wilson, 1975a for a history of the concept), it has now been recognized in representatives of all vertebrate classes and many invertebrates as well, in particular amongst the Crustacea. Territories in phylogenetically primitive animals appear to be associated primarily with the monopoly of food and refuges within the habitat (for example, mantis shrimps, coral fish). Their function, however, especially amongst higher vertebrates, has been augmented to encompass breeding sites in addition to areas where females and young may be resident until the latters' independence. The ultimate of the pure breeding territory has been reached in the lek, a temporary territory solely for mating and found amongst birds, such as ruff, mannikins, birds of paradise and grouse (Gilliard, 1962; Diamond and Terborgh, 1968; Ellison, 1971) as well as mammals such as the Uganda kob and hammer-headed bat (Buechner, 1961). Probably the commonest form of territory is, however, the familial breeding territory, either held temporarily for a single breeding season or permanently. This area must offer sufficient resources to support either the mother alone (as is typical of most mammals), or both parents (which is more typical of birds), as well as the young (until they disperse). An extension of the familial breeding territory is the group territory. In this situation, the young, at least those of one sex, do not disperse on maturity but remain with their parents, the space then being defended by all group members. These usually consist of several successive generations and therefore show a high degree of relatedness. This situation is the one most commonly met with amongst the primates.

Intrinsic to the concept of territoriality is that of defence. Living areas which are not defended are termed home ranges (Burt, 1943). In the case of a territory, the area concerned is actively defended by the holder, usually by means of a variety of aggressive displays ranging from ones with low energy expenditure, like odour marks and vocalizations such as bird song, on the one hand, to high-energy expenditure, such as escalated fighting, on the other. Territoriality is therefore a complex phenomenon, the basic function of which depends on the resource or resources being defended, these being as simple as an area in which to copulate or as complex as an area containing sufficient food, refuges and mating opportunities for an entire group.

Although the primary function of a territory may be the monopolization of resources, territories also serve to space populations as well as performing a function in selection. Since territories are contested and space or resources are limited, not all individuals may be capable of holding and maintaining a territory and thereby directly or indirectly influencing their future reproductive success. Territoriality can thus be considered as a mechanism by which individuals are selected for fitness, only successful territory holders being able to pass on their genes to future generations thus fixing territoriality as a behavioural strategy in the genome. The correlation between efficient territory holding, quality of territory and reproductive success has been especially well demonstrated for several species of birds for example, dickcissel (Zimmerman, 1971) and great tits (Krebs, 1971).

Certain species of mammals use marks of various kinds to indicate their occupancy of a particular space. The majority of territorial mammals are macrosmatic and territory occupancy is indicated effectively by means of odour, either in the form of elimination products or by means of the secretions from special glands being placed at points along the territory border or within the territory itself. In the majority of the Prosimia and the Callithricidae, olfactory territorial marking is still well represented. In the case of microsmatic mammals such as the Old World monkeys, certain of the New World monkeys and the great apes, however, other means of indicating territory occupancy have evolved. The most spectacular of these are the vocal choruses of gibbons and howler monkeys, in which all members of the group participate (Carpenter, 1940). Visual signals are more rarely encountered, probably owing to the fact that the majority of primate species are forest and bush dwellers, habitats where visual signals would not have carrying power beyond a few metres; these do not appear to have been selected for. In man, one of the few primates to have adapted to life in open areas, territory demarcation has become almost entirely visual and the borders of territories are marked by barriers such as fences and walls, and augmented by 'keep off' signs and locks.

8.3 TERRITORIALITY IN MAN

Territoriality in humans is a complex, often nebulous concept made even more perplexing by the unique orientations of the different disciplines which touch it. In its extreme and simple forms, human territoriality is easy to identify and can be discussed as a simple extension of territoriality in animals. Analysis at this level is mostly space related and sees national boundaries, property ownership or street gang 'turfs' as examples of resource defence in humans.

Early devotees of this point of view strongly advocated the animal–human continuum, and focused their attention primarily on the active defence of a fixed spatial territory (see Lorenz, 1969; Eibl-Eibesfeldt, 1970). Other contributors, notably Hall (1959); Parr (1965); and Stea (1965) stressed the 'spatial claim' component which must precede any readiness to defend a territory. Their main attention is on territorial display rather than on fixed boundaries and it provides a suitable background for the development of two related concepts: temporary territories and mobile territories. Temporary territories are occupied for a time only, while mobile territories are volumes of space individuals 'carry around them' wherever they go. Such mobile territories can further be classified into different types of space according to the degree of territorial infringement an individual will tolerate by outsiders. Summarizing theoretical work on proxemics (the name given to the body of theory on human space utilization), we may thus speak of 'social space', 'informal space', 'personal space' and 'intimate space', each denoting distances individuals tend to put between themselves and others according to the situation and the degree of closeness they feel towards these others (for extensive reviews of the anthropology of space, see Goffman, 1972; Hall, 1976).

The concept of a mobile territory liberates the theory from the confines of fixed demarcated boundaries but does little to extend it beyond overt defence behaviour or display. This presents several problems. First, it is not clear if humans, under normal conditions, expend a major portion of their energy in constant territorial defence or display. Secondly, normal humans adapt well to extreme personal space variation (i.e. congestion on a train or hospitalization in a public ward). Thirdly, the daily course of human affairs requires a routine transgression of personal boundaries which is mostly accomplished with no conscious hostility by the parties involved. Finally, humans readily leave fragments of their territories undefended for lengthy periods of time (for example, a car in a parking lot or an apartment during the day) their ownership being indicated by 'proxy' displays such as locks, bars and signs.

Some of these concerns are partially addressed by the concept of 'body language'. This is based on a theory of unconscious, non-verbal communication and holds that all individuals constantly transmit involuntary messages to conspecifics through their body posture, facial features and other visual cues. These advertise internal constituents such as mood, health, status, and readiness to fight and obviate the need for overt territorial display (Fast, 1972).

An alternative, though not exclusive solution is proposed by Altman and

Haythorn (1967), who suggest that: 'A territory is defined in terms of the degree of consistent and mutually exclusive use of particular chairs, beds, sides of tables etc.' (see also Altman, 1970). Altman and Haythorn thus dispense with the primacy of territorial defence in favour of a theory that considers territorial possessions as objects that are intuitively recognized and accepted by all. A claim, for example, a bed, does not have to be defended again and again because it is implicitly recognized as belonging to the person who claims it and because other individuals tend to shun objects such as chairs, beds, clothes or even cars which belong to others.

The attractiveness of Altman's formulation lies in its concordance with the realities most people intuitively recognize. It affords a gestalt of human coexistence in which, under normal conditions, people need not endlessly patrol the boundaries of their territories or display aggressively at any approaching stranger. Such conditions exist on communal rookeries of many bird species but are not suggestive of the human strategy.

There is an additional aspect which singles out human territoriality from that of all other animals, and here too Altman's approach makes good sense. Humans are unique in their ability to visualize abstract territorial violations. They can reason a sequence of occurrences and conjure an infinite set of potential territorial calamities long before they are likely to happen. To maintain a defensive attitude under such conditions would demand either a solitary existence or a complete devotion to total paranoia. The fact that people are generally able to feel safe within the confines of their private universe, and not compelled to actively defend their possessions at all times, flies in the face of the defensive territoriality alternative.

Under the conditions of a relatively easygoing territorial system in which one's real estate properties, resources and personal safety are not constantly threatened, a reformulation of the territoriality question is in place. Assuming that people are not on an ever ready alert to defend their rights, they run the opposite risk of becoming overconfident. Some mechanism must exist, therefore, to steer an appropriate course between the extremes of carelessness and paranoid suspicion and to elicit defensive behaviour only in the presence of certain stimuli and conditions. Rather than investigate the extent of human territoriality we must study the mechanisms which put people on their guard, in other words, study what makes people afraid.

The second part of this chapter is devoted to this question by way of a preliminary investigation in a quasi-experimental study on the territorial components of anxiety responses in humans.

8.4 EXPERIMENTAL DESIGN AND METHODOLOGY

To test people's responses to the various aspects of territorial, personal and property threat, a survey was conducted on a randomly chosen representative sample of adults in Israel. The study constituted a part of a larger Public Opinion questionnaire in which respondents were questioned on various topics in face to face interviews with trained interviewers. In total, 1200 people were interviewed in the study and 1183 valid responses were obtained.

The study comprised a set of 14 situational propositions, ordered in pairs and printed on a card. Participants were asked to look at the card, choose from each pair the more anxiety provoking statement of the two and mark it on the card. After they had ranked all 7 pairs, respondents were requested to go over the statements again and point out the most anxiety provoking statement in the whole series.

The statements or propositions portray various unpleasant situations which any individual in urban society might encounter. By design, every situation contains three menace elements (or variables): physical threat; territorial threat; resource threat. In every pair of statements one variable was held as constant as possible while two were varied as much as possible. Items were chosen on the basis of assessments made by 10 innocent respondents who were asked to rank 30 items on these 3 scales and were not told what the purpose of the study was. Their assessments were correlated and averaged to derive the overall rating of each of the items on a two level ordinal scale (low and high). The 14 best fitting items on the score were then chosen for the study. Care was taken in pairing the statements to control for the range of severity between the items in every pair and amongst the variables within each item.

Table 8.1 lists the items chosen in the order of their presentation to respondents with the rater assessments for each proposition on the three experimental variables. Three types of territories are represented – a static one, 'home', a mobile one, 'car' and a personal distance type variable (statements 3b and 7a). Classification of physical threat is more difficult, since it comprises a level component (low to high threat) and an overtness element (implicit threat in statements 1a and 7b, and explicit threat in statements 3b, 4a, 6b and 7b). Resource threat refers primarily to actual property loss or damage, with its associated effort and costs.

Before analysis the data were cross-checked to determine the effects of intervening variables related to cultural norms, such as level of education, religious affiliations, security consciousness (measured as the number of security measures taken to prevent burglary) and ethnic background. None of these were shown to have any significant effect on the results obtained.

The study investigates whether men and women view territory, possessions and personal threat in the same way, which would indicate an underlying species concept of what constitutes danger, or whether these variables have different emphasis for the two sexes. It attempts to relate the findings to human social structure and behaviour in the urban environment.

Table 8.1
Ranking of paired items testing respondents' anxiety level

		Rater assessment		
	Statement	*Physical threat*	*Territorial threat*	*Resource threat*
1a.	You wake up in the morning and discover that a thief was in the house at night. Nothing was stolen.	HIGH	HIGH	LOW
1b.	You come home and find that while you were gone a thief was in the house. An expensive watch was stolen.	LOW	HIGH	HIGH
2a.	You come home and find that while you were gone, a thief was in the house. Nothing was stolen.	LOW	HIGH	LOW
2b.	You come home and find that while you were away, a water pipe burst. Neighbours came to help and caused damage.	LOW	LOW	HIGH
3a.	A nasty neighbour comes in, shouts about neighbourhood tax matters, and refuses to vacate the premises.	LOW	HIGH	LOW
3b.	Someone nasty latches on to you in the street and threatens you physically.	HIGH	LOW	LOW
4a.	You have a minor car accident at a traffic light. The other driver behaves violently.	HIGH	LOW	LOW
4b.	You come to the car in the morning and find that one of its bumpers was smashed during the night.	LOW	LOW	HIGH
5a.	You come to the car in the morning and find that someone maliciously made a long scratch on one side.	LOW	HIGH	HIGH
5b.	You come to the car and find someone has let the air out of all the tyres.	LOW	LOW	LOW
6a.	You come to your car in the morning and find that someone had broken in and upset everything inside.	LOW	HIGH	HIGH
6b.	You arrive at your car just as someone is about to break into it. No damage has yet been done.	LOW*	HIGH	LOW
7a.	You go home in the dark and are followed by two suspicious looking characters.	HIGH	LOW	LOW
7b.	You look out the window and think there are two suspicious characters lurking outside.	MEDIUM/ HIGH*	HIGH	HIGH

*These items received intermediate rankings. Item 6b was classified low/medium on Physical Threat, and item 7b was classified medium/high on this variable.

8.5 RESULTS

8.5.1 Threat perceptions

We analysed the choice of what people considered the most hazardous item in each pair, first for the whole population and then by gender. Table 8.2 presents the distribution of items chosen within each pair in percentages, together with significance levels for the overall difference between choices for the population and the difference between genders.

The findings indicate that, in the vast majority of cases, both men and women perceive various anxiety provoking situations in essentially the same way. Any differences in the perception of what constitutes 'danger' between the sexes must therefore reflect testable gender specific orientations.

Situations perceived as presenting actual physical threat, whether implicit or explicit, are scored higher by both sexes than ones not containing this variable, independent of the territory or resource loss content of the situation (1a vs. 1b, 3b vs. 3a, 7a vs. 7b). Both men and women rank resource loss or damage in conjunction with territory invasion higher than territory invasion alone (2a vs. 2b, 5a vs. 5b). Actual territory invasion, however, even though no damage was done, ranks higher than threatened invasion (6a vs. 6b). Only in the case of statement 4 was no significant difference found in the overall response of both sexes to the imagined situation.

It appears from these findings that there is an underlying uniformity in what urban Man considers to be a cause for anxiety. Threat of personal damage ranks highest followed by damage to property or resources, invasion of territory or personal space being the lowest of the priorities.

Although members of both sexes regard the imagined situations as anxiety provoking in essentially the same way, there are also significant quantitative differences between the sexes. Women respond to situations involving physical threat significantly more strongly than men do (1a vs. 1b, 3b vs. 3a, 7a vs. 7b, 4a. vs. 4b). It is the female response to the physical threat situation in statement 4 rather than property damage alone which influences the non-significance of the choice differences observed for this imagined situation in the overall analysis. Women find the implied physical threat variable significantly more anxiety provoking than resource damage alone, while men regard this juxtaposition of variables from the opposite point of view.

8.5.2 Threat ranking

In order to determine what men and women considered to be the most anxiety provoking situation of the 14 listed, after adjustment for the number of males and females in the sample, the item scored as 'most hazardous' was weighted to reflect a comparable level of severity by multiplying the absolute ranking of each item within a pair by its ranking on the 'worst item' list:

weighted rank = ranking within pair x ranking as 'worst item'

The weighted scores thus allow a comparative ranking of the items on a gradient from the most to the least anxiety provoking situation, and these are shown in Table 8.3.

The most striking difference between men and women is their ranking of the most anxiety provoking situation of all. For women it is one consisting, in its variables, entirely of implied rather than explicit physical threat with no territorial or resource connotations whatsoever (statement 7a). For men, however, the greatest hazard was considered a situation in which both territorial threat and physical threat were scored as 'high' (statement 1a). Men scored the situation women found the most hazardous only fifth on their list, while women found the situation men considered most hazardous their second most anxiety provoking choice. The importance of the physical threat component can be implied from the finding that the same situational constellation, differing only in the fact that the territory owner was absent i.e., there was no physical threat involved (statement 2a), was ranked last by women and twelfth by men.

With the exception of statement 1b, which was ranked fourth, women find situations in which they are threatened physically, either explicitly or implicitly, the most anxiety provoking ones. Situations in which there are resource losses or damage (ranks 4 to 11) elicit more anxiety responses than those where a territorial component is present in the absence of physical or resource threat (ranks 8 to 14). For men, resource loss in the absence of physical threat scores higher (ranks 3 to 8) than it does for women, while again, variable constellations with a high territorial threat component but a low one for physical and resource threat are ranked lowest (ranks 9 to 14).

The importance of a physical threat component to women's anxiety responses can be implied from the more gradual reduction of the weighted male scores compared with the female ones. Female scores drop rapidly from rank 4 onwards while scores for males show a more gradual decline. Women appear to consider threats to resources and/or territory of far less importance than do men. This difference in score drop between the sexes can be related to the finding that more men than women score situations in which resource loss is involved, in either actual cost or effort involved, as the 'most anxiety provoking' of the situations presented. Invasion of the territory alone scores the lowest for both sexes. One anomaly here is that, for women, invasion of the mobile territory – car (statement 6a) – is considered far more anxiety provoking than invasion of the static territory – house (statement 2a) – while, for men, the two are almost equal in rank.

Table 8.2
Distribution of more hazardous items within paired statements by sex. In percentages for each pair with significance scores for differences between sexes and between items in each pair

		Pairwise perceived threat (%)			*Significance of differences*	
	Statement	*Males*	*Females*	*Population overall*	*Male/ female*	*Between items*
1a.	You wake up in the morning and discover that a thief was in the house at night, nothing was stolen.	53.7	64.1	59.1	< 0.01	< 0.001
1b.	You come home and find that while you were gone a thief was in the house. An expensive watch was stolen.	46.3	35.9	40.5		
2a.	You come home and find that while you were gone, a thief was in the house. Nothing was stolen.	42.8	41.7	42.1	N.S.	< 0.001
2b.	You come home and find that while you were away, a water pipe burst. Neighbours came to help and caused damage.	57.2	58.3	57.5		
3a.	A nasty neighbour comes in, shouts about neighbourhood tax matters, and refuses to vacate the premises.	37.0	22.9	25.8	< 0.001	< 0.000
3b.	Someone nasty latches on to you in the street and threatens you physically.	63.0	77.1	69.9		
4a.	You have a minor car accident at a traffic light. The other driver behaves violently.	48.9	53.8	51.2	< 0.05	N.S.
4b.	You come to the car in the morning and find that one of its bumpers was smashed during the night.	51.1	46.2	48.7		

5a.	You come to the car in the morning and find that someone maliciously made a long scratch on one side.	66.9	68.8	67.4	N.S.	< 0.000
5b.	You come to the car and find someone has let the air out of all the tyres.	33.1	31.2	32.6		
6a.	You come to your car in the morning and find that someone had broken in and upset everything inside.	67.0	65.9	66.5	N.S.	< 0.000
6b.	You arrive at your car just as someone is about to break into it. No damage has yet been done.	33.0	34.1	33.5		
7a.	You go home in the dark and are followed by two suspicious looking characters.	64.0	77.5	70.2	< 0.001	< 0.000
7b.	You look out the window and think there are two suspicious characters lurking outside.	36.0	22.5	29.1		

Table 8.3
Weighted, standardized gender specific rank scores in a descending order

		Standardized scores		
Rank	*Item*	*Men*	*Women*	*Item*
1	1a	1526	1890	7a
2	3b	871	1820	1a
3	1b	696	864	3b
4	2b	605	429	1b
5	7a	470	250	7b
6	7b	261	153	2b
7	4a	200	128	4a
8	4b	168	77	6a
9	5b	108	66	3a
10	3a	80	50	4b
11	5a	60	32	5a
12	2a	39	18	5b
13	6a	16	12	6b
14	6b	16	4	2a

8.6 DISCUSSION

The findings of this investigation have thrown a new light on how urban humans conceive of objects and spaces that they profess to 'own' as components of a territory. This study was designed to test Altman's definition of a human territory, in which he states that, in contrast to other primates, humans are less overtly territorial than hitherto suggested and that a flexible mitigation of territoriality by mechanisms of mutual trust is a necessary part of human social strategy.

Despite the implicit conceptualization of a territory as the exclusive ownership of a particular space, which is reflected in sayings such as 'an Englishman's home is his castle', the findings of this study have indicated that, at least for the urban Israeli population, such concepts are not of primary importance in regulating human spacing and relationships. Imagined fear of territorial invasion was found to have a low precedence on the scale of situations provoking anxiety whereas, if Man was primarily a territorial species living in a familial territory, as do the majority of birds and mammals, threat of territorial invasion would be expected to provoke high anxiety levels.

One hypothesis that can be put forward to explain this anomaly from a sociobiological point of view is that, in contrast to lower vertebrates, man does not need a spatially cohesive territory in order to reproduce successfully. There has been no evolutionary pressure for the selection of traits predisposing towards exclusivity of ownership of a particular spatial area. Our findings indicate, however, that although territoriality itself may not be well established in humans of either gender, males, especially, consider a threat to their resources as being highly anxiety provoking. This suggests that for males, in contrast to females, 'success' or 'status' are more closely associated with possession of resources rather than space alone and this success or status may influence the quality rather than the quantity of females he can attract and the offspring a male can produce (for a review see Grammer, 1989). For the urban Israeli man, therefore, ownership of resources is an important factor and imagined loss or violation of those resources are highly anxiety provoking situations, while simple invasion of the living space by a stranger is less so.

Although situations in which physical threat and resource loss are, in general, more anxiety provoking than territorial invasion alone, one indication that territoriality may be present in human males is the finding that by far the majority of men in this survey considered that invasion of their territory (house) while they were in it and unaware (statement 1a) was the most anxiety provoking of all the alternatives they were presented with. With the exception of statement 7b (ranked sixth), this was the only situation presented in the questionnaire where territory invasion occurred during occupancy of the territory and has implications, therefore, for physical threat and active counteraction of that threat as well, the effects of which might override those of territory alone. In statement 7b, however, the threat is only an implied one and the territory owner is aware of the potential invaders. This situation proved far less anxiety provoking than the former, where territory invasion was a *fait accompli* and the possibility of physical threat was present, even though the threat was not carried out. This suggests that territorial invasion when the owner of the territory is present and relatively helpless may have a completely different connotation to the same invasion in the absence of the territory owner (see statement 2a, ranked twelfth). Whether this is a more widespread phenomenon amongst vertebrates is not known, since we have no means of measuring, either objectively or subjectively, an animal's territorial responses towards a territory from which it has been removed. To determine whether these traits are common to *Homo sapiens* in general, however, or whether they are peculiar to the urban environment, the study needs to be conducted on populations from other races and cultures.

Probably the most striking finding of this study is that women appear to be almost non-territorial and to have little anxiety with respect to resource loss, as it pertains to property. By far the most anxiety provoking situation for the urban human female is physical threat, either explicit or implied, especially in the absence of any supportive help. The two most anxiety provoking situations which score as worst items far above all the others, however, are ones involving probable inter-sexual confrontations and possible sexual violation (rape situations). These can be considered a form of resource loss, since a situation in which a woman could get pregnant against her will with no further material support from the impregnating male, represents a high cost factor, not only from the point of view of possible physical injury and the biological investment in the child, but also with regard to the sociological and sociobiological implications of the outcome. Although the mother can replicate her genes in this particular child, the child itself can represent a block to further successful reproduction. The presence of an illegitimate child in urban society reduces status and, with it, the possibility of making a good marriage (i.e., subsequent reproduction with more than one offspring in a qualitatively good environment) or may even disrupt an existing marriage. Considering how widespread both geographically and culturally the negative connotations associated with non-marital reproduction are, the extreme fear of potential rape situations recorded in our data may reflect a more basic biological trend rather than a purely cultural one.

This preliminary study on urban man's concepts of what is anxiety provoking and what is not has raised a host of questions and hypotheses. Without doubt, situations in which physical threat is involved are considered by both men and women to be the most anxiety provoking of all, although the reasons for this anxiety may differ between the sexes. Loss of resources, as indicated by personal possessions, ranks second in the anxiety-provoking scale while invasion of personal living space, if neither of the two variables above are involved, is of low significance. This could be taken to mean that urban man is non-territorial; the concept of territory here, however, may not be equivalent to the concept of the same name in lower vertebrates: an area defended in space and time. Our investigations have indicated that territory, as far as urban man is concerned, may be a fragmented rather than a cohesive entity, each of these fragments containing different types of resources. Strong feelings of anxiety are aroused by violation of space or possessions in the presence of the owner but less so in his or her absence. For urban man, the concept of territory appears to be more than a coherent spatial one, as it is in most other vertebrates, and may be expanded to constitute spatially scattered sites where it is the resources that these sites contain that are of primary importance, rather than the sites themselves. Evidence in support of this hypothesis is that violation or loss of resources rank higher than actual spatial invasion in the absence of the subject on the scale of most anxiety provoking situations.

Animal territories always include a resource or resources, even if the resource in question is restricted to mating opportunities. The whole purpose of territoriality is to defend the resources an area contains rather than the area itself. We hypothesize that the way of life of urban man, which necessitates the spatial scattering of resources, has resulted in the adoption of a non-spatially cohesive territory and this has posed problems with regard to its defence. An average urban territory could contain sleeping, eating, working and breeding sites, with their own specific resources and spatial requisites, each of which the subject only visits at intervals, usually by means of a territorial extension, the car. Our study has revealed strong defensive behaviour (as its mirror image, anxiety) with respect to the site the subject momentarily inhabits, but less marked defence responses towards temporarily vacated areas and the resources they contain. This makes sociobiological sense from the cost/benefit point of view. The costs entailed to actively defend all sections of a spatially scattered set of resources would far exceed the benefits of attempting to retain those resources in the face of opposition, especially when the resources are relatively easily renewable or replaceable. Sites are defended passively in the absence of the owner by means of visual signals such as fences, 'keep out' signs and locks, but these are not completely effective. It is interesting to speculate on the whole concept of property insurance in this regard, which exists in the expectation that such resources will be plundered in the owner's absence and spreads the actual cost of such resource loss in small instalments over a long period of time. Insurance and modern urban man arose contemporaneously, indicating a possible connection between urbanity, expected territorial invasions with resource loss and the non-spatially cohesive territory hypothesis presented here. This study has shown that even the imagined loss of resources is still an anxiety provoking situation, indicating that protectiveness towards property (i.e., territorial behaviour), with all the connotations this has for status and success, still plays a major role in urban man's behavioural repertoire.

Appendix

Unweighted distribution of the most hazardous item in percentages by sex

Statement	*Most hazardous item Unweighted distribution* Men	Women	Overall
1a. You wake up in the morning and discover that a thief was in the house at night. Nothing was stolen.	19.7	23.4	21.1
1b. You come home and find that while you were gone a thief was in the house. An expensive watch was stolen.	10.2	6.9	8.5
2a. You come home and find that while you were gone, a thief was in the house. Nothing was stolen.	2.5	0.7	1.6
2b. You come home and find that while you were away, a water pipe burst. Neighbours came to help and caused damage.	9.6	3.4	6.5

3a. A nasty neighbour comes in, shouts about neighbourhood tax matters, and refuses to vacate the premises.	2.8	1.6	2.2
3b. Someone nasty latches on to you in the street and threatens you physically.	11.5	13.6	12.6
4a. You have a minor car accident at a traffic light. The other driver behaves violently.	4.3	3.3	3.8
4b. You come to the car in the morning and find that one of its bumpers was smashed during the night.	4.2	1.4	2.8
5a. You come to the car in the morning and find that someone maliciously made a long scratch on one side.	2.8	1.3	2.0
5b. You come to the car and find someone has let the air out of all the tyres.	3.0	0.9	1.7
6a. You come to your car in the morning and find that someone had broken in and upset everything inside.	1.1	1.8	1.5
6b. You arrive at your car just as someone is about to break into it. No damage has yet been done.	1.3	0.5	1.9
7a. You go home in the dark and are followed by two suspicious looking characters.	8.5	24.8	16.8
7b. You look out the window and think there are two suspicious characters lurking outside.	5.1	5.3	5.2

PART THREE

'Primitive' Warfare

CHAPTER NINE

Origin and evolution of 'primitive' warfare

J.M.G. van der Dennen

CONFLICT AS STRUGGLE

9.1 SOCIAL DARWINISM

In this chapter biological theories of the origin of warfare in human evolution and/or biological theories of primitive warfare will be reviewed and criticized. Biological theories are considered to be those which derive their main concepts from evolutionary biology, ethology and related disciplines. The theories reviewed range from the period of Social Darwinism to contemporary neo-evolutionist anthropology and modern sociobiology. Most biological-evolutionary theories of primitive war and the origin of war have their roots in Social Darwinist thinking (although they are reluctant to acknowledge this heritage), so it seems only fair to begin with a brief *tour d'horizon* of Social Darwinism. Social Darwinism may be regarded as a distinct historical period ranging from approximately mid-nineteenth century to about mid-twentieth, but this is rather uninteresting. It is more interesting to regard it as a confluence of ideas, a syndrome of doctrines and recurrent themes. As the main intellectual inputs we shall consider: racialism; evolutionism/selectionism; instinctivism (and functionalism).

9.1.1 Racialism

At the end of the eighteenth and the beginning of the nineteenth centuries, a number of philologists and historians started the theory of Aryanism (and later on, Teutonism and Nordicism). Sorokin (1928) explains that though some of them understood that the Aryans were a linguistic group, they often mixed it, nevertheless, with the Aryan race, and in this way facilitated the appearance of a purely racial interpretation of history. The most influential were the racial theories of Gobineau, Gumplowicz, and Chamberlain. Gobineau's (1853) thesis was that the 'inequality of races is sufficient to explain the entire enchainment of the destinies of peoples'. There are the inferior and the superior races, and only the latter are able to attain true civilization. Chamberlain (1899) adds little, except by trying to show that the most superior race is the white, particularly the Aryan 'race'.

The principles of Gumplowicz's (1883) theory are as follows: first, the theory of polygenesis, or the multiple origin of mankind, developed by Gobineau 30 years before. Second, the assumption of an inherent and deadly hatred and animosity in the relationship of one racial group to another, resulting in an inevitable and deadly struggle between the groups (*Rassenkampf*). Third, the assumption that only through such a struggle has any enlargement of the social group, or any consolidation of two or more groups into one social body, been possible. Fourth, the victorious group, having conquered its victim, pitilessly exploits it, turning it into slaves or subjects. For the sake of successfully controlling them it enacts laws, and in this way we have 1. the origin of the state, 2. the origin of law, and 3. the origin of stratification and inequality.

Once the means of warfare became developed, it was only natural that war would be a constant feature of humanity. With war comes the possibility of the conquest of one group by another and the development of a strong state to ensure a means for domination. Disciples and followers of Gumplowicz were, among others, Small, Ward, Kochanowsky, Oppenheimer, de Savorgnan, and Ratzenhofer. Ratzenhofer (1893) condenses his theory to a single proposition: 'The contact of two hordes produces rage and terror. They throw themselves upon one another in a fight to exterminate, or else they avoid contact'.

'Until now', Novikow (1912) comments, 'it was believed that men fought their fellows in order to obtain food, women, wealth, the profits derived from the possession of the government, or in order to impose a religion or a type of culture. In all these circumstances war is a means to an end. The new theorists proclaim that this is all wrong. Men must of necessity massacre one another because of polygeny. Savage carnage is a law of nature, operating through *fatality*'. Novikow has three objections against the racialist doctrine:

1. Until now there have been no race wars, for the simple reason that the races have not been conscious of their individuality.
2. When the wars of political domination took place between two linguistic groups, they became race wars by chance.
3. The Swedes, the Danes, and the Germans are Teutons. That has not prevented them from fighting one another furiously. While conversely, on numerous occasions the first contact of two very different races has been peaceful (like that between Welsh and Tehuelche in Patagonia, 1865).

Note that this racialist intellectual input to the movement or period called Social Darwinism has very little to do with Darwin. The *Zeitgeist* was apparently such, however, that it could easily be incorporated into mainstream Social Darwinism.

9.1.2 Evolutionism/selectionism: the struggle for existence

Conflict and struggle were long ago declared a fundamental law of the universe, of life and of man's existence; and hence the source of all change and progress. Even the theory of the 'survival of the fittest' was outlined not later than the fifth century BC (Empedocles, Heraclitus), and may also be found in the Zend-Avesta (Sorokin, 1928). In the nineteenth century a great impetus to the idea was given by Spencer, Darwin, Wallace and Huxley. Darwin (1859 *et seq.*) took the idea of a struggle for existence from Malthus (1798) (and the term from Spencer), but hardly defined this basic concept. Later authors have interpreted it in their own way. There are authors who talk of the struggle for existence among atoms, molecules, organisms, human beings and societies (Novikow, 1896; Tarde, 1897). Some other authors use the term only in application to living beings, but in a very general and loose sense (Bagehot, 1884; Nicolai, 1919).

The essence of Malthus' (1798) hypothesis is that population tends to increase faster than the means of subsistence and that this increase is checked by wars, epidemics and famines, to which Malthus subsequently added 'moral restraint', meaning deferred marriage and sexual abstinence. He regarded warfare in the earlier ages of the world as 'the great business of mankind', and considered as one of the first causes and most powerful impulses of war 'undoubtedly an insufficiency of room and food; and greatly as the circumstances of mankind have changed since it first began, the same cause still continues to operate and to produce, though in smaller degree, the same effects'. Before Malthus many authors indicated the demographic factor as one of the principal causes of war. Malthus generalized the theories into a law where war functions as one of the effective checks on population. Since that time, this idea has become quite common in various formulations. Novikow (1896; 1912) distinguished four principal types in the evolution of the struggle for existence among human beings: physiological, economic, political and intellectual. According to him, in the course of time the ruder forms of struggle are superseded by milder ones: 'No grim fatality obliges us to massacre one another eternally like wild beasts . . . The Darwinian law in no wise prevents the whole of humanity from joining in a federation in which peace will reign . . . All the theories based on that alleged fatality are pure phantasmagorias absolutely devoid of all positive reality' (Novikow, 1912). Also Vaccaro (1886 *et seq.*) envisioned progress from ruthless extermination at the earliest stages of human evolution to the disappearance of war in the future. These ideas were welcomed by many authors such as Tarde (1899), Kropotkin (1902), Kovalevsky (1910), Ferri (1895), de Molinari (1898), Ferrero (1898), Nicolai (1919), Sumner and Keller (1927), among many others. Steinmetz (1907; 1929) formulated '*Das Gesetz der abnehmende Kriegsverluste*' (the law of diminishing war casualties). Sorokin (1928) outlined the fallacies in this kind of reasoning.

Darwin himself contributed little to what has come to be called Social Darwinism, (Schellenberg, 1982) – and what more aptly would be called Social Spencerism. Other writers were far less reticent in tracing the implications of Darwinism for society. One whom Darwin cited with general approval was his fellow Englishman, Herbert Spencer, who was also influenced by Malthus. Spencer was an even more thoroughgoing evolutionist than Darwin, extending this idea to the far reaches of human society and, indeed, the whole universe. Spencer also talked about the importance of the 'survival of the fittest' (a phrase which Darwin borrowed from him before the *Origin of Species* appeared). Such evolution he viewed as a cosmic law of nature, applicable alike to the inorganic (physical), organic (biological) and superorganic (or sociocultural) realms.

On the superorganic level, societies may be seen as significant integrations which continually increase in coherence and heterogeneity. This view of human society is parallel to the conception of biological organisms as emergent wholes. Societies, like individual organisms, develop their integrities or unities through a struggle for survival, a struggle which in large pits society against society. Competition for survival is therefore characteristic of both individual organisms and societies, and in this struggle are forged the characteristic forms taken by animal species and human societies. Fear is endemic in the uncertainties of early forms of human society, which leads to religious and political forms of social control. Especially prominent is a military form of social organization. With the extending scope of social organization, however, more emphasis can be given to peaceful pursuits (Schellenberg, 1982; cf. Service, 1975; Hofstadter, 1955).

Spencer was the principal proponent of the group-selection thesis. In his *The Study of Sociology* (1873) he states: 'Warfare among men, like warfare among animals, has had a large share in raising their organizations to a higher stage. The following are some of the various ways in which it has worked. In the first place, it has had the effect of continually extirpating races which, for some reason or other, were least fitted to cope with the conditions of existence they were subject to. The killing-off of relatively feeble tribes, or tribes relatively wanting in endurance, or courage, or sagacity, or power of co-operation, must have tended ever to maintain, and occasionally to increase, the amounts of life-preserving powers possessed by men. Beyond this average advance caused by destruction of the least-developed races and the least-developed individuals, there has been an average advance caused by inheritance of those further developments due to functional activity . . . A no less important benefit bequeathed by war, has been the formation of large societies. By force alone were small nomadic hordes welded into large tribes; by force alone were large tribes welded into small nations; by force alone have small nations been welded into large nations.'

Inter- and intra-group selection go on to the present day, but, Spencer asserts, in civilized warfare the intra-group selection has become negative and retrogressive, eliminating the best elements of the population. Furthermore, warfare is at variance with industrial development. As an institution it has become dysfunctional.

Spencer was also one of the first to discuss what we call today 'ethnocentrism' or the phenomenon of ingroup–outgroup differentiation. In his *Principles of Ethics* (1892–3) he wrote: 'Rude tribes and . . . civilized societies . . . have had continually to carry on an external self-defence and internal co-operation – external antagonism and internal friendship. Hence their members have acquired two different sets of sentiments and ideas, adjusted to these two kinds of activity'. The theme of ethnocentrism was later elaborated by Sumner (1906; 1911), who coined the term. Spencer was also the founder of functionalism in sociology and anthropology.

Two years after Darwin's *The Descent of Man*, appeared the first significant work of biologically derived speculation to break Spencer's monopoly in that field: Bagehot's (1884) *Physics and Politics*. Bagehot – the first avowed Social Darwinist according to Service (1975) – attempted to reconstruct the pattern of growth of political civilization in the manner of evolutionary ethnologists like Lubbock and Tylor, from whom he drew some of his data. There is no doubt of the predominance of natural selection in early human history, Bagehot asserted: 'The strongest killed out the weakest as they could'. He felt that warlike competition among societies in early times would select for those with the best leadership and most obedient populace. Hence his much quoted adage: 'the tamest are the strongest'. Progress, habitually thought of as a normal fact in human society, is actually a rare occurrence among peoples. Of the existence of progress in the military art there can be no doubt, nor of its corollary that the most advanced will destroy the weaker, that the more compact will eliminate the scattered, and that the more civilized are the more compact (Hofstadter, 1955).

Continuing Spencer's functionalist line of thought was the American sociologist Sumner. 'It is the competition of life', Sumner (1911) asserts, 'which makes war, and that is why war always has existed and always will. It is the condition of human existence.' The foundation of human society, said Sumner (1911; Sumner and Keller, 1927), is the man/land ratio. Conflict over the means of subsistence is the underlying fact which shapes the nature of human society. When population presses upon the land supply, earth-hunger arises, races of men move across the face of the world, militarism and imperialism flourish, and conflict rages. Where man are few and soil is abundant, the struggle for existence is less savage: 'Wherever there is no war, there we find that there is no crowding'. Sumner emphasized group factors (including the binding power of folkways and mores) more strongly than did Spencer, and Sumner was less optimistic about the direction of evolutionary change (Hofstadter, 1955; Schellenberg, 1982).

Among the other theories which composed the anthropo-racial school in sociology were the selectionist theories of de Lapouge and Ammon, and the hereditarist school of Galton and Pearson, which led to a huge eugenics movement especially in the Anglosaxon countries (see Hofstadter, 1955).

Following Spencer's suggestion, de Lapouge (1896 *et seq.*) contended that the selection caused by war and other forms of social selection is essentially negative and deleterious. This led him to the formulation of his law of the quicker destruction of the more perfect racial elements. Ammon (1893 *et seq.*) agrees in essence with this law of decay, as did Ferrero (1898), de Molinari (1898), Novikow (1912), Nicolai (1919), and many others. The argument runs as follows: armies, as a general rule, are composed of the 'best blood' of the population. During a war, it is the army which suffers losses most. This means that war exterminates the best blood of a nation in greater proportion than its poorer blood. This means that war facilitates a survival of the unfit, favours a propagation of the poorer blood and in this way is a factor of negative selection and racial degeneration (Sorokin, 1928). Vaccaro (1886) stressed another form of this: the Roman rule *parcere subjectes et debellare superbos* (spare the submissive and demolish the proud men) has been a general rule of almost all wars. Therefore: 'since the submissive, to the exclusion of the brave and upright men, beget children, the traits of baseness and servility become fixed in the race'. In this way war selection has exterminated millions of the best individuals.

Hundreds of studies were dedicated to the problem whether selection due to war was negative, neutral, or positive. Some of the authors went so far in an evaluation of the negative selection of war that they made it responsible for the decay of nations (Seeck, 1910; Jordan, 1907). A number of authors indicate that, even at the present time, war's selection is far from being as negative as supposed. After all, 'the great fact remains that somehow man has evolved, and he has fought, presumably, half of the time. If warfare is so deleterious it may be asked: how did he get where he is?' (Woods and Baltzly, 1915; Sumner, 1911).

Gini (1921) and Savorgnan (1926) add to these considerations a new one. If, in regard to men, war's negative selection is true, its harm is compensated for through the positive selection of females due to war. Steinmetz (1907; 1929) brings out two reasons in his endeavour to show that even if war selection is in some degree negative, this harm is counterbalanced by war's positive effects. Following the opinion of Plutarch, Polybius, Aristotle,

Machiavelli, Vico, and of many others, he claims that the peacetime selection is negative also. Peace leads to vice, loss of virility and to survival of the people who are far from being the best blood of the nation. Peaceful competition leads to a regressive selection too. Therefore it is questionable which of these two negative selections is more harmful and regressive: 'War that shatters her slain / And peace that grinds them as grain'. Above all, war is the supreme instrument of group-selection. It is the only test serving this purpose, and the only test which is adequate because it tests at once all forces of the belligerent groups: their physical power, their intelligence, their sociality and their morality.

Steinmetz stated that war is the usual business of primitive tribes; that 'the savages, probably after the very first stage, were bloodthirsty and waged their wars in a most cruel way and with horrible losses of human life' (Steinmetz, 1907). From a biological standpoint, he continues, aggressiveness has been a condition necessary for progress. Without it, man could not emerge from his animal state because he would be exterminated by other species. Without war an upward movement within humanity would not be possible because any means of finding out which social group is superior and which is inferior would be absent. A long or eternal peace would make man an exclusively egotistical creature, without virility, courage, altruism, or bravery. Such a man would be entirely effeminated and corrupted to the very heart of his nature (Sorokin, 1928).

Why and how did the eighteenth century Rousseauian image of the 'noble savage', uncorrupted by the evils of civilization, change into the image of the 'brutal, violent, ape-like savage'? First of all the new conflict-model of nature and human society, introduced by Darwinian, Spencerian and Haeckelian thinking, also introduced the slogans of 'struggle for life' and 'survival of the fittest', which were taken quite literally, and soon became interpreted in terms of selection for the strongest and most violently aggressive individuals, groups, classes, ethnies, peoples, races and nations: the Agent of Progress. These notions combined with the pre-nineteenth century idea of a '*scala naturae*' (or Great Chain of Being), a natural, fixed and linear hierarchy of species with man, the Crown of Creation, at the top, and a similar hierarchy of races with the white, Caucasian race superior to all others and the final, ultimate stage of evolution and civilization. These views were translated in the new evolutionary framework as orthogenesis toward predetermined goals. Secondly, by the same logic, preliterate cultures and primitive peoples were considered to be at the lowest rung of the orthogenetic ladder, and by implication living fossils: wild, brutal, savage, base, crude, low, backward, uncivilized, and intellectually, morally and technologically inferior. In short, our 'contemporary ancestors', as they were called, were contemptible creatures to be civilized by the White Man's pacifying mission. Similarly, the simians were considered to be everything we, the civilized and superior were not: debased, despicable, ferocious, voracious, shamelessly promiscuous, lewd and lascivious creatures. Thirdly, if primitive people were our contemporary ancestors, would not their appearance and conduct be similar to our real ancestors, the 'missing link' between man and the apes: the apeman, whose fossil remains were slowly but gradually to be uncovered? One of the first to describe man's alleged ancestors as violent, ferocious and bloodthirsty creatures was Friedrich Albert Lange in 1866. He depicted them as prehistorical brutes who bashed each other's skulls with clubs in order to devour the raw brains of their hapless competitors (Corbey, 1988).

The authoritative and influential French archaeologist, Gabriel de Mortillet, completes the now-stereotypical and well-known image of the 'missing link', which was to dominate the thinking about our origins till far into the twentieth century, and which was revived by the sanguinary slaughterhouse phantasmagorias of Dart and Ardrey and their disciples. The apeman, de Mortillet writes in 1883 in his *Le Préhistorique – antiquité de l'homme*, was '*colère, violent et bataillard*': a brute, prone to fits of violent, furious and uncontrolled rage, without the faculty of speech, naked, still a semi-animal with apelike features and morally and intellectually a moron or worse. In this same tradition developed the explanation of the biocriminologist '*avant la lettre*', Cesare Lombroso, of the criminal as a living atavism, a remnant from prehistorical times when we were still violent, bestial, brutal and barbarian semi-apes (see also Corbey, 1988 for more examples).

There was nothing in Darwinism, as Hofstadter (1955) points out, that inevitably made it an apology for competition or force. Kropotkin's interpretation of Darwinism was as logical as Sumner's. After the Franco-Prussian War Darwinism was for the first time invoked as an explanation of the facts of battle. 'The greatest authority of all the advocates of war is Darwin', explained Max Nordau in the *North American Review* in 1889. 'Since the theory of evolution has been promulgated, they can cover their natural barbarism with the name of Darwin and proclaim the sanguinary instincts of their inmost hearts as the last word of science'.

It would nevertheless be easy to exaggerate the significance of Darwin for race theory or militarism, as Hofstadter observes. Neither the philosophy of force nor doctrines of *Machtpolitik* had to wait upon Darwin to make their appearance. Nor was racism strictly a post-Darwinian phenomenon (as Gobineau testifies). Still, Darwinism was put in the service of the imperial urge. Although Darwinism was not the primary source of the belligerent ideology and dogmatic racism of the late nineteenth century, it did become a new instrument in the hands of theorists of race and struggle. The likeness of the Darwinian portrait of nature as a field of battle appealed to the prevailing conceptions of a militant age (Hofstadter, 1955), in which the struggle for existence was translated as permanent and bloody war, nature was conceived of as 'red in tooth and claw', and in which selection was interpreted as violent elimination of the weak. (See van der Dennen (1977) for an extensive review of the Apologists of War in European history.) In the public's mind, by the turn of the century, Darwin's word 'fitness' had already lost its biological meaning of reproductive success and gradually came to imply physical strength or individual survival. If fitness is incorrectly interpreted to mean strength, then survival of the fittest means survival of the strongest rather than propagation of those genes which confer the most adaptive advantage (Barash and Lipton, 1985). Nasmyth (1916) and Perry (1918) launched a formidable assault upon Social-Darwinism and all its works. 'Perry's *Present Conflict of Ideals* was the most substantial of all the refutations of the Darwinized ethics and sociology that had culminated in the monstrosities attributed to von Bernhardi and Nietzsche. The whole evolutionary dogma, the Darwin-Spencer legacy of progress, the glib optimism of John Fiske, the warnings of Benjamin Kidd, the natural-selection economics of Thomas Nixon Carver – all fell under Professor Perry's axe' (Hofstadter, 1955). Like William James before him, Perry pointed out the essential circularity of the Darwinian sociology, in which power and strength are defined in terms of survival, and survival is, in turn, explained by power and strength.

9.1.3 Instinctivism

Instinctivism was a relative latecomer in the intellectual inputs that constitute the syndrome of Social-Darwinism. Darwin (1873) in *The Expression of the Emotions in Man and Animals* suggested an evolutionary interpretation of human emotions, which led many to base them on man's inheritance from his animal ancestry. This phylogenetic theory led to the preparation of numerous catalogues of (sometimes thousands of) human instincts (e.g., Bernard, 1924). William James (1890; 1910) was one of the first to postulate an instinct of pugnacity: '. . . modern man inherits all the innate pugnacity and all love of glory of his ancestors' (1910). James was soon followed by McDougall (1915, 1927) – 'The instinct of pugnacity has played a part second to none in the evolution of social organization' – and many others. The war instinct is sometimes regarded as being similar to the fighting instinct, as in the writings of Nicolai (1919), but in other cases the two are regarded as something quite different (e.g. Woods and Baltzly, 1915). Some sociologists indicate a herd instinct as indirectly responsible for the existence of war (Trotter, 1916; MacCurdy, 1918), or a sexual androgenic instinct (van Bemmelen, 1928); Weule (1916) envisaged '*rein physiologische Kraftüberschusz*' as the root cause of war. Other formulations are more or less variations on the themes of man's ineradicable warlike urges or combativeness or bellicosity, etc. Even relatively recent writers as Richardson (1960) and Falls (1961) use such instinctivist terminology. Thus Freud's formulation of the externalized *Todestrieb* or *Thanatos* underlying war already had an instinctivist context.

Many psychoanalysts still believe in the existence of destructive instincts underlying war; for example, as Strachey (1957) says: ' . . . it must be remembered that the destructive instincts which, when all is said and done, are the greatest single cause of war, are instincts and that they are impossible to eradicate altogether, greatly though they may be modified'. A similar view seems to underlie Durbin and Bowlby's (1939) theory of war. They state: 'War occurs because fighting is a fundamental tendency in human things – a form of behaviour called forth by certain simple situations in animals, children, human groups, and whole nations . . . nations *can* fight only because they are able to release the explosive stores of transformed aggression, but they *do* fight for any of a larger number of reasons'.

Sociologists and psychologists like Steinmetz (1907, 1929), Thorndike (1913) and many others indicate several varied instincts responsible for war, regarding it either as an outcome or as a drive for rejuvenation stimulated by a superabundance of the social bonds imposed by a social life and various social rules which finally repress the source of life itself; or as a form of relaxation from those conventional rules which, through their drudgery, monotony, and repression, tend to turn man into an automaton; or as an outlet for the satisfaction of the innate drives of anger, *Wanderlust*, the military spirit, courage, the spirit of adventure *Grausamkeit*, *Uraggressivität*, and so forth and so on (Sorokin, 1928; van der Dennen, 1977).

Instinct explanations of warfare lost their appeal in the 1930s (presumably under the influence of cultural relativism in anthropology and behaviourism in psychology, schools of thought which emphasized the plasticity and environmental determination of human behaviour). They were revived by Lorenz, Ardrey, Alcock, and others (cf. Flügel, 1955; Malmberg, 1983). The universal law of the struggle for existence was revived by sociobiological and biopolitical theory: relationships between states 'are treated as an inevitable outcome of the "struggle for survival" to which all living organisms are presumably condemned' (Somit, 1972).

CONFLICT AS COMPETITION

9.2 HAMILTON: THE EVOLUTION OF CRUELTY

According to Hamilton (1971), warfare was a natural development from the evolutionary trends taking place in the hominid stock. He finds it only too easy to imagine that the genes that reared cruelty out of the primate's aggressive drive have been favoured by natural selection in the hominid line. For the selection of cruelty, indeed, it is unnecessary even to consider inclusive fitness, except insofar as this may have been involved in the speeding-up of progress in mental and linguistic ability. In short, Hamilton states that vicious and warlike tendencies are natural in man and were formerly (at least) adaptive. In a similar vein, Lopreato (1984) subsumes primitive war and other forms of violence under the behavioural predisposition of self-enhancement, particularly the urge to victimize.

9.2.1 Criticism

Anyone familiar with the kaleidoscopic picture of the agonistic behavioural repertoires of more than 100 primate species and subspecies might wonder what the primate's aggressive drive might be. Furthermore, it is not necessary to invoke an aggressive drive as a source of warfare (cf. van der Dennen, 1986). I find myself much more in agreement with Young (1975), who states: 'War is not a maladaptive human trait arising out of a previously adaptive animal instinct of aggression. It is a highly evolved aspect of human political organization which has proved its viability by becoming more highly organized and more murderous as cultural evolution has advanced. It is not derived from the day-to-day aggression balanced by submission that serves an adaptive role in the societies of nonhuman primates. Its source has been the previously untapped potentialities resident in the norm of the reaction of the human genotype. Among these potentialities has been the aggressive potential characteristic of all animal species, but this potential has no more forced the evolution of war than some mythical killer instinct has.'

9.3 TERRITORIALITY AND PRIMITIVE WAR

In ethological theory, aggression, as Pettman (1975) says, is inextricably interwoven with the concept of territorial defence. In anthropology, territorial divisions of human groups have been claimed to be virtually universal (Carr-Saunders, 1922; Malinowski, 1941; Wynne-Edwards, 1962). No wonder that it was an ethologist to propose territoriality as an explanation of human warfare. 'In order to understand what makes us go to war', Tinbergen (1968) contends, 'we have to recognize that man behaves very much like a group-territorial species'. As a social, hunting primate, man must originally have been organized on the principle of group territories. Thus Tinbergen, having implicated group-territoriality in the evolution of human warfare, goes on to delineate other preconditions; the upsetting of the balance between aggression and fear (to which he adds the somewhat arcane assertion: 'and this is what causes war'), is due to at least three other consequences of cultural evolution – the invention of long-range weapons which make killing easy, sophisticated indoctrination, increased population pressure and other factors. In later articles (Tinbergen, 1976, 1981) he developed his thesis of group territorialism as the root of war more fully.

Reynolds (1966) also relates warfare to territoriality, though he regards territoriality as a relatively recent phenomenon, from the time that man became sedentary: '. . . now that communities were attached to particular territories as an ecological necessity, the advent of other tribes, still nomadic, on their land would be viewed with hostility; or, when a community grew too large for its land, a subcommunity might break off and search for new land on which to hunt or to cultivate or to graze its domestic animals. In these circumstances territoriality and inter-group aggression began'. This appears to be a popular view in the literature on the subject, for example, Carneiro, 1970; van Sommers, 1972; Galtung, 1973; Vine, 1973; Harris, 1975; Van den Berghe, 1978; among many others. Ardrey (1966) set out to demonstrate that man defends territory by instinct (the so-called territorial imperative). Another blatant example of naïve biologism is Alcock's (1972) theory: 'War is an overt action resulting from man's innate aggressiveness. Like other social animals, groups of men defend their home territories and aggressively seek to own or control larger territories . . . '.

9.3.1 Criticism

According to Wilson (1975a), it is reasonable to conclude that territoriality, in the broad sense of any area occupied more or less exclusively by an animal or group of animals through overt defence or advertisement, is a general trait of hunter-gatherer societies. In a review of the evidence, Wilmsen (1973) found that these relatively primitive societies do not differ basically in their strategy of land tenure from many mammalian species. One might well object that Tinbergen's theory is not really about 'group-territoriality' (it does not seem to play a large role either as a necessary or as a sufficient condition), rather his theory is a modest psychological one, which specifies some psychological preconditions of the origin of predatory warfare. Furthermore, it is not only territorial defence which has to be explained – obviously territorial defence itself does not constitute war ('it takes two to tango') – but the reasons and motives of territorial transgressions and violations, territorial conquest, etc. Territorial defence is immaterial without its correlative: offence. Wright's (1965) conclusion that 'wars seldom have the object of territorial aggression or defence until the pastoral or agricultural stages of culture are reached, when they become a major cause of war' may still be considered valid.

9.4 HUNTING AND WARFARE: CARNIVOROUS PSYCHOLOGY THEORY

In the following collection of theories, hunting is supposed to be the master behaviour pattern of the human species. Man evolved as a hunter (so the argument goes); he spent over 99% of his species' history as a hunter, and he spread over the entire habitable area of the world as a hunter (Laughlin,

1968; Freeman, 1973; Nelson, 1974). The predatory adaptation achieved by the Australopithecinae probably involved 'a behavioural transition from a retreating to an attacking pattern' (Freedman and Roe, 1958) during which, as Freeman (1973) contends, 'carnivorous curiosity and aggression have been added to the inquisitiveness and dominance striving of the ape'. With these considerations in mind, Washburn and his associates (Washburn, 1959; Hamburg, 1963, Washburn and Avis, 1958; Washburn and Howell, 1960; Washburn and DeVore, 1961; Washburn and Hamburg, 1968; Washburn and Lancaster, 1968) postulate a special learning disposition in the human animal for hunting and killing, with its own intrinsic source of satisfaction, pleasure and lust. This learning disposition would have been selected for during the – very long, in evolutionary terms – period during which the protohominids as well as *Homo sapiens* lived as nomadic hunters in small tribal communities. The hunting and killing of prey animals facilitated, in this view, the transposition to the hunting and killing of conspecifics – and even torture and the (vicarious) enjoyment of cruelty. The authors also point to the popularity of war: '. . . until recently war was viewed in much the same way as hunting. Other human beings were simply the most dangerous game. War has been far too important in human history for it to be other than pleasurable for the males involved' (Washburn and Lancaster, 1968). This complex of factors the authors call man's carnivorous psychology.

This theory of Washburn and associates was already anticipated by William James (1910) in his *Principles of Psychology*, and by Frobenius (1914), who formulated the hypothesis that war originated from the man-hunt. 'If evolution and the survival of the fittest be true at all', James (1910) wrote, 'the destruction of prey and human rivals must have been among the most important of man's primitive functions, the fighting and the chasing instincts must have become ingrained . . . It is just because human blood-thirstiness is such a primitive part of us that it is hard to eradicate, especially where a fight or a hunt is promised as part of the fun'. In a similar vein, Pfeiffer (1972) writes: 'As long as man lived in the wilderness, the excitement and glamour of the hunt had meaning in the context of survival, in promoting aggression against prey and predators. Man deprived of hunting as a major source of prestige, deprived of wild species as a major focus of aggression, began playing the most dangerous game of all. Men began to go after other men as if their peers were the only creatures clever enough to make hunting really interesting. So war, the cruelest and most elaborate and most human form of hunting, became one of the most appealing ways of expressing aggression'. To the extent that intelligence, cooperation, tool use, bipedalism, inventiveness, physical stamina, and aggressiveness all enhanced the chances of success between warring groups, Corning (1975) argued, intergroup selection would have served to reinforce selection pressures already at work with respect to big-game hunting and other aspects of protohominid social life. Rather than opposing genic selection, war-based group selection might merely have accelerated the trend (see also Corning, 1983).

9.4.1 The 'killer ape' hypothesis

In a series of 39 papers published between 1949 and 1965, Dart reviewed the archaeological evidence of our hominid-becoming-carnivore ancestors found at Makapansgat. On the basis of fractured skulls and bone fragments of baboons and other Australopithecines, Dart (1953) concluded that ' . . . man's predecessors differed from living apes in being confirmed killers'. Killers, that is, of both prey animals and conspecifics, in spite of the fact that their brains were still relatively small. Yet, Dart asserts, 'this microcephalic mental equipment was demonstrably more than adequate for their crude, omnivorous, cannibalistic, bone-club wielding, jaw-bone clieving, Samsonian phase of human emergence . . . [the so-called "osteodontokeratic culture"] . . . The loathsome cruelty of mankind to man forms one of his inescapable, characteristic and differentiative features; it is explicable only in terms of his cannibalistic origins' (Dart, 1953). This is, as Wilson (1975a) aptly remarked, 'very dubious anthropology, ethology, and genetics'.

Dart's view inspired the first of Ardrey's series of popular books. The basic tenet of Ardrey's (1961) *African Genesis* is that contemporary man is a descendant of a race of 'terrestrial, flesh-eating killer apes', and that this fact *an sich* explains the aggressiveness and warlikeness of modern man: man is a predator, with all the pleasure and lust in killing springing from his predatory nature. The African ancestor of modern man was a rather primitive primate, Australopithecus, who out of sheer physical weakness – not being equipped by nature with claws, fangs, hooves, horns, or agility in locomotion – began to use weapons. In Ardrey's line of thought tools are identified with weapons, and these weapons – indeed, all human culture – result from man's predatory nature. Culture is a product of the use of weapons: 'The weapon . . . had fathered man'.

9.4.2 Melotti: a redirection hypothesis

Melotti's (1984b, 1986b) interpretation of the evolutionary origins of human aggressiveness is the following: owing to their original lack of disposition for interspecific aggression, our hominid ancestors, in their transition from anthropoid life to systematic hunting, were obliged to derive the biological basis necessary for the new way of life from the potential for intraspecific aggressiveness that is common to all primates. In effect, human hunting (like hunting behaviour in chimpanzees) appears to be much more aggressive than the predation practised by carnivores, and this seems to confirm the idea that human interspecific aggression is the result of the redirection of an originally intraspecific disposition against extraspecific targets. Middle and Upper Palaeolithic hunting probably selected for increasingly aggressive individuals: for, to face big mammals, such a vulnerable being as man needed a great endowment of aggressiveness. When, at the end of the last glacial era (about 10000 years ago), big-game hunting became impossible in a large part of the then inhabited world, man lost the main interspecific outlet for his increased aggressiveness. Thus, this drive, which was already intraspecific at its origins, was again directed at its former target: fellow members of the human species.

9.4.3 Criticism

Apart from the fact that carnivorous psychology theory is marred by confusion among the concepts predatory, carnivorous and hunting, what is the evidence to substantiate these claims? First, as Fromm (1973) emphasized, the idea that hunting produces pleasure in torture is an unsubstantiated and most implausible statement. Hunting, as practised by primitive hunters at least, is not an aggressive activity at all (for example see Turnbull, 1965). There is also no evidence for the assumption that primitive hunters were motivated by sadistic or destructive impulses. On the contrary, there is evidence that they had an affectionate feeling for the killed animals and possibly a feeling of guilt for the kill (Mahringer, 1952; Lewinsohn, 1954; Rensberger, 1977), as was already noted by Frazer in the nineteenth century. Fromm especially criticizes the notion that man has a drive for and pleasure in killing, although, as Nelson (1974) remarks, 'Fromm's interpretation that the pleasure was not in the killing but in the development of hunting skills is no more compelling than Washburn's'. Finally, what body of information we have about contemporary primitive hunter-gatherers does not indicate that hunting is conducive to destructiveness, cruelty, or warfare (Fromm, 1973; Sahlins, 1960; Turnbull, 1965; Service, 1966; among many others).

It has now become apparent that Dart's and Ardrey's sanguinary and phantasmagoric imaginations have more (horror-)literary than scientific qualities: 'First, the question of the incidence of interpersonal violence among australopithecines remains an open one. Secondly, it would appear that the Makapansgat fossils are not the tools and weapons of an ancient culture, but the leftovers of many carnivore meals' (Leakey, 1981). Furthermore, the fact that the 'killer' prehominid *Australopithecus africanus* was a 'ferocious carnivore' and that this accounts for modern man's aggressiveness is not at all convincing. There is hardly any evidence that carnivores exhibit more, or more intense, intra-specific aggression than other mammals (Rapoport, 1965). Herbivores, too, show agonistic behaviour, as two fighting bulls demonstrate (Eibl-Eibesfeldt, 1963). 'Today the vast majority of experts familiar with the fossil evidence agree that while man's ancestors appear to have been hunters who killed animals and consumed quantities of meat, there is no suggestion that they were driven by a blood lust any more than any other predator. More important, there is certainly no hint that they killed each other more than does any animal species known today' (Rensberger, 1977). Recent discoveries of hominid fossils suggest that Ardrey's 'killer apes' may not even be man's ancestors at all, 'but rather an unsuccessful side branch in the line of human evolution' (Nelson, 1974; cf. Rensberger, 1977).

Furthermore, Steinmetz (1896), Helmuth (1973) and others have made clear that the bases of human anthropophagy are most probably of a religious, ritual and ceremonial nature, exocannibalism developing out of endocannibalism. Finally, of course, if such biological freaks as killer apes had ever existed, they would have exterminated each other long before they would have given rise to modern man.

Melotti's notion of aggressive drive, here envisaged to be some kind of energetic reservoir, is in all probability not warranted, and is obsolete. There is no evidence that big-game hunting specifically requires aggression ('a great endowment of aggressiveness') as a prime or even co-motivation. The hypothesis that Palaeolithic hunting selected for increasing aggression is therefore very dubious. Finally, the aggression–warfare linkage is highly problematic (van der Dennen, 1986). The redirection hypothesis advanced by Melotti thus seems to have a very weak factual basis.

Diametrically opposed to the foregoing theories, Scott (1974 *et seq.*; cf. also Simeons, 1960) proposed a picture of early man as essentially a timid, relatively unaggressive 'fear-biter'. Scott holds that early man was not basically adapted as a predator and that he has only become one secondarily by the use of tools. Early men and women must have been few in number and relatively weak and defenseless; they may have been able to survive only by being extremely timid and wary of danger and, secondarily, by pretending to be brave.

9.5 HUMAN BRAIN EVOLUTION AND WARFARE

The idea that warfare influenced human brain evolution was proposed by Darwin (if not before) and again by Keith (1947). They meant, however, warfare in historical times and saw the effects of warfare in terms of refinements of the human brain. Alexander and Tinkle (1968) and Alexander (1971) were the first to propose that the evolution of the novel substance of the hominid brain was driven by warfare. Bigelow (1969 *et seq.*) presents the most comprehensive case and Pitt (1978) reexamines the arguments and evidence.

Bigelow (1969 *et seq.*) covers much of the ground of Alexander and Tinkle, showing how the needs of warfare would foster the duality of man's nature, and how well it accounts for the evolution of intelligence and the trebling in volume of man's brain. Warfare need not be a continuous series of year-in, year-out skirmishes in order to exert a strong selective pressure. A war once a generation can radically alter gene frequencies, especially if the losing females propagate the victors' genes. Human warfare, Bigelow emphasizes, is an organized activity. Success in war has always been due primarily to cooperation. Cooperation required communication and efficient brains, emotional restraint as well as intelligence. Since success in intergroup competition was determined mainly by capacities for cooperation and intelligent self-control, the result of the selective process was definitely not an increasingly uncontrollable 'aggression instinct' (Bigelow, 1972). What Bigelow does not adequately explain is why warfare should have evolved at such an early date, even though he remarks that it is most unlikely that intelligent human groups would have starved peacefully while the other groups were getting the game first. According to Pitt (1978), 'Warfare was the logical culmination of escalating territorial competition between groups as population densities increased. The turning point from uneasy peace to war probably came at moments when a natural calamity, such as drought, caused starvation to become a serious threat to some of the groups'. Hominid populations must have reached critical densities at fairly frequent intervals through the Pleistocene, and perhaps the late Pliocene. Several scenarios can be pictured in such a situation:

1. peaceful coexistence of the groups;
2. peaceful competition between groups with the losers starving;
3. violent conflict between individuals;
4. scrambling competition;
5. violent group conflict, i.e., warfare.

Warfare would be the best alternative for the group that practised it successfully, assuming it to have been within their biological reach. Assuming that different groups tended towards one or another of these strategies, in varying degrees, it is easy to see that the warmongers would be the most successful and could indeed overrun any group attempting to practise one of the other strategies. Plainly it will be the warmongers whose genes are represented in the next generation. Indeed, the only possible competitive strategy for survival in competition with a group practising warfare, is warfare itself, either defensive or offensive. Warfare, therefore, would be an evolutionary successful strategy, and would tend to spread itself, if it once began.

It is interesting to note that Pitt is one of the few evolutionary theorists who does not agree with the underlying assumption of other theorists that the principle characteristic which an animal needs to carry it over the threshold of warfare is an unusual degree of aggression.

9.5.1 Weapons, intelligence, and warfare

One does not necessarily need the Freudian phallic symbolism to acknowledge that the males of the human species are fascinated by weapons. Andreski (1954) and Tinbergen (1976) pointed out that weapons fostered the concept that it is advantageous to kill the enemy, since a dead enemy does not return to fight again. Baer and McEachron (1982) and McEachron and Baer (1982) made it clear that the development of weapons lowered the costs of attacking while increasing the costs of being attacked. Thus, there was a selection pressure to develop 'pre-emptive strike' or attack before being attacked behaviour. There is a long-standing principle in evolution that the greatest natural competitor to any animal is another member of the same species. This is because conspecifics share almost the exact same requirements and, when resources are limited, must contest for the same resources. Within groups, the dominance system and genetic interrelatedness tend to control and modulate aggressive competition. Between conspecific groups, it is quite a different story. When one group encounters another over a limited resource, each group has a number of options. If troop A is using a limited resource and troop B arrives, troop A can (a) avoid troop B by retreating, (b) try to ignore troop B, (c) cooperate with troop B, or (d) compete with troop B. If the resource is easily available, it might pay troop A to retreat and avoid any possibility of conflict. However, in the long evolutionary run, this is a self-defeating strategy. Sharing a resource (options b and c) is likely to occur when it is not very limited or extremely difficult to defend. If the resource is really limited, sharing is very unlikely. If conflict is inevitable, it makes better evolutionary sense for the troops to determine ownership of the resources as groups, rather than having both conflict and decreased inclusive fitness.

Once started, selection for conflict and weapons technology rapidly gained momentum by leading into a positive feedback system: better weapons led to increased levels of group conflict. Conflict selected for (among other things) enhanced mental capacity in the form of increased learning capacity, improved communications, the emergence of the ability to plan, have foresight, improve technology etc. This increased mental capacity in turn not only created better weapons through an improving technology, but made the group a better fighting unit and thus a more dangerous adversary. These factors in turn increased the selective pressure for conflict – and the cycle began again. The feedback system would have had other effects as well. Conflict tends to select for better organized groups.

Hamilton (1975) presented a 'stepping-stone' model of hominid intergroup conflict. In his model, he proposed that a hominid group could expand into the territories of defeated neighbouring groups, enlarge in size and consolidate its position and then expand again and so on. This model points out two other interesting aspects of intergroup warfare. First, warfare may gain the victorious group territory and resources that may in themselves have selective value. This would promote conflict in addition to, and somewhat independent of, the positive feedback system. Secondly, the capture and integration of a limited number of 'enemy' females might be used to prevent inbreeding depression in the now-closed hominid groups, providing yet a third possible adaptive value for warfare (McEachron and Baer, 1982). Alexander (1971) proposed that intergroup conflict would select for greatly increased capacity to recognize relatives, friends and enemies. According to McEachron and Baer, weapons would increase xenophobia, fear and antagonism toward strangers. Even after intergroup conflict ceased to be selective, the xenophobic attitude would remain in the form of emotional tendencies and reinforcement. These emotional tendencies would then initiate conflicts on their own.

9.6 HUMAN SOCIALITY AND WARFARE

Alexander (1974 *et seq.*) has pointed out that there is no automatic or universal benefit from group living. Indeed, the opposite is true; there are automatic and universal detriments, namely increased intensity of competition for resources, including mates, and increased likelihood of disease and parasite transmission. What, then, are the benefits of group living that offset its automatic detriments? There are only three:

1. susceptibility to predation may be lowered;
2. the nature of food sources may make splintering off unprofitable;
3. there may be an extreme localization of some resource.

To explain primate and hominid groups only the causative factor of predation remains. And once predation from other species relaxed, early man became his own predator: 'When man developed his weapons, culture and population sizes to levels that essentially erased the significance of predators of other species, he simultaneously created a new predator: groups and coalitions within his own species' (Alexander, 1974).

In a subsequent publication, Alexander (1979) proposed the so-called 'Balance-of-Power Hypothesis'. This hypothesis contends that at some early point in our history the actual function of human groups – their significance for their individual members – was protection from the predatory effects of other human groups. The premise is that the necessary and sufficient forces to explain the maintenance of every kind and size of human group above the nuclear family, extant today and throughout all but the earliest portions of human history, were (a) war, or intergroup competition and aggression, and (b) the maintenance of balances of power between such groups. This argument would roughly divide early human history into three periods of sociality as follows:

1. Small, polygynous, probably multi-male bands that stayed together for protection against large predators.
2. Small, polygynous, multi-male bands that stayed together both for protection against large predators (probably through aggressive defence) and in order to bring down large game.
3. Increasingly large, polygynous, multi-male bands that stayed together largely or entirely because of the threat of other, similar, nearby groups of humans.

Thus, Eaton (1978) suggests that early humans may have spent a very long time during which their social behaviour was largely structured by both defence against predators and competition with them. The complex behaviours required for such activities could have 'primed' or preadapted humans for their later evolution in hostile intraspecific groups.

Andreski (1954) and Lorenz (1966) – the principal proponent of the instinct-cum-selection thesis – had already advocated a similar notion: '. . . it is more than probable that the destructive intensity of the aggression drive, still a hereditary evil of mankind, is the consequence of a process of intraspecific selection which worked on our forefathers for roughly 40 000 years, that is, throughout the Early Stone Age. When man had reached the stage of having weapons, clothing and social organization, so overcoming the dangers of starving, freezing and being eaten by wild animals, and these dangers ceased to be the essential factors influencing selection, an evil intra-specific selection must have set in. The factor influencing selection was now the wars waged between hostile neighbouring tribes' (Lorenz, 1966).

Lorenz, basing himself on a hydraulic model of aggression, regards human hypertrophied aggression as an anomaly compared to other species. Indeed, he seems unusually murderous, for his development of weapons came so fast that he has not yet evolved the biological mechanisms of restraint, the built-in inhibitions so common in the ritualized aggression of other species. He also identifies 'militant enthusiasm', originally evolved as a form of communal defence, as an easily appealed to and easily surfacing psychological constellation in modern man. Militant enthusiasm is considered to be a true autonomous instinct.

9.6.1 Criticism

Hostile neighbouring hordes may well be, as Montagu (1974) suggests, 'the invention of nineteenth century antiquarians and their modern counterparts', and he continues: 'As for the allegation of hostility between neighbouring prehistoric populations, there is not the least evidence of anything of the sort having existed. This by no means rules out the possibility that such hostilities may occasionally have occurred. If such hostilities did occur, it is extremely unlikely that they were frequent. Neighbouring populations in prehistoric times would have been few and far between, and when they met it is no more likely that they greeted each other with hostility than do gatherer-hunter peoples today' (Montagu, 1974; Shepard, 1973; Melotti, Chapter 12). Referring to contemporary hunter-gatherers, Service (1966) states: 'The birth–death ratio in hunting-gathering societies is such that it would be rare for population pressure to cause some part of the population to fight others for territorial acquisition. Even if such a circumstance occurred it would not lead to much of a battle. The stronger, more numerous group would simply prevail, probably even without a battle, if hunting rights or rights to some gathering spot were demanded'. Alexander (1979) counters with the view that there is not an iota of evidence to support the idea that aggression and competition have *not* been central in human evolution. However, since he does not specify what the counter-evidence should look like, this remains a rather gratuitous statement.

9.7 ECOLOGICAL/DEMOGRAPHIC AND OTHER FUNCTIONALIST THEORIES

As was shown in the first paragraphs of this chapter, Spencer, Sumner and Keller founded the functionalist ecological orientation in sociology and anthropology. The foundation of human society, said Sumner, is the man/land ratio. The ecological approach to the study of war thus has a long history.

In his classical study on the evolution of primitive war, Davie (1929) lists population growth as one of the basic causes of the competition of life which makes war. He states: 'Since war is so fundamental a phenomenon, its explanation must be sought in the basic conditions of life. One such life condition is land, for it is from the land that all means of subsistence are ultimately drawn. Man has had to struggle for a living, however, for there is no "boon of nature" or "banquet of life"'. Also Bernard (1944) acknowledged overpopulation or the pressure of population as the 'only one very significant biological cause of war in a general and fundamental sense'. War, according to Bernard, is also a social invention, for which most peoples have models which they can imitate and for which all peoples have abundant materials. Any people that has reached the hunting stage of development possesses the weapons at hand with which to enter upon war. The early warriors were hunters who merely turned the weapons of the chase against other men and used the tactics of hunting game in ambushing and slaughtering human beings. It was easier for them to kill and rob than to find new ways of procuring a new food supply by peaceful means.

Andreski's (1954 *et seq.*) most general assumption is the recognition of the fact that the struggle for wealth, power, prestige and glory (for invidious values, i.e., desirable things of life) is the constant feature of the life of humanity (cf. Hobbes, 1651). Killing one another could not have remained one of the chief occupations of men if there was no surplus of men available. And it was the natural tendency of the population to grow beyond the means of subsistence that assured the permanence of bloody struggle. The recognition of this fact enables him to advance a hypothesis about the origin of war. As nothing of that sort exists among the mammals this institution must be a creation of culture. It probably came into existence when the advance in material culture enabled man to defend himself better against the beasts which preyed on him, and thus to disturb the natural balance which keeps the numbers of any species stationary in the long run. After the beasts had been subdued, another man became the chief obstacle in the search for food and mutual killing began (Andreski, 1954). In a later version (Andreski, 1964) he succinctly states that 'given the propensities of human nature, the tendency of the population to grow beyond the resources has ensured the ubiquity of wars, although not every single instance of war had this factor as an immediate cause'.

The neo-evolutionist school of American cultural anthropology begun by White (1949), primarily interested in the ecological relationships of man and environment, originally emphasized the level of socioeconomic development necessary before organized warfare can be conducted. White himself viewed warfare as a consequence rather than a cause of evolutionary development (cf. also Gorer, 1938; Newcomb, 1960). A key aspect of the neo-evolutionist theory is that 'high-energy societies replace low-energy societies', thus 'long range trends toward higher levels of productivity are related to intergroup hostility' (Harris, 1971; 1975). 'Primitive warfare arose as part of a complex system that prevented human populations from exceeding the carrying capacity of their habitats' (Harris, 1972). Or, in Vayda's (1968 *et seq.*) terminology: war functions to adjust the man/land ratio. Harris (1978) discussed and criticized the 'war as solidarity', 'war as play', 'war as human nature' and 'war as politics' explanations of primitive warfare. He rejects them all, noting especially that political expansion cannot explain warfare among band and village societies because most such societies do not engage in political expansion. Why then, do they practise warfare? While bands and villages do not conquer each other's land the way states do, they nonetheless destroy settlements and rout each other from portions of the habitat that they would otherwise jointly exploit. Raids, routs and the destruction of settlements tend to increase the average distance between settlements and thereby lower the overall regional density of populations. One of the most important benefits of this dispersion – a benefit shared by both victor and vanquished – is the creation of 'no man's lands' in areas normally providing game animals, fish, wild fruits, firewood, and other resources. Because the threat of ambush renders them too dangerous for such purposes, these no man's lands play an important role in the overall ecosystem as preserves of plant and animal species that might otherwise be permanently depleted by human activity. The dispersal of populations and the creation of ecologically vital no man's lands are very considerable benefits which derive from intergroup hostilities among band and village peoples despite the costs of combat.

9.7.1 Population control, preferential female infanticide, the male supremacist complex and primitive war

A theory of population control in primitive society involving the effects of preferential female infanticide and warfare has been developed by Divale (1970 *et seq.*), Harris (1971 *et seq.*) and Divale and Harris (1976). The theory holds that every human society must take steps to control population growth. This is especially the case in primitive societies because their simpler technology places greater limits on their ability to expand food energy production. The manner in which most primitive societies regulate their population is postulated as follows: 'Infanticide is practised on both males and females for a variety of reasons. However, since there is a general preference to have a boy as the first child and since the ratio of males and females at birth is almost equal, many more girl infants get killed. In terms of population control this is significant because excess females are eliminated before they reproduce. The effect of selective female infanticide is that many more boys reach maturity than do girls and a shortage of marriageable women exists among young adults. The women shortage leads to adultery, rape and wife-stealing which in turn lead to frequent disputes over women. Deaths which result from these disputes lead to blood-revenge feuding and warfare in which the excess male population is eliminated. In the childhood generation boys greatly outnumber girls because of female infanticide. But in the adult generation the ratio between the sexes is balanced because males die in warfare, However, even though the adult ratio is about equal, a relative women shortage still exists due to the practice of polygyny – the older and more influential males have several wives, leaving younger males wifeless. The constant warfare of these societies creates a constant need for warriors and it is this need which causes the cultural preference for a boy as the first child which begins the process in the first place.

The root of this entire system is a culturally produced women shortage. Female infanticide and polygyny create a shortage of women which leads to wars which in turn favour male infants etc. The cycle is continuous and each generation creates conditions which perpetuate the process in succeeding generations. The next effect is the control of excess population' (Divale, 1974). There seems, according to the same author, to be a direct correlation between population density and the forms of violence (feuding, raiding, open battle) that occur.

It is not suggested that war caused female infanticide, or that the practice of female infanticide caused war. Rather, it is proposed 'that without reproductive pressure neither warfare nor female infanticide would have become widespread and that the conjunction of the two represents a savage but uniquely effective solution to the Malthusian dilemma' (Harris, 1978). Furthermore, warfare functions in this system to sustain the so-called male supremacist complex (social practices such as patrilocality, polygyny, marriage by capture, brideprice, postmarital sex restrictions on women, sexual hierarchy with female subordination, sexual privileges for fierce warriors, male machismo, masculine displays, dangerous and competitive sports and martial skills, militancy, the warrior cult; and in general war-linked, male-centered institutions, prerogatives and ideologies – because the survival of the group is contingent upon the rearing of combat-ready males) and thereby provide the practical exigencies and ideological imperatives for postpartum cultural selection against female infants (Divale and Harris, 1976). The males have a price to pay for all this: they constitute the bulk of the victims of warfare in preindustrial societies (Harris, 1975).

Finally, the theory explains that peoples like the Yanomamö, Dani, Maring and Maori attribute their wars to disputes over women. Although, Harris (1975) warns, such explanations cannot be accepted at face value; the belligerents who lose their lives in armed combat seldom accurately understand why they do so. 'Excessive warfare is an ecological trap into which humanity has probably fallen again and again' (Harris, 1975). He concludes that 'war has been part of an adaptive strategy associated with particular technological, demographic, and ecological conditions', but that this does not require us to 'invoke imaginary killer instincts or inscrutable or capricious motives to understand why armed combat has been so common in human history' (Harris, 1974).

9.7.2 Criticism

The assumption of the authors that the primary function of warfare is to encourage female infanticide and hence population limitation is ingenious but probably unnecessary. Alternative interpretations of preferential female infanticide have been proposed by Alexander (1974) and Dickemann (1979). The supporting evidence for the Harris-and-Divale argument consists of sex-ratio statistics and census data from indigenous bands and tribal societies around the world. The validity of their statistical material and tests has been severely challenged by Harrison (1973), Hirschfeld, Howe and Levin (1972) and Bates and Lees (1979). 'There is no reason to conclude that population regulation itself requires warfare, which would be an extremely costly and risky means of promoting female infanticide' (Bates and Lees, 1979). 'It does not appear that we can accept Divale's infanticide–marriage alliance–

polygyny–warfare syndrome as having as universal an applicability as his data suggests might be the case' (Harrison, 1973). Harris and Divale assume that overbreeding is endemic to primitive populations, which seems unlikely from ethnological evidence and unsupportable from ethnography: not all or even most primitive societies are or were expanding (Nettleship, Givens and Nettleship, 1975). Furthermore, they wonder, how could it be that 'warfare began as part of an ecologically adaptive system of population control if the people fighting fought for prestige, or – more troublesome – for brides with whom to breed more children to add to the crowding?' 'To say that the function of primitive warfare is to limit population sounds almost as odd as saying that the function of hot weather is to increase beer consumption or that we have famines to keep population down', Howell (1975) remarks, and he adds that a great deal of the confusion in the contemporary arguments could be eliminated by consistently distinguishing effect and purpose instead of ambiguously employing the term function.

9.7.3 Multifunctionalist theory

Vayda (1961 *et seq.*) has challenged the view of warfare as a 'safety-valve' institution (Wedgwood, 1930; Wright, 1965; Murphy, 1957 *et seq.*; among many others), and has argued that the adaptive value of warfare is not limited to the promotion of solidarity. In accordance with this expectation, several evolutionary anthropologists have sought ecological functions for warfare in populations that occupy certain types of ecological niches. Vayda suggests that different predictions must be made for expanding and nonexpanding populations. For expanding populations, warfare is hypothesized to serve the function not only of reallocating resources but also of killing enough people to reduce population pressure on resources. He speculates that in nonexpanding populations warfare might serve other purposes no less important ecologically, for example, dispersal of groups in finite territories to optimally exploit the ecological niche, or the capture of women and children in small populations with unbalanced sex ratios. Most of these other adaptive functions do not require large-scale killing or even particularly lethal warfare.

Studies attempting to show that warfare serves an adaptive purpose, or is ecologically functional, abound nowadays (e.g. Wedgwood, 1930; Newcomb, 1950; Vayda, 1961 *et seq.*; Leeds, 1963; Lathrap, 1968; Krapf-Askari, 1972; Graham, 1975; Morey and Marwitt, 1975, among many others).

In a subsequent article, Vayda (1967) proposed that the functions of primitive war may be the maintaining of one or more variables (for example, man/land ratio) or activities in a certain state or within a certain range of states. He grossly designates these variables as psychological, socio-political, economic and demographic.

1. Functions in the regulation of psychological variables: the notion that primitive war may operate so as to keep such variables as anxiety, tension and hostility from exceeding certain limits is implicit in the statements of anthropologists who speak of primitive wars as 'flight-from-grief' devices (Turney-High, 1949), as 'enabling people to give expression to anger caused by a disturbance of the internal harmony' (Wedgwood, 1930), and as serving to divert intrasocietal hostility onto substitute objects (Whiting, 1944; Murphy, 1957).
2. Functions in the regulation of socio-political variables: a characteristic hypothesis about stateless societies that lack a central government with penal jurisdiction over the separate local groups is the familiar hypothesis of deterrence or preventive war. According to this hypothesis, warfare undertaken by a group to avenge an insult, theft, non-payment of bride price, abduction, rape, poaching, trespass, wounding, killing, or some other offence committed against its members deters members of other groups from committing further offences.
3. Functions in the regulation of economic and demographic variables: according to hypotheses presented by various writers, war breaks out when the inequalities between groups in their possession of or access to certain economic goods or resources reach a certain magnitude. Such hypotheses are functional ones if they go on to state that the effect of warfare is to reduce the inequalities to a point where they do not exceed a proper or acceptable level. Similar to these hypotheses are those about the regulation of demographic variables. In this case, the resources redistributed as a result of war are human beings. The redistribution of land as a result of primitive war can, of course, equally be seen as a redistribution of people upon the land, a process involving a victorious group's movement into a vanquished and dispossessed enemy's former territory. Should there be no place to which the vanquished can flee, the answer to problems of local population pressure may be heavy battle mortality.

Comparable multifunctional theories have been provided by Leeds (1963), Vayda and Leeds (1961) and Leeds and Vayda (1965). Leeds asserts that individual motivations in war are irrelevant to the socio-cultural demands and responses (what he calls the social 'motivations').

In a number of recent anthropological studies of warfare, different grades of violence have been distinguished, separate causes have been sought for fighting at each grade, and, in some cases, escalations from grade to grade have been noted (e.g. Warner, 1931; Otterbein, 1968; Chagnon, 1967 *et seq.*). Vayda (1971; 1974) describes a multiphase war process operating among the Maring of eastern New Guinea. The significant features of this process include the following:

1. The later phases of the process that involve heavy mortality and sometimes leads to territorial conquests cannot occur unless preceded by periods of weeks or months marked by rather ritualized hostilities in which mortality is low.
2. Escalations from phase to phase in the war process are not inevitable.
3. The causes of entry into war are not the same as the causes of escalation from one phase to another of the war process.

Evidence from case studies (Vayda, 1970; 1971) raises serious questions about the validity of cross-cultural or cross-societal studies that depend on the fixed assignment of the warfare of various societies to one or another of a limited number of such categories as revenge warfare. The case studies point to the possibility that the ethnographic reports on which the assignments to the categories are based may be describing the causes of only the first phases of war processes; fighting for blood revenge, magical trophies, or sacrificial victims can become something else if there is escalation to the later phases.

9.7.4 Criticism

As Hallpike (1973) has pointed out, Vayda's claims do not seem, sometimes, to be supported even by his own data. The obvious fact that warfare among the Maring patently did not have any noticeable effect in redistributing land drives Vayda to state that even so, their system of warfare might have had such effects if things had been different. All that Vayda establishes is that 'in some cases warfare may be the means by which groups which are short of primary forest for new gardens may acquire other people's. No one would suggest that in some cases, people may not fight over land, or anything else which they fancy and which is in short supply, but this does not explain why the Marings fought' (Hallpike, 1973).

Vayda is even more prone than Harris to reify adaptive systems and to express them in equilibrating functional terminology. He explains the complex system of pre-contact Maori fighting as a system of population growth, dispersal and access to resources which 'while they did not *know* about the system, the Maori were still *motivated* to follow. It is implied that the "motivation" somehow stemmed from the functional nature of the system' (Nettleship *et al.*, 1975).

9.7.5 General criticism of functionalist theories

The most thorough criticism of functionalist theories of primitive warfare has been provided by Hallpike (1973), from which the following is epitomized.

By its very nature, functionalist analysis cannot explain the genesis or emergence of primitive warfare. As Vayda (1968) himself states, the object of functional analysis is a 'demonstration of how things work rather than an explanation of why they exist or how they have come to be'.

The concept of function is used to mean almost everything from operation to effect and purpose. It is also often confused with adaptation, social cohesion, equilibrium, etc. An adaptive institution may be dysfunctional, while a functional institution may be maladaptive.

Functionalist arguments often depend upon the arbitrary selection of a particular organizational level, or group, in the society, with reference to which warfare or any other practice can then be said to be functional. A case in point is Chagnon's (1967) hypothesis: 'The hypothesis I put forward here is that a militant ideology and the warfare it entails function to preserve the sovereignty of independent villages in a milieu of chronic warfare'. Now, disregarding the essential circularity of the hypothesis, 'while it can be argued that it is adaptive for any one village to engage in warfare, and be generally ferocious, in a situation where everyone else is equally ferocious, it does not follow that it is adaptive for that group of villages to engage in constant raiding and feuding among themselves – they would be much better off in terms of material prosperity if they lived at peace . . . The Yanomamö, like the Tauade, and other acephalous societies, engage in warfare because among other reasons they cannot stop, not because they necessarily as a culture derive any benefit from fighting. In the absence of any central authority they are condemned to fight forever, other conditions remaining the same, since for any group to cease defending itself would be suicidal'.

Functional analysis may be conducive to dismiss the motives of the individuals concerned as irrelevant. 'Warfare', White (1949) claimed, 'is a struggle between systems, not individuals. Its explanation is therefore social or cultural, not psychological'. Following White's lead, Newcomb (1950) claimed that all warfare is 'motivated by economic need, and the biological competition of societies'. Now the claim that all warfare is always motivated by economic need is, in terms of the consensus of ethnographic facts, merely ridiculous, and needs no special refutation here. It turns out, however, that what Newcomb means by motivated is not what most people would take the word to mean. In this way Newcomb can dismiss all the evidence of ethnographers of Plains Indian warfare upon the reasons why the Indians valued warfare so highly, on the grounds that 'the motivation of the individual is not the cause of warfare, it is rather the method by which a cultural irritation or need is satisfied' (Newcomb, 1950).

One feels inclined to ask at this point if a culture can scratch itself. Newcomb and White are both in a philosophical muddle, which leads them to suppose that a social system can have needs, motives and frustrations unknown to its members. When Newcomb claims therefore that all warfare is economically motivated, he means that the motives of real people are irrelevant, and that warfare is 'a function of socio-cultural systems, and individuals are . . . no more than the means through which these systems attain their ends . . . It does not matter for what reason the individual thinks he is fighting, and dying, as long as he is satisfying the needs and imperatives of his culture' (Newcomb, 1950). Leeds (1963) and Harris (1975) have made similar claims. Of course, the claim that a culture is an integrated kind of 'being' with distinct needs which have to be satisfied is to indulge in fantasy. When one reads such vain attempts to explain primitive warfare by appeals to its ecological effects or functions, one realizes that function has frequently the covert significance of 'what a twentieth century materialist rationalist intellectual from Europe or America thinks is a sensible allocation of labour and resources'. In consequence, many such functionalist explanations of primitive warfare have a strong ethnocentric bias. 'That primitives could have genuine, even if sometimes mistaken or inadequate, reasons for these beliefs and behaviours is dismissed as the fallacy of taking native explanations at face value' (Hallpike, 1979).

Finally, functionalists are almost always able to show that everything which people do has some advantageous aspects for someone, so that diametrically opposite situations will be described as adaptively – or functionally – advantageous. 'For example, if a group of tribes habitually lives at peace, it will be shown that there are certain conditions which make this possible, whose function will then have been demonstrated to contribute to the maintenance of peace. But, should the tribes concerned habitually have lived in a state of chronic anarchy and violence, the functionalist is not discouraged. He may either say that each tribe is a separate society, and that warfare contributes to the solidarity of each so-called society, or, that it eliminates weaklings and contributes to the vigour of the group, besides keeping down the surplus population, and supplying protein if they are cannibals' (Hallpike, 1973).

9.8.1 *Cherchez la ressource*: materialism

Lately, theorists of ecological, ethological, sociobiological, and Marxist perspectives and/or signature have found themselves happily united, *bien etonnés de se trouver ensemble*, in a common emphasis on primitive war as a strategy to secure scarce and vital or strategic resources, such as land, protein, women, etc. Simply *cherchez la ressource* and you will find the basic cause/deeper reason/ultimate explanation/the in-the-last-analysis rationale of that war. Though not all these scholars would call themselves such, I will, for the sake of convenience, call them the materialists or the school of materialism. The basic tenets of materialism have recently been eloquently exposed by Ferguson (1984): materialism contrasts with theories that explain war as generated by certain values, social structures and so forth, in the absence of any material rationale. Such factors do affect the conduct of war and thresholds of violence, but a materialist theory directed at explaining the occurrence of war must hold these factors to be secondary and not regularly capable of generating and sustaining war patterns in themselves, if that theory is to be subject to falsification. The presence of war in an area and, probably, the presence of other forms of group competition–conflict as well, adds a new unnatural dimension to the effective environment. War is one of those hazards, like disease or exposure to difficult climatic conditions, that affects the total costs of living in and exploiting a given area.

Ferguson's version of materialism is a very reasonable one. He states: '. . . the focus on land and game has created an over-simplified picture of ecological explanations. Whatever significance environmental phenomena have is a result of their interaction with a society of a given form. The salient environmental condition in any case may be something other than a scarce resource . . . War is never a simple function of the natural environment' (Ferguson, 1984).

There are, however, proponents and advocates of this school who may be considered to be vulgar materialists. War – *any* war, whether primitive or modern – is waged for the sake of securing material resources: territory, protein, women, pigs, cattle, mineral deposits, oil, etc., and that's all there is to it (e.g. Eibl-Eibesfeldt, 1986; p. 530). Implicit assumptions of this and related approaches seem to be:

1. Primitive war is a homogeneous and unitary phenomenon, on which simple propositions can be made. There is little consideration of kinds or types of, or evolutionary stages in, primitive war.
2. War is universal and ubiquitous.
3. There exists a clear-cut demarcation between peace and war.
4. The motivation is always of the material-economic kind, reflecting a truncated and vulgar materialist realistic-conflict paradigm.
5. War was a prime mover of sociocultural evolution, an 'agent of progress'.
6. The theories are based on a fundamentally Hobbesian worldview and a basically cataclysmic paradigm: organisms are simply dictated to by evolutionary imperatives, instinctoid 'whisperings within', or other forces beyond their control.

In his 1988 article, Chagnon attempts to show how several forms of violence in a tribal society are interrelated, and he describes his theory of violent conflict among primitive peoples in which homicide, blood revenge and warfare are manifestations of individual conflicts of interest over material and reproductive resources.

Chagnon observes that many anthropologists tend to treat warfare as a phenomenon that occurs independently of other forms of violence in the same group. However, duels may lead to deaths which, in turn, may lead to community fissioning and then to retaliatory killings by members of the two now-independent communities. As a result many restrict the search for the causes of war to issues over which whole groups might contest – such as access to rich land, productive hunting regions, and scarce resources – and, hence view primitive warfare as reducible solely to contests over scarce or dwindling material resources. Such views fail to take into account the developmental sequences of conflicts and the multiplicity of causes, especially sexual jealousy, accusations of sorcery and revenge killing, in each step of conflict escalation. If, as Clausewitz suggested, (modern) warfare is the conduct of politics by other means, in the tribal world warfare is *ipso facto* the extension of kinship obligations by violence because the political system is organized by kinship.

9.8.2 Meyer: endemic vs. instrumental war

Collective violence between groups seems in the most primitive stage to be caused by the respective mythologies: they produce values and norms which result in collective acts. The strategic concept in these mythologies is power: actors strive for power in order to improve their positions. An immanent analysis of such socio-cultural systems must take the close interrelation of empirical and transempirical processes into account. Power, being essentially a quality of the transempirical realm nevertheless is also the decisive quality of empirical social life. These psycho-cultural processes are interrelated with the social level, where the subsequent blood feuds support the tendency towards an incessant cycle of violent interaction. The state of war between such societies may thus be characterized as endemic war – a relation between societies where war seems to be an end in itself rather than an instrument. The decisive breakthrough in the evolution of war as in general socio-cultural evolution came about with the instrumentalization of collective violence, promoted by development of specialized warrior societies, technical inventions, mainly weapons, and social inventions, mainly the idea of incorporating defeated social groups.

Fear as the universal motive behind primitive warfare has been most eloquently exposed by Meyer: 'Die Manifestationen kollektiver Gewalt erscheinen vielmehr einer Atmosphäre der Furcht vor dem Fremden zu entspringen, die, den Angriff des Fremden antizipierend, diesem zuvorzukommen trachtet' (Meyer, 1981). Such fear-inspired preemptive attack is embedded in the extreme ethnocentrism of primitive societies. For a further elaboration see Meyer (Chapter 11).

9.8.3 Durham: genic theory and the evolution of warfare

Ideally, according to Durham (1976), the adaptiveness of primitive warfare would be ascertained by a rigorous test of three competing hypotheses:

1. cultural traditions of warfare in primitive societies evolved independently of the ability of human beings to survive and reproduce;
2. cultural traditions of primitive warfare evolved by the selective retention of traits that enhance the inclusive fitness of individual human beings;
3. cultural traditions of primitive warfare evolved by some process of group selection which commonly favoured the altruistic tendencies of some warriors.

Durham then develops a model, based on the assumption of genotypic selfishness, which predicts the occurrence of 'intergroup aggression' (Durham's equivalent of warfare) both defensive and offensive, under certain well-defined conditions. Theoretically, direct intergroup aggression would exist as a cultural tradition only where the participating individuals each derive net intrademic fitness advantages. Thus, warfare would exist as a cultural tradition only where social and environmental conditions result in continuous or recurrent net benefits to the aggressors.

A cultural pattern of direct intergroup aggression could be the result of selection in at least three instances:

1. vengeance;
2. failure of reciprocity;
3. resource competition.

Durham focuses on resource competition. He emphasizes at the start that even where we may attribute human aggression to resource competition, relations of kinship and friendship may have an important effect on that behaviour. It may well be that the human ability to recognize – and therefore avoid harming – kin and friends is what has allowed the evolution of deadly conflict.

In principle there are two ways in which participation in group aggression can have net individual fitness benefits. First, when human groups compete for limited resources, successful warriors may themselves directly benefit from the fitness value of resources defended or acquired. The requirements here are two: 1. The spoils must not be shared throughout the deme but must be shared within the group or subgroup of aggressors, so that 2. the fitness-value of resources gained by each participant must exceed his or her accumulated costs. Secondly, intergroup aggressiveness can evolve by selection (i.e., biological and cultural selection) even when all of the warriors do not derive direct resource benefits from the conflict, so long as their fitness costs are more than compensated by other benefits from within the group. The requirements here are 1. at least one important figure in the group must secure resource benefits, and 2. the benefactor(s) must provide other goods or services (or both) on which the other participants' fitnesss depend. In theory, direct aggressiveness is only to be expected, therefore, in habitats where dependable or predictable high quality resources are in short supply.

In contrast, more peaceful relations between groups are expected where:

1. the resource demands show little overlap;
2. the combined demand for any common resource(s) is regulated below the levels of supply by other factors (disease, predation, or parasitism, for example);
3. the spatial and temporal distribution of a scarce resource favours either high mobility (precluding high frequency contact of distinct groups) or reciprocal sharing of unpredictable patches.

In environments where an alternative (if somewhat less desirable) supply of a scarce resource is available, the process of selective retention could actually favour reduced competition between groups, either through specialization of the techniques of resource harvest (cultural character displacement) or migration. For instance, in many arctic areas, the absence of intergroup warfare appears to correlate with high regional variability in food supply.

Durham's model has several important theoretical implications. First, it counters the view that primitive warfare constitutes a 'theoretical embarrassment to a discipline (i.e. anthropology) which has tended to believe that human societies are functionally integrated systems, well adapted to their environments' (Hallpike, 1973). The model describes conditions in which intergroup aggression can be highly adaptive in terms of basic survival and reproduction. Secondly, it challenges a common presumption that primitive war has always evolved for some transcendent group-level function requiring individual sacrifice (as suggested, for example, by Harris and Divale). At least in circumstances of resource competition, it is possible for aggressive intergroup behaviour to have real benefits for participating individuals. Furthermore, selection can favour widespread member participation in collective aggression even though each individual behaves in his or her own, distinct, genetic self-interest. Finally, the general model has several implications for the dynamics of intergroup aggression. Selection would mould a warfare strategy that maximizes the participants' net gains. Thus, Durham's modest conclusion is that at least some cases of intergroup aggression can be seen as a behavioural adaptation to conditions of competition for limiting resources.

Durham's theory has been more or less adopted by Wilson (1978). In addition, Wilson suggests that a principle of animal sociobiology, still only partly tested, is that in times of plenty and in the absence of effective predators females tend to become a density-dependent factor limiting population growth. Wilson defines war as 'the violent rupture of the intricate and powerful fabric of the territorial taboos observed by social groups'. The force behind most warlike policies is ethnocentrism, the irrationally exaggerated allegiance of individuals to their kin and fellow tribesmen. Primitive men cleaved their universe into friends and enemies and responded with quick, deep emotion to even the mildest threat emanating from outside the arbitrary boundary. The practice of war is a straightforward example of a hypertrophied biological predisposition. The evolution of warfare was an autocatalytic reaction that could not be halted by any people, because to attempt to reverse the process unilaterally was to fall victim. A new mode of natural selection was operating at the level of entire societies.

9.8.4 Criticism

Basing himself on models of the evolution of individual aggressive behaviour, Durham is almost forced to regard warfare as 'intergroup aggression' which is unfortunate terminology (van der Dennen, 1986), while the adoption of a realistic conflict paradigm leads him somewhat astray. The bulk of examples of primitive warfare does not evidence competition over scarce resources (e.g. Malinowski, 1941; Wright, 1965; Broch and Galtung, 1966; Service, 1966; Harrison, 1973; Harris, 1978). In the words of Wright: 'Neither territorial conquest nor seizure of slaves nor plunder of economic goods is characteristic of primitive warfare'. Most primitive warfare seems, at least in our Western eyes, rather pointless. It is this conspicuous absence of economic motives that ought to be explained. In many cases the total losses exceed whatever benefits could possibly be gained. In other words, there is a discrepancy between Durham's ultimate explanation and the proximate motives of primitive peoples. Furthermore, it is not clear from Durham's model why it is virtually only human beings that make war, or why intergroup aggression is not more widespread in the animal world. Nor does it explain why warfare is predominantly a male business, or why disculpation ritual (indicating a profound ambivalence of the warrior *vis-à-vis* his victims) is so universal.

9.8.5 Eibl-Eibesfeldt: cultural filter superimposition and preadaptations

According to Eibl-Eibesfeldt (1977), man, like other organisms, has inhibitions against killing as part of a biological filter of norms. Yet he kills conspecifics on a large scale. How does this come about? Man tends to form closed groups. Cultural peculiarities tend to diverge rapidly, and the varieties of culture behave as if they were different species (Erikson's (1966) 'pseudospeciation'). Others are not considered to count as full members of mankind, or even as human beings at all. By cultural definition, intraspecific aggression gets shifted to the level of interspecific aggression, which is destructive in the animal kingdom as well. Facilitated by communical barriers and by armament which kills quickly, and often at a distance, man shuts himself off against all appeals normally releasing the fighting inhibitions which are subjectively experienced as pity. Upon the biological filter of norms which inhibits killing, is superimposed a cultural filter of norms commanding killing of the enemy. This leads to a conflict of norms, bad conscience, guilt and ambivalence, as already noted by Freud (1913). It takes quite a lot of indoctrination and coercion to bring people to fight each other. Unfortunately war had functions to fulfil, it is not to be considered an evolutionary cul-de-sac or pathology: 'Es handelt sich nicht um eine funktionslose Entgleisung, sondern um eine spezifisch menschliche Form der Zwischengruppen-Aggression mit deren Hilfe Menschengruppen um Land und Naturgüter konkurrieren' (Eibl-Eibesfeldt, 1975).

In later publications (Eibl-Eibesfeldt, 1979; 1982; 1984), he argued that different levels of selection are discernible in man, and that the group is indeed an important unit of selection in *Homo sapiens*. He identifies a number of preadaptions which evolved by individual selection, but which make group selection feasible: 1. maternal care and the individualized bond, leading to xenophobia; and 2. mechanisms of individual bonding. Eibl-Eibesfeldt suggests that human indoctrinability and the inclination to polarize values are specific traits difficult to explain via selection at the level of the individual. The emotional basis of this response has its roots in family defence, but cultural evolution led to the development of warfare ethics which cause individuals to act against their self-interest. At the tribal level, the costs for the young individual male are extremely high. Thus in nonliterate tribal cultures, indoctrination of heroic virtues creates a readiness for self-sacrifice for the group. This often goes hand in hand with the training for obedience.

Campbell (1972), in a revision of an earlier paper (Campbell, 1965), notes that: 'I now believe that these self-sacrificial dispositions, including especially the willingness to risk death in warfare, are in man a product of social indoctrination, which is counter to rather than supported by genetically transmitted behavioural dispositions'. However, if there were genetic differences in indoctrinability, and if warfare were the main selective system operating, then there would be genetic selection against indoctrinability. Probably, Campbell argues, the overall adaptive advantage for indoctrinability, group identification and fear of ostracism is strong enough to overweigh the negative selection produced when the most indoctrinable incur greater fatality rates in wartime. There is probably positive selection for heroic bluff that persists as long as successful but turns into cowardly retreat when the odds become overwhelming.

9.8.6 Kin selection and the evolution of warfare

According to contemporary evolutionary theory, so-called group selection is not typically expected; in general, social cooperation will develop only when the actor's own kin ultimately benefit through reciprocity – and this factor can be expected to limit the size of groups to that in which individuals have a reasonable probability of recognition or role reversal in future encounters (Masters, 1983; Wilson, 1975a). There are, however, some restrictive circumstances in which this individualistic calculus does not hold. Where a number of small populations of extended kin are relatively isolated from each other, conflicts between groups can and will occur if resources are insufficient to support all of them. In that case, a behaviour that benefits the group – even at a cost to the actor – can be favoured by natural selection whenever competition within the group is harmful to both the individual's inclusive fitness and the group's collective interest. Under these circumstances (which Alexander (1979) describes as a 'balance of power' between competing bands of extended kin), group selection would expand the sphere of social cooperation. As Benjamin Franklin put it 'We must all hang together, or assuredly we shall all hang separately' (Masters, 1983). Furthermore, in such situations, kin selection might virtually encompass the group, and selection between groups might amount to a special case of kin selection (Corning, 1975). Willhoite (1980) has shown how this 'balance of power' hypothesis can explain the progressive evolution of hominids through the stages of the band, the sovereign village, the chiefdom, and finally the early state. At each level, as communities expand to include more distant kin or non-related kin-groups, the role of coercive institutions also expanded (Masters, 1983).

The possibility that endemic warfare and genetic usurpation could be an effective force in group selection was clearly recognized by Darwin (1871). He saw that not only can group selection reinforce individual selection, but it can oppose it – and sometimes prevail, especially if the size of the breeding unit is small and the average kinship correspondingly close. By adding the additional postulate of a threshold effect, according to Wilson (1975a), it is possible to explain why the process has operated exclusively in human evolution: the capacity to consciously ponder the significance of adjacent social groups and to deal with them in an intelligent, organized fashion. The only combinations of genes able to confer superior fitness in contention with genocidal aggressors would be those that produce either a more effective technique of aggression or else the capacity to preempt genocide by some form of pacific manœuvering. Either probably entails mental and cultural advances. In addition to being autocatalytic, such evolution has the interesting property of requiring a selection episode only very occasionally in order to proceed as swiftly as individual-level selection. By current theory, genocide or genosorption strongly favouring the aggressor need take place only once every few generations to direct evolution. This alone could push truly altruistic genes to a high frequency within the bands. Furthermore, it is to be expected that some isolated cultures will escape the process for generations at a time, in effect reverting temporarily to what ethnographers classify as a 'pacific state' (Wilson, 1975a).

Warfare is viewed by Wilson as a significant general factor in human evolution. Anthropologists, however, have not shown any consensus in support of such an interpretation. For example, Montagu (1976) counters with the view that 'up to some 12000 years ago war played an insignificant role' in evolution, and that in the last 12000 years war has become either biologically irrelevant or dysgenic.

Building on Alexander's (1971, 1979) balance of power hypothesis, McEachron and Baer's (1982) hypothesis on the evolution of weapons, and especially the principle of kin selection (Hamilton, 1964, 1975), Shaw (1985) and Shaw and Wong (1987, 1988) present a model of kin selection and the evolution of human warfare.

They propose that inclusive fitness considerations (an ultimate cause) have combined with competition over scarce resources (environment), intergroup conflict and weapon development (changing environment), to:

1. reinforce humanity's propensity to band together in groups of genetically related individuals;
2. predispose group members to act in concert for their own well-being, first and foremost;
3. promote xenophobia, fear and antagonism among genetically related individuals towards strangers.

They interpret these responses as 'emerging' or reinforcing proximate causes which shaped the structure of social behaviour in hunter/gatherer groups for 99% of humanity's existence.

Their model rests on three premises:

1. that individuals have evolved not only to be egoistic, but to be nepotistically altruistic;
2. that individuals, and individuals in nucleus ethnic groups, are predisposed to mobilize for resource competition in ways that will enhance inclusive fitness and reproductive potential;
3. that a link exists between ethnic mobilization for competition over scarce resources and the idea that intergroup conflict/warfare has been positively functional in humanity's evolution.

The axiom of inclusive fitness is crucial to understanding the importance of ethnicity in the expression of humanity's propensity for warfare for five reasons. First, it implies that individuals judge net benefits of engaging in competition not only in terms of direct private gain but also in terms of indirect gain associated with the well-being of genetically related individuals. It thus provides a social rationale for related individuals banding together to pursue competition (nepotistic altruism). Secondly, inclusive fitness considerations militate against freeriders. Thirdly, inclusive fitness considerations reduce problems of unequal distribution of the spoils of conflict/warfare. Fourthly, inclusive fitness considerations enhance the process of selecting a group leader. Finally, the axiom of inclusive fitness allows to postulate how death can be tolerated in conflict/warfare situations.

In accordance with Alexander's 'balance of power hypothesis', Shaw and Wong stress the point that in the past one million years or so an increasing proportion of man's 'hostile environment' has been other nucleus ethnic groups engaged in resource competition. While the unit of selection remains that of the gene and their individual carriers, intergroup conflict has rendered groups of ever-expanding size and internal structure effective forces of selection. According to this idea, expansion of nucleus ethnic groups through intermarriage, or their expansion via amalgamation with other nucleus ethnic groups, was motivated by the fact that other groups were doing so. Failure to maintain a balance of power (initially in terms of numbers only), would inevitably mean the domination of one group by a larger group and, consequently, unequal access to fitness enhancing resources. From this perspective, large scale agriculture and an increasingly elaborate division of labour follow as concomitant developments. The underlying momentum of such developments is group selection (to maintain the balance of power), which, in turn, is a consequence of genetic selection.

9.8.7 The evolution of war and its cognitive foundations: the theory of Tooby and Cosmides (1988)

Coalitional aggression evolved because it allowed participants in such coalitions to promote their fitness by gaining access to disputed reproduction enhancing resources that would otherwise be denied to them. Far fewer species manifest coalitional aggression than would be expected on the basis of the actual distribution of social conditions that would favour its evolution. Tooby and Cosmides propose that the distinctive and frequently surprising features of war stem from an underemphasized dimension: cooperation. Although a fight is an aggressive conflict between two individuals and involves no cooperation, a war is an aggressive conflict between two coalitions of individuals and would not be possible unless each coalition were able to coalesce, function and sustain itself as a group of cooperating individuals. They suggest that a detailed analysis of the evolutionary dynamics of cooperation in the context of coalitional aggression may explain:

1. adaptive obstacles in the evolution of coalitional aggression;
2. why war is so rare among animal species;
3. why, nevertheless, it is so easy to generate conditions in which human males find initiating warfare so psychologically appealing.

Each coalition member has impact on the coalition 1. by regulating the level of his own direct participation in the joint action, and 2. by the actions he undertakes to enforce the risk contract on the other coalition members. These two dimensions of regulating direct participation and enforcement have important and sometimes surprising properties. For example, the optimum level of direct participation is extremely sensitive to the probability of success, and the relationship between these variables may help explain why males will engage so readily in warfare when they are confident of success. It can be shown that given:

1. certainty of victory;
2. the assurance of a random distribution of risk of death among participants;
3. the assurance of a relatively 'fair' allocation of the benefits of victory;
4. efficiency in the utilization of reproductive resources on a zero-sum basis,

selection will favour participation in the coalitional aggression regardless of the existence or even the level of mortality (within broad limits).

Within a polygynous system with certain formal properties (for example, access to females being the limiting resource for male reproduction; male labour being comparatively unimportant to female reproduction etc.), the death of some members of a coalition will not decrease the average reproduction of the members of the coalition, because the reproductive resources and opportunities within the coalition, or gained as the result of victory, will simply be reallocated among the survivors. As long as the members of the coalition do not lose reproductive resources, the level of deaths among the males will not influence the average success of the coalition members. Each individual who dies loses, but each survivor gains to the same extent, and provided the participants do not know in advance who will live and who will die, but rather that the risk is distributed randomly, and provided they are assured of success (as in, for example, a much larger group attacking a much smaller one), the collective decision of the coalition to go to war will benefit its members (in the currency of fitness).

Natural selection weighs decisions on the basis of their average consequences to individuals, summed over evolutionary time; consequently, these factors explain why males can so easily be induced to go to war, despite its lethal effects on many of them. This zero-sum nature of within-coalition reproductive reallocation cushions successful coalitions from most of the negative fitness consequences that would seem to necessarily follow from the decision to initiate warfare. Because evolved psychological mechanisms will be shaped by the average result of a decision, the finding that average fitness is enhanced by the decision to embark on a successful war provides a powerful explanation for the existence of strong pro-war emotions (given the necessary conditions).

Coalitions of males, when they assess the relevant variables indicating that they are larger or more formidable than any local competing coalitions, should appear to manifest an eagerness and satisfaction in initiating warfare and an obliviousness or insensitivity to the risk they run as individuals, in terms of their individual somatic welfare.

This approach also predicts the striking asymmetry that exists between males and females in coalitional aggression: females are rarely limited by access to males, so that the net production of a coalition of females would drop in direct proportion to the number of females killed. In a curious fashion, males may be so ready to engage in coalitional aggression because it is reproductively safer for them to do so. Females have more to lose, and less to gain, and such differences in consequences should be reflected in psychological sex differences in attitudes towards coalition formation and coalition-based aggression (cf. also Symons, 1979). In short, war is not simply a response to resource scarcity: when times are good, and male productivity irrelevant, war may be very advantageous.

CHAPTER TEN

The Inuit and the evolution of limited group conflict

C. Irwin

10.1 INTRODUCTION

From an examination of the archaeological and ethnographic record Bigelow (1969) and Eibl-Eibesfeldt (1979) draw the conclusion that the human behaviour we call war is and has been a universal human trait perhaps as old as *Homo sapiens.** Bigelow (1969), in his book *The Dawn Warriors*, goes so far as to suggest that many of man's so called advanced facultative skills evolved specifically to enhance man's warring skills, as war, Bigelow suggests, was adaptive. The advent of nuclear weapons may have irrevocably changed the cost to benefit ratio of warring behaviour. So long as there were some winners then war may have been adaptive but when there are only losers the behaviour of war ceases to have any adaptive advantages. How will this change in the cost to benefit ratio of warring behaviour be played out in evolutionary terms? Will man become extinct? Perhaps.

If it is the expressed desire of mankind not to become extinct, at least not in the near biological future, then how can this warring animal be modified without paying the costs of selection by nuclear conflict? If it is impossible to alter the genetic hubris of human nature by genetic engineering (we don't know how) or natural selection (we don't have time) the only course open to us, for the salvation of our genotype, may be through the cultural manipulation of human behaviour. This possibility for saving mankind, by rational and cultural intervention, at first appears to be an obvious solution that

*For an additional review see Alexander (1979), for cross-cultural studies see Otterbein (1970) and Wheeler, V. (1974) and, for a study of hunter/gatherers, see Ember (1978) who regards the anthropological view that hunter/gatherers are peaceful and non-competitive as a 'myth'.

should be promptly embraced. Unfortunately when we take even the briefest look at human history we quickly discover that man has continually been pleading with his fellow man to be nicer than he is. Philosophers and kings have repeatedly tried various political solutions and rational appeals to project man into a state of peace, at least since the times of Ancient Greece, without success. If the cultural modification of human nature, to curb an innate propensity for group conflict, is impossible, then evolutionary logic would suggest that we should get the process of selection over and done with as soon as possible. Why wait? Press the button now.

This conclusion is built on the premise that human salvation, by cultural means, is as impractical as changing the genetic basis of human antisocial behaviour. This chapter is written to document the fact that the cultural restriction of group conflict is indeed possible, as at least one society, on at least one occasion by some process of sociocultural evolution, has done it. The Inuit* of the central Canadian Arctic did sometimes commit murder but the killing of a fellow tribesman and a member of another tribe cannot be distinguished in the Inuit culture. It will be argued that the behaviour we call war was not a part of their social repertoire. The very term 'war' cannot be translated into their language Inuktitut. This rare event, in the annals of human biocultural evolution, deserves our close attention.

10.2 INTERTRIBAL CONFLICT IN THE CENTRAL ARCTIC: MURDER, FEUD OR WAR?†

Before I can begin to make the case that the Inuit of the central Canadian Arctic, unlike many of the peoples that surrounded them, did not engage in the social behaviour war it will be necessary to briefly discuss what is being defined by the term war in this chapter. Gwynne Dyer, in his book *War* (1985), wishes to suggest that prehistoric hunter gatherers did not engage in warfare as they did not have a division of labour society that produced a warrior class or a chief that would lead them into battle. For Dyer war is a cultural invention that merely has to be uninvented if mankind is to be saved. However the prehistoric hunter gatherers Dyer discusses did possess clearly

*Although most of the original ethnography in this chapter relates to the Inuit tribe that are known as the Netsilingmiut much of what is said here would apply to the tribes that surround them and perhaps to most of the Inuit in general who are the Eskimo that occupy North America from east of the Mackenzie Delta to west and north of Labrador, exclusive of these regions.

† As this study required information on historical intertribal conflicts a methodology was employed in which community elders confirmed and elaborated previously collected ethnographies. This field work was undertaken in Arctic settlements on the west coast of Hudson Bay during 1976, 1980 and 1982. For more detail on methodology see Irwin (1985a).

defined territories,* peopled by clearly defined groups,† that, as Dyer admits, frequently engaged in organized intergroup killing.‡ For the purposes of this chapter this behaviour will be defined as war. The socially and technically complex behaviour we associate with war today is, in this view, a cultural elaboration of a primitive behaviour. Bigelow (1969) and Eibl-Eibesfeldt (1979) would probably agree with this perspective. By the criteria defined here let us now take a look at the Inuit of the central Canadian Arctic, and the peoples that surrounded them, and decide if they engaged in war or existed, excepting murder, in a state of relative peace.

In Alaska the coastal Eskimo and their inland Indian neighbours were in an almost continual state of war that was frequently genocidal (Petroff, 1884). In Weyer's account of such a conflict it should be noted that the women and children were killed. 'The attacking Indians stuffed the door with brush wood and set fire to it. Entrapped, the Eskimo women were shot by the Indians from the smoke hole, and most of those not killed by arrows suffocated in the smoke.' These Indians were later killed 'to the last man' in a counter attack (Weyer, 1932). Nelson (1899) attributes a similar level of fighting to Eskimo groups although these encounters fell short of total genocide, 'they killed all they could of the males of the opposing side, even including infants, to prevent them from growing up as enemies. The dead were thrown in heaps and left. The females were commonly spared from death, but were taken as slaves.'

The Eskimo of the Bering Strait region of Alaska would appear to have been culturally adapted to the behaviour of war. The Unalit called the warrior who planned the attack *mu-gokh-ch-ta* and the enemy was termed *um-i-kis-tu-ga*, 'one who is angry with me' (Nelson, 1899). The young men of the region:

> regularly trained for war, hardening themselves in all manner of athletic exercises, dieting themselves, and often obliged to fast in order to habituate themselves to great hardship, or making journeys on foot for many days in succession as a test of endurance. Not only were the different tribes constantly at feud among themselves: they did not hesitate to enter upon combats with Indians or white men when these ventured into their territory. Fighting was carried on as a rule with bow and arrow, but they

*Inuit tribes name themselves after geographic locations so that the Netsilingmiut, for example, are not, as is popularly believed, 'The People of the Seal', but are in fact 'The People of Seal Lake', as this tribe live close to a large tidal lake that contains seals.

†These groups or tribes can be defined in many different ways. One of the most useful perspectives that can be used here is that of tribes as dialect groups (Birdsell, 1972) that were substantially endogamous (Irwin, 1985a, 1987).

‡ Burch and Correll (1971) define war as conflict between the members of a number of extended families from two or more regional groups and feuding as hostilities limited to the members of two extended families. These definitions may be used here.

> had also special inventions of their own; among the most notable were breastplates of walrus tusk, proof against arrows, or great saw-toothed clubs designed to crush the skull of an enemy. (Rasmussen, 1927)

Warring behaviour does not seem to be so culturally well developed as we move east along the coast of the North Slope of Alaska and into the Mackenzie Delta region. Here, as on the Yukon, Eskimo/Indian conflict extended up the river system (Rasmussen, 1927). However it is worth pointing out that Weyer (1932) does not consider these encounters to be war campaigns as such (Stefansson, 1914) but rather murderous skirmishes (Rink, 1887). I will not attempt to adjudicate this point of definition here but this equivocation may be significant as it is probably symptomatic of a real reduction in the level of conflict when compared to South West, West and North West Alaska.

In comparison to the Eskimo of Alaska and the Mackenzie Delta the Inuit of the central Arctic had little contact with the Indian population to the south. However, what contact they did have may frequently have been merciless and one sided as typified by the encounter on the Coppermine river at what is now called Bloody Falls.

> By the time the Indians had made themselves thus completely frightful, it was near one o'clock in the morning of the seventeenth; when finding all the Esqimaux quiet in their tents, they rushed forth from their ambuscade, and fell on the poor unsuspecting creatures, unperceived till close at the very eves of their tents, when they soon began the bloody massacre, while I stood neuter in the rear.
>
> In a few seconds the horrible scene commenced; it was shocking beyond description; the poor unhappy victims were surprised in the midst of their sleep, and had neither time nor power to make any resistance: men, women, and children, in all upward of twenty, ran out of their tents stark naked, and endeavoured to make their escape: but the Indians having possession of the land-side, to no place could they fly for shelter. One alternative only remained, that of jumping into the river: but, as none of them attempted it, they all fell a sacrifice to Indian barbarity!
>
> The shrieks and groans of the poor expiring wretches were truly dreadful; and my horror was much increased at seeing a young girl, seemingly about eighteen years of age, killed so near me, that when the first spear was stuck into her side she fell down at my feet, and twisted round my legs, so that it was with difficulty that I could disengage myself from her dying grasp. As two Indian men pursued this unfortunate victim, I solicited very hard for her life; but the murderers made no reply till they had stuck both their spears through her body, and transfixed her to the ground. They then looked me sternly in the face, and began to ridicule me,

> by asking if I wanted an Esquimaux wife; and paid not the smallest regard to the shrieks and agony of the poor wretch, who was twining round their spears like an eel! Indeed, after receiving much abusive language from them on the occasion, I was at length obliged to desire that they would be more expeditious in dispatching their victim out of her misery, otherwise I should be obliged, out of pity, to assist in the friendly office of putting an end to the existence of a fellow creature who was so cruelly wounded. On this request being made, one of the Indians hastily drew his spear from the place where it was first lodged, and pierced it through her breast near the heart. The love of life, however, even in this most miserable state, was so predominant, that though this might justly be called the most merciful act that could be done for the poor creature, it seemed to be unwelcome, for though much exhausted by pain and loss of blood, she made several efforts to ward off the friendly blow. My situation and the terror of my mind at beholding this butchery, cannot easily be conceived, much less described; though I summed up all the fortitude I was master of on the occasion, it was with difficulty that I could refrain from tears; and I am confident that my features must have feelingly expressed how sincerely I was affected at the barbarous scene I then witnessed; even at this hour I cannot reflect on the transactions of that horrid day without shedding tears. (Hearne, 1795)

Encounters of this kind were so effective, from an Indian point of view, they may have been partly responsible for the creation of a mythology amongst the Inuit that living trees were dangerous and it was unsafe to camp in the Indians habitat, the forest (Rasmussen, 1927; Birket-Smith, 1929; Weyer, 1932).

Where contact between the Inuit and the Indians was more frequent conflict was probably less one sided. On the south west coast of Hudson Bay the Caribou Eskimo were probably always hostile toward the Indians, the Cree and Chipewyan (Rasmussen, 1927; Birket-Smith, 1929; Weyer, 1932). In northern Quebec and Labrador the Ungava District Hudson Bay post built a palisade around their houses in 1831 'to prevent the intrusion of the natives, Indians and Eskimos, who were lately at war with each other that the rancorous feeling had not subsided and might break out afresh at any moment without warning' (Turner, 1894; Weyer, 1932).

In the Central Arctic the blood feud probably replaces war as the most deadly form of social conflict* (Boas, 1888; Weyer, 1932). Boas tells us:

> Real wars or fights between settlements, I believe, have never happened, but contests have always been confined to single families. The last instance

*One explorer, Klutschak (1881), suggests that the Netsilingmiut had conquered their neighbours the Ukussiksalingmiut. However I have not been able to find any recollection of such a war amongst the members of either of these tribes. Boas (1888) also finds no record of such a conflict and makes an effort to refute Klutschak.

> of a feud which has come to my knowledge occurred about seventy years ago. At that time a great number of Eskimo lived at Niutang, in Kingnait Fjord, and many men of this settlement had been murdered by a Qinguamio of Anarnitung. For this reason the men of Niutang united in a sledge journey to Anarnitung to revenge the death of their companions. They hid themselves behind the ground ice and killed the returning hunter with their arrows. All hostilities have probably been of a similar character. (Boas, 1888)

It is worth noting that this conflict resulted in the final killing of only one person, the murderer, and not his relatives, or other members of his tribe. This kind of action could just as easily have taken place within a tribe as the execution of a murderer by the victim's closest male relative was the normal course of action for all Inuit (Saabye, 1818; Lyon, 1824; Hall, 1864; Gilder, 1881; Klutschak, 1881; Turner, 1887; 1894; Boas, 1888; Nelson, 1899; Hawkes, 1916; Stefansson, 1921; Jenness, 1922; Rasmussen, 1927; Birket-Smith, 1929; Cadzow, 1929; see Weyer, 1932 for a review), even within an extended family. This so called blood feud, by Inuit standards, was probably nothing more than an emicly correct dispensation of justice. If, as Boas suggests, this example is typical of intergroup conflict amongst the Inuit of the Central Arctic then it may not be possible to distinguish this kind of human killing from murder or execution in either semantic or practised behavioural terms. If this distinction cannot be made then it may be reasonable to conclude that the Inuit did not participate in the behaviour we call war. In an attempt to resolve this question it will firstly be necessary to complete a cognitive analysis of the Inuit in terms used to describe the taking of human life and secondly make some estimate of the incidence and extent of murder, both within and between tribes. Finally, if it is concluded that the Inuit did not practise war, then did this state of relative peace represent some lost state of nature or were cultural mechanisms instituted to limit the escalation of deadly conflict?

In Greenland, where there are no Indians, intertribal relations may have been for the most part peaceful (Nansen, 1893; Weyer, 1932). Perhaps it was limited to murder and feud (Saabye, 1818) much as it appears to have been in the Central Arctic. Nansen tell us that the Greenlanders find war 'incomprehensible and repulsive, a thing for which their language has no word'. As before, knowledge of numbers killed, what groups they belonged to and descriptions of cultural mechanisms instituted to limit intertribal hostility would be welcome. However this chapter is limited to an analysis of these questions with respect to the Inuit of the Central Arctic.

The question of whether the taking of human life in the Central Arctic should be classified as murder, feud or war can be answered from both the perspective of the Inuit culture (emic) and from the perspective of the ethnographer (etic). As these terms exist in the English language it is clearly possible to distinguish between these forms of killing in our culture. If these semantic discriminations cannot be made in Inuktitut then some of these behaviours may not exist in an emic sense and are probably less likely to exist in a real behavioural, etic, sense. Inuit* terms for the taking of human life are given in Table 10.1.

Table 10.1
Inuit cognates for the taking of human life

English	*Eskimo (Inuktitut) and literal translation*	
Murder	Inuaktok Human – he or she took life	
Revenge killing or execution	Akeyauok Back – he or she – to him or her	
Two people fight	Unatuktook Fight – them (two)	
Many people fight	Unatuktoon Fight – them (many)	
War	None	
Many people fighting and killing each other or mass murder	Unatuktoon Inuakgrotioon Fight – them (many) – Human – they took life (many)	
Baby abandonment	Nutaraarluk Baby	Iksingnaoktauyoq Leave it – he or she – doing it
Baby freezing	Nutaraarluk Baby	Qiqititauyoq Freeze it – he or she – doing it
Suicide	Inminik Self	Pitariok He or she took life

*This analysis uses Netsilingmiut terms. However, as the dialects of the Inuit tribes are very similar (Irwin, 1985a, 1987) this analysis is probably cognitively accurate for most Inuit in general. For more information on the methodology used for this semantic analysis see Irwin (1985a).

A number of observations can be made on this collection of terms. Inuktitut (the Eskimo language) contains a term for murder, *inuaktok*. This behaviour is considered to be wrong (blameworthy) by the Inuit, it is in most cultures. Like English there is also a term for execution, *akeyauok*. This behaviour is not considered blameworthy. Under some circumstances execution may even be praiseworthy in the Inuit culture if revenging a murder requires a brave act. However revenge killing or execution, *akeyauok*, can only be used with respect to specific acts of murder and the associated murderer(s). In discussions about war the Inuit use the term *unatuktoon*, 'many people fighting'. However this term does not necessarily imply the taking of human life unless a form of the term for murder is added, *unatuktoon inuakgrotioon*. Given this semantic construction it is not possible for the Inuit to associate praiseworthy behaviour with human killing on a large scale as the closest they can get to a term for war in their language is *unatuktoon inuakgrotioon*, 'mass murder'. To make this point clearer it is worth noting that some forms of human killing, that are blameworthy in the English language and culture, are not blameworthy in the Inuit culture. Infanticide, generally female infanticide, is not described by a term that would be the equivalent of 'baby murder'. The Inuit use the terms baby abandonment, *nuaraarluk iksingnaoktauyoq*, and baby freezing, *nuaraarluk qiqititauyoq*, and these behaviours are not considered blameworthy by the Inuit* (Irwin, 1985). We may conclude, therefore, that like the Eskimo of Greenland (Nansen, 1893) the Inuit of the Central Arctic appear to have no word in their language that is the cognitive equivalent of war, with its associated licence to take human life. It should be pointed out that the Inuit word for murder is based on the Inuit word for human, *inuk*. Nearly all Eskimo are some form of Inuit. Indians, however, are not, they are Iqeelik, lice people. It follows, therefore, that killing an Indian may not have been murder in a strict cognitive sense although killing another Inuk, whatever their tribe, excepting execution, always was murder.

To decide whether or not war took place in some etic sense it would be helpful to know, first, if murder between members of different tribes was significantly more frequent than murder between members of the same tribe. Secondly, it would be helpful to know how many individuals were killed in the most bloody intertribal conflicts: 50, 20, 10, 5 or less?

Murder may not have been an infrequent event amongst the Inuit. Of 36 myths collected from my mother and father-in-law, approximately half deal with the wrongness of murder and the inevitable justice of execution at the

*It should be pointed out that these babies were not considered human, in a metaphysical sense, as these babies would not have been given a name/soul and could be cognitively discriminated as, *nutaraarluk atikungitok*: baby – name – he or she – without.

hands of the closest male relative. Rasmussen's (1931) collection of myths shows a similar balance. The myth 'Quqtaq' is typical (Irwin, field notes):

> When Quqtaq was little more than a baby his father was killed by a man and that man took Quqtaq's mother for a wife. Quqtaq grew up with the man who had killed his father and this stepfather looked after Quqtaq well even though Quqtaq was not his own son.
>
> Every morning, while Quqtaq was still growing up, he would leap out of bed and then jump through the door. He would jump through the snow door even when the igloo was closed up. Sometimes he would jump through the snow blocks, even when they were not the door!
>
> The man who had killed Quqtaq's father and who had married Quqtaq's mother, was getting worried. He would stare at Quqtaq when he leapt out of bed and jumped through the igloo wall or jumped through the closed door. This made the man suspicious and one time he even said, 'Quqtaq, I think you want to kill me'.
>
> But Quqtaq replied, 'No, I do not wish to kill you because you have looked after me, I am growing up and I eat well because of you. Why would I want to kill you?'
>
> One morning Quqtaq jumped out of his bed, leapt through the door, and when he jumped back in he grabbed hold of a knife, so fast, no one saw him. Then he stabbed the man who had married his mother and he pulled the knife out of the body, covered in blood.
>
> When Quqtaq killed his mother's second husband, she exclaimed, 'Quqtaq, do you have a mind? Quqtaq, do you have a heart? Why did you kill this man?'
>
> Quqtaq answered, 'I do have a mind and I do have a heart. But I remember, when I was so very little, trying to stop myself from crying, when this man killed my father. I will never forget the time when this man killed my real father. As I have told you, I remember the time when I struggled to hold back the tears. So I have killed this man who has been my stepfather'.
>
> Kako, 1980

In contrast to this collection of murderous tales, that frequently involved the killing of close relatives (for example jealous brothers), it is very difficult to find any mythology that can be said to describe intertribal conflicts that may bear some resemblance to war. Boas (1888), Rasmussen (1931) and my informants all refer to a mythical people, the Tunrit, who were very large and strong, half Inuit and half white, that could move enormous rocks and lived in stone houses (the Inuit of the Central Arctic only use tents and snow houses, igloos). The remains of these dwellings can still be found. Boas (1888) and Rasmussen (1931) tell us that the Tunrit ran away and left their land after one of them had accidentally killed an Inuk boy or one of their dogs. Samik told Rasmussen (1931) that the 'Tunrit were strong, were easily frightened, were easily put to flight, their lust to kill, we hear nothing of'. Boas (1888) also describes what he considers to be the only tradition of a real fight between tribes. It also involves a conflict between newcomers and people who lived in stone houses who rolled boulders down on their enemy. In the accounts I have heard there was loss of life in the conflict that led to the Tunrit deserting their land (field notes). Perhaps these different tales are all variations of the same story.

If we review the first and second hand accounts of murder we find evidence that suggests the ratio of tribal to intertribal killing may have been biased toward the taking of human life within the tribe. Boas does not provide many detailed accounts of murder. Rasmussen (1931) gives a detailed account of his meeting with the 'outlaw', Ingivalitaq, who had murdered his hunting companion and points out that of the 60 families of the Netsilingmiut tribe included in his census the male heads of the four families living at the remote camp on Belliot Strait were murderers 'or at any rate men who had kidnapped their wives'. Balikci (1970) gives several historical accounts of murder amongst the Netsilingmiut and their neighbours (see below), but Steenhoven (1959) provides what may be the most authoritative descriptions of seven murders that he is able to attribute to members of this tribe in recent times; Sumisertok in about 1900, Sivatkaluk also in about 1900, Krabviojark in 1910, Atuwir in 1913, Amaroalik in 1921, Arnasluk in 1935 and Angotaojajok in 1940. However all of these murders involved fellow camp members that were related and therefore probably of the same tribe.

Some additional sense of the level of human conflict in the Central Arctic may be gained from an assessment of how many people were killed in the most deadly of these human exchanges. In Boas's single account of an intertribal conflict only the avenged murderer was killed. However Balikci (1970) provides a detailed account of what may have been the most bloody exchange involving Netsilingmiut. Balikci's informant, Irkrowaktoq, places the incident before the arrival of Sir John Ross's ship in 1829 and describes how Kujaqsaq and some of his relatives from around Netsilik travelled to Arvilikjuak to avenge the murder of his father, Ugak. In this account many Arvilik men are killed but we are not told how many. In three accounts, which Balikci tells us are fragments of the same incident, Rasmussen's (1931) informants Samik, Manelaq and Tieksaq name three persons killed by Kujaqsaq who is described as being 'eager to kill people' and having arrows that were directed by magic words. Itimangnerk, the brother of Balikci's informant, told Steenhoven (1959) that this kind of incident would be resolved when one side 'suffered, say, four casualties or wounded'. My informants were generally not sure how many people were killed in such conflicts. The most specific answer I was given to this question was 'four, or three, or two. Four was the most. The losers would be those with three or four dead'.

Both Inuit mythology, and the accounts of killing that can be recalled, do not suggest that the rates of intertribal killing exceeded the rates of killing within the tribe. It may therefore be reasonable to conclude that intertribal conflict amongst the Inuit of the Central Arctic, was limited to murder, and revenge killing, or execution, as is suggested by their semantics. These conflicts generally involved individuals but on at least one occasion before 1829 a revenge conflict produced as many as 'four casualties or wounded' (Itimangnerk to Steenhoven, 1959) which isn't very many when compared to the 'upward of 20' Inuit killed by the Indians at Bloody Falls (Hearne, 1795). Let us now explore what may have been the evolutionary causes for this state of limited intertribal hostility.

10.3 ALTERNATE THEORIES FOR THE EVOLUTION OF LIMITED GROUP CONFLICT: ULTIMATE CAUSE

Many social scientists fall into the trap of characterizing prehistoric, hunter gatherer man in either Rousseauian 'noble savage' terms or Hobbesian 'brutish' terms so that they can develop a thesis of civilization that is a fall from the Garden of Eden (for example, Dyer, 1985 above) or a struggle to create an Eden where there never was one. Neither picture of human nature is quite correct as sociobiology suggests that humans have a capacity for altruism, egoism and reciprocity depending on the coefficient of relatedness of the parties involved and/or the cost to benefit ratio of the social behaviour being engaged in (Hamilton, 1964; Trivers, 1971). As far as our hunter gatherer ancestors are concerned the limits of reciprocity and the beginnings of group conflict would seem to be the tribe or dialect group that was substantially endogamous (Irwin, 1985, 1986). If the humans in these population structures stopped making war then sociobiological theory suggests that something happened to the coefficient of relatedness, or cost benefit ratio, or both. Let us briefly consider these alternatives that are tabulated in Table 10.2.

10.4 COEFFICIENTS OF RELATEDNESS

Hamilton first suggested that the theory of kin selection (Hamilton, 1964) could explain human group behaviour in genetically structured tribal populations in his 1975 paper 'Innate Social Aptitudes of Man: An Approach From Evolutionary Biology'. Although this paper is perhaps the first to explain human intertribal relations in population genetic terms, Hamilton fails to realize that relative differences in coefficients of relatedness can frequently be better predictors of social behaviour than simple similarities in coefficients of relatedness (West Eberhard, 1975). Because of this oversight

Table 10.2
Summary of biological costs and benefits (ultimate cause) for Inuit/Inuit and Indian/Inuit war

Benefits of Inuit/Inuit war	*Costs of Inuit/Inuit war*
More Wives	Very few hunters could reliably feed more than one wife and her children
More Land	Hunters cannot exploit new territory as efficiently as old territory
	Hunters as warriors were a valuable and limited resource
Benefits of Indian/Inuit war	*Costs of Indian/Inuit war*
More Wives	Indian and Inuit genotypes would be degraded
More Land	Indians did not possess a technology to exploit the Inuit territory

Hamilton is brought to the false conclusion that in some demicly structured human populations:

> . . . hardly any extra hostility is expected to members of neighboring demes. . . . a sea-shore phase of hominid evolution, if it occurred, should have been particularly harmonious. (Hamilton, 1975)

In my paper, 'A Study in the Evolution of Ethnocentrism' (Irwin, 1987, also see Irwin, 1985a), this error in theory is corrected with a mathematical model that predicts that social behaviour can be polarized at all population boundaries where there is some variation in the coefficient of relationship between the adjacent demes. In other words rivalry between closely related human populations is as predictable a phenomenon as sibling rivalry. Brothers have been known to kill each other and closely related tribes sometimes do go to war. So the fact that the Inuit tribes of the central Canadian Arctic are closely related to each other, because they are arranged in a linear stepping stone population structure (Irwin, 1985a, 1986) cannot, in and of itself, explain their state of relative peace.*

*It should be noted, however, that inertia in the evolutionary process could produce a situation in which the proximate mechanism of group identification and behaviour could produce population structures with unusually low levels of intergroup conflict. For a discussion of this issue see 'Badging' in section 10.7.4.

It is worth noting that the intensity of hostility between adjacent Eskimo tribes in Alaska is much less bloody than Indian/Eskimo conflict in the same region. This fact can probably be explained in terms of the relative differences in the coefficients of relatedness of these different groups. However, if the Alaskan Eskimo were not at war with the Indians of the region then these Eskimo might have been expected to increase the intensity of their own intertribal conflicts. Why hasn't this happened in the Central Arctic? Why didn't the Inuit tribes, who had very little contact with the Indian population to their south, fight it out for possession of the land?

10.5 COSTS AND BENEFITS OF WAR TO THE INUIT

Weyer provides the following suggestion as to why the Inuit tribes of the Central Arctic did not engage in warfare:

> A further hindrance may lie in the fact that both factions are so poverty-stricken and harassed by the common hardships of their antagonistic and ungenerous habitat that intertribal fighting but poorly repays even the victor for the time and energy it consumes. (Weyer, 1932)

This thesis may very well be correct but it begs the question as to what the potential benefits of war for the Inuit might have been and what costs made the prosecution of war an unprofitable venture for them. From a biological perspective the benefits of war can be understood in terms of securing additional resources for the purposes of reproduction and survival. I shall organize my analysis along these lines.

10.5.1 Reproductive resources, costs and benefits

With respect to reproduction, females that are approaching or who have reached the age of reproductive maturity could be considered a valuable resource. This may be particularly true for the Inuit of the Central Arctic who practised high rates of female infanticide (Rasmussen, 1931; Weyer, 1932; Balikci, 1967; Freeman, 1971; Irwin, 1985a). However male foragers, who provided most of the meat for their wives and children, may have been a more critical limiting resource than reproductively mature females (Irwin, 1985a, 1989). It should be pointed out here that most marriages amongst the Inuit were monogamous (Rasmussen, 1931; Irwin, 1985a) as the adult population was balanced by high rates of male mortality (due to hunting accidents) equal to the rates of female infanticide; so, although one of the more frequent motives for murder amongst the Inuit was to steal a wife from another man this was never done on a mass scale, involving war, as only a few exceptional hunters were ever able to support more than one wife and their children (Irwin, 1985a, 1989). By way of contrast it is worth noting that in the kinder climate of Western and Southern Alaska, where rates of female infanticide are much lower (Weyer, 1932; Irwin, 1985a), the capture of females was a common cause of intertribal war (Petroff, 1884, Nelson, 1899; Weyer, 1932). Therefore, it may be reasonable to speculate that the frequency of polygamous relationships would have been much higher amongst these Alaskan Eskimo when compared to the Inuit of the Central Arctic.

10.5.2 Survival resources, costs and benefits

With respect to survival, a valuable resource would have been productive hunting grounds. The distribution of the large mammals that are the principle food source for the Inuit, is quite uneven across the Arctic (patchy). For example, concentrations of marine mammals are often found at prominent head lands or narrow straits that generate strong currents and open water (for example, Point Barrow, Belliot Strait, Igloolik) and fish are caught in abundance at the mouths of the largest rivers (for example, Backs River and Coppermine River). A monopolistic occupation of these sites could provide an Inuit tribe with a valuable and stable food source that might well be worth going to war over. However, the earlier review of the ethnographic record suggests that the Inuit did not fight wars over these 'oases of protein' that are scattered across their habitat. If this is indeed the case then biological theory suggests that the costs of perpetrating such wars must have outweighed the rich benefits that might have come to the victors. I will now attempt to review what some of these costs could have been.

Territorial specialization

A territory that is very distant or separated from the 'home' territory by a geographic barrier, such as a mountain range, glacier field or wide expanse of open water will clearly be of less value than an easily accessible territory. However, if an Inuit tribe develop a subsistence technology that is specialized to exploit the ecological characteristics of their own habitat then even an accessible adjacent territory may be of only marginal value if its ecological profile is different. The coarse grain environmental change that occurs between the lands occupied by the Indians and Eskimo could be such a case in point. If the Indians had been able to survive the Arctic winters north of the tree line, then wars between these two megapopulations may have been more bloody and decisive than they were. The Indians did not possess the Inuit technologies that use animal fats and oils for heat, snow houses for mobile insulated dwellings and all the tools required for the implementation of these technologies, such as specialized clothing, spears, snow knives, sleds etc. Without the benefit of these inventions Indian expeditions beyond the tree line were limited to the brief Arctic summer.

In a similar way, highly specialized forms of hunting, focused on specific animals, at specific times and places, may have helped reduce conflict between the Inuit tribes of the Central Arctic. This thesis was addressed in my field work by asking elderly hunters which tribes were best at different hunting techniques. In spite of a certain degree of ethnocentrism in some of the responses, in that they may have tended to exaggerate the hunting abilities of their own tribe, most of the respondents agreed that the Netsilingmiut were the best at hunting seals through winter ice (some of their dogs are trained to detect breathing holes buried beneath the snow). This would be expected as the Netsilingmiut live in the middle of the Northwest Passage where, in some years, the sea ice does not break up even in the summer months. By way of contrast the Igloolingmiut, who live by a turbulent strait in Northern Hudson Bay, were considered to be most proficient at hunting sea mammals in open water (their kayaks are large for extended use at sea) and the inland Inuit, of the Barren Grounds, were considered most expert in the various techniques of hunting caribou (their sleds were long and slender which sacrificed manœuverability for speed). It seems possible, therefore, that the Inuit tribes that occupied the Central Arctic possessed technologies that were, to a considerable degree, adapted to their particular environmental niche. If they moved to their neighbours' territory they may have increased the risk of starvation as many of their skills and much of their local knowledge may not have been applicable to the new habitat.

Although hunting specialization may have made warring over land less profitable for the Inuit, it should be noted that the introduction of the rifle, around the beginning of this century, turned hunting specialists into hunting generalists. This revolution in Inuit hunting methods may have played a significant role in the migration of a large number of Netsilingmiut families south to the west coast of Hudson Bay in the early 1920s (Rasmussen, 1931; Irwin, 1985a). It may be speculated that if the introduction of the rifle and this migration had not been accompanied by the introduction of intertribal law, in the form of the Royal Canadian Mounted Police, then the migration of the Netsilingmiut to Hudson Bay may not have been so swift and unopposed.

Population control

From the perspective of sociobiology, with its emphasis on inclusive fitness, population control is an unlikely evolutionary imperative, as it probably requires selection at a group level (for a discussion of this argument with respect to humans see Bates and Lees, 1979). Current orthodoxy in sociobiology considers group selection to be an improbable biological event

(Williams, 1966; Wilson, 1975a). I have argued elsewhere (Irwin, 1985a) that Eskimo female infanticide was an adaptive manipulation of the sex ratio that increased reproductive success. Therefore, to suggest that female infanticide and high male mortality replaced war as a means of population control amongst the Inuit (for example see Weyer, 1932) would appear to be making a biologically flawed and outdated argument. This line of explanation will not be used here.

Male hunters as a limiting resource

Active male hunters, that were able to consistently harvest protein resources, were a valuable and possibly limiting resource for the Inuit of the Central Arctic. As already mentioned, this may have been one of the causes of Eskimo female infanticide (Irwin, 1985a). Anecdotal data is also available that suggests that the scarcity of good hunters may have been an important element in the limitation of Inuit intertribal conflict. For example, revenge execution was sometimes discouraged in order to bring an end to a potential string of reciprocal, feuding murders (Steenhoven, 1959; Balikci, 1970). However if an execution could not be avoided, because the murderous crime had been particularly onerous or because the murderer was a continued threat, then:

> . . . relatives could take revenge. But if there were young men involved in the revenge, good hunters who bring in food, then others accompanied him because they like to see him alive rather than killed. When someone says that he wants to take revenge, then all will accompany that man. And if that man is a good hunter, the others accompany him for his protection. (Itimangnerk, in Steenhoven, 1959)

Rates of female infanticide and male mortality are temperature dependent phenomena, across the Arctic, from Alaska to Greenland (Irwin, 1985a). Extensive intertribal hostility in the warmer parts of Alaska and a relative absence of warfare amongst the Inuit of the Central Arctic at first suggests that Eskimo intergroup violence may also be a temperature dependent phenomenon. However, the ethnographic record suggests that warfare did not rise to Alaskan levels of conflict in the kinder climate of Greenland. If these data are correct then this observation presents a puzzle for the thesis being developed here and I should make an effort to speculate on a possible explanation. It should be noted that the Eskimo of Greenland reached their habitat by migrating to it from the harsher environment of Arctic Canada (Crawford and Enciso, 1982). If they had already evolved a cultural capacity to limit group conflict before they reached Greenland then perhaps they simply have not lost this adaptation by virtue of some process of cultural inertia. It could also be speculated that the natives of Greenland genuinely found the reintroduction of warfare abhorrent and rationally chose to maintain their evolved cultural capacity for limiting group conflict. This suggestion, however, probably takes an overly optimistic view of human nature and so this thesis should not be seriously considered until all other possible explanations have been exhausted. Unfortunately this interesting question is beyond the scope of this chapter.

10.6 COSTS AND BENEFITS OF WAR TO THE INDIANS

With respect to reproductively mature females, as a resource to be gained from warring behaviour, it should be noted that the Indians and Inuit of the region did not regard each other as human. Another quote from Hearne's account of the massacre at Bloody Falls may be instructive on this point:

> The brutish manner in which these savages used the bodies they had so cruelly bereaved of life was so shocking, that it would be indecent to describe it; particularly their curiosity in examining, and the remarks they made, on the formation of the women: which, they pretended to say, differed materially from that of their own. For my own part I must acknowledge, that however favourable the opportunity for determining that point might have been, yet my thoughts at the time were too much agitated to admit of any such remarks; and I firmly believe, that had there actually been as much difference between them as there is said to be between the Hottentots and those of Europe, it would not have been in my power to have marked the distinction. I have reason to think, however, that there is no ground for the assertion; and really believe that the declaration of the Indians on this occasion, was utterly void of truth, and proceeded only from the implacable hatred they bore the whole tribe of people of whom I am speaking. (Hearne, 1795)

Although there was some trafficking in females, between the Indians and Eskimo in South West Alaska (Petroff, 1884; Weyer, 1932) and perhaps occasionally in the Mackenzie Delta (Rasmussen, 1927; Weyer, 1932), this behaviour was the exception rather than the rule. It may be important to note that this genetic migration only took place where Inuit extended their territory into the relatively warmer Indian habitat, below the tree line. The Inuit of the Central Arctic are genetically adapted to their harsh environment with short limbs, improved circulation, improved metabolism and small hands and feet, to reduce heat loss and thus prevent frost bite (Folk, 1966; Hanna, 1968; Moran, 1979). Intermarriage could have destroyed these genetic adaptations and proved detrimental to both the Indian and Inuit populations with a resultant loss of fitness due to outbreeding depression (Shields, 1982a, 1982b).

With respect to land as a resource I have already discussed the reasons why the Indians did not make war with the Inuit to take their territory away from them. The Indians simply did not possess the technology to inhabit the Arctic environment north of the tree line on a permanent basis. However the Inuit might have been able to adapt some of their technologies to live in the Indian habitat, particularly along the coast line, where their fishing and sea mammal hunting skills could have been put to productive use. In the context of these observations Indian/Eskimo conflict may be interpreted as an Indian behaviour directed principally at preventing the Inuit from migrating south of the tree line and not as a means to acquire new hunting and fishing grounds. The Inuit believed they were taking a life threatening risk if they camped amongst the trees for any lengthy period of time (Rasmussen, 1927; Birket-Smith, 1929; Weyer, 1932). This fact suggests that the Indian practice of making war against the Inuit of the Central Arctic, whenever the opportunity allowed, was most successful in keeping the Inuit out of Indian territory.

10.7 CULTURAL ADAPTATIONS FOR THE EVOLUTION OF LIMITED GROUP CONFLICT: PROXIMATE CAUSE

If humans have evolved an innate capacity for group conflict then the Inuit described here may have had to develop cultural mechanisms to overcome phylogenetic inertia in order to restrain tribalistic hostility. As with questions of ultimate cause these cultural proximate mechanisms (summarized in Table 10.3) would have to deal with the evolutionary problems of limited resources and differences in coefficients of relationship. I will begin with the cultural management of limited resources.

10.7.1 Sexual and marriage relationships

Of 58 Netsilingmiut families surveyed by Rasmussen (1931) in the Central Arctic in 1923, 54 were monogamous, 1 was polyandrous and 3 were polygamous. The Inuit did not have any cultural prohibitions on adult males or females from having more than one partner. Marriage relationships were often very pragmatic. In Inuit society it was possible for a young man to be married to an older widow if she needed a man to hunt and provide for her and her children, in return he needed a woman to make his Arctic clothing. Such a marriage may appear to have more to do with survival than reproduction. However, in this kind of economic relationship the young man's sexual needs may have been satisfied by a relationship with the wife of one of his hunting companions. As all the parties to such an economic/sexual relationship would probably be close relatives then it may not be necessary to look beyond some form of cultural kingroup selection as an evolutionary explanation for these behaviours. Given this kind of cultural flexibility in

Table 10.3
Summary of cultural adaptations (cultural proximate mechanisms) for limiting Inuit conflict

	Limited resource	*Cultural adaptation*	*Effect*
1.	Reproductively mature females	Monogamy favoured but polyandry and wife exchange permitted	Reduces risk of not having access to a female
	Productive hunters	Monogamy favoured but polygamy and wife exchange permitted	Reduces risk of not having access to a provider
2.	Land and game	Hunting territory held in common	Reduces risk of not having a hunting territory
	Limiting conflict	*Cultural adaptation*	*Effect*
3a.	Language	No cognates available to conceptualize and discuss war	Planning war is difficult
3b.	Political organization	No institutions for social organization above the extended family	Organizing war is difficult
3c.	Socialization of nonaggression	Child rearing practices manipulate infants, rewards, affection, and punishment, aggression	Develops self control of aggression
3d.	Magic	Murder by magical means is ineffective	Murder rate reduced
3e.	Threat of retribution	Execution of murderer is socially encouraged	Costs of murder increased
		Mythology teaches the high costs of murder	Apparent costs of murder increased
3f.	Formal conflict	Ritualization of conflict limits combatants	Death rate reduced
4a.	Kinship	Intermarriage increased ties between tribal subpopulations	Coefficient of relationship increased

Table 10.3 continued

Limiting conflict	*Cultural adaptation*	*Effect*
4b. Pseudokinship	Itlureet (ritual cousins) increased ties between potentially antagonistic tribes	Apparent coefficient of relationship increased
4c. Badging	Increased cultural similarity reduced innate capacity for intertribal hostility	Apparent coefficient of relationship increased

marriage and sexual relationships the need to kill in order to obtain a husband or wife, or simply to have sexual intercourse, was greatly reduced. Needless to say, given the imperatives of reproductive success, this cultural flexibility and tolerance for different sexual arrangements frequently failed, individuals sometimes became jealous and, occasionally, murdered so they were able to monopolize the favours of a partner. Although it may be impossible to prove, it may well be the case that sexual tolerance and flexible marriage relationships increased the reproductive success of Inuit kingroups and possibly reduced rates of murder.

10.7.2 Property rights

I have probably just suggested, in so many words, that competition for reproductively mature females and productive male hunters may have been reduced by partially treating them as a common resource within the kingroup. This conclusion may be given additional support from an analysis of the way in which the Inuit regard property rights in general. Within the tribe the Netsilik, and the tribes adjacent to them, shared most of their material resources (Irwin, field notes) for the practical reason that:

> A greedy person is not smart because a greedy person will end up with nothing some day and then he will end up with no help.
>
> Kako, 1980

Land, animals that had not been killed and even fish traps were held in common; the Inuit considered it wrong to 'possess' such property (Weyer, 1932; Birket-Smith, 1959). Indeed the denial of free access to these resources has, on at least one occasion, been a cause for murder (Balikci, 1970). With only one exception all my informants felt fish traps would not be a cause for intertribal conflict. According to Hawkes (1916) the Eskimo of Labrador:

> . . . may occupy a fishing station in summer year after year undisputed, but it does not give them any special right to it. Anyone else is free to come and enjoy its benefits and, according to Eskimo ethics, they would move away before they would start a dispute about it. (Hawkes, 1916)

However, as Weyer points out, quite a different situation is found amongst the warring Eskimo of Alaska:

> . . . certain duties accompanying the accumulation of much property by a single person are peculiar to the Eskimos of the Bering Sea. Furthermore, personal rights are more specifically defined here. Among the Unalit, thus, the most productive places for setting seal and salmon nets are sometimes regarded as being privately owned. (Weyer, 1932)

Within the tribe the Inuit of the Central Arctic further promoted sharing with the institutionalization of different kinds of partnerships. For example, seal meat sharing (Van de Velde, 1956; Balikci, 1970; Trivers, 1971), name sharing gift exchange (Guemple, 1966; Irwin, 1981), wife sharing (Guemple, 1961) and various forms of pseudo kinship or kin extension (Irwin, 1985a and Burch, 1975 for Alaska); for a review see *Alliance In Eskimo Society* (Guemple, 1971). Although these partnerships rarely extended beyond the tribe it might be reasonable to suggest that the extensive sharing of resources within these groups that went some way to creating a just, egalitarian society, may have taken pressure off individuals and small groups to look beyond the tribe for the necessities of survival and reproduction.

10.7.3 Limiting conflicts

Given the conflicts of interest implicit in the nature of human sociality it is to be expected that institutionalized communalism was far from being a flawless mechanism for social conflict control. When the cultural mechanisms just described failed then other cultural mechanisms were deployed to contain and limit the destructive effects of human conflict. Some of the cultural proximate mechanisms for limiting antisocial behaviour may be described as outlined below.

Language

It is difficult to know to what extent language may be an epiphenomenon that simply provides a cognitive map for evolved human behaviours and to what extent language may be an instigator of behaviour. Either way it has been noted that there is no cognitive equivalent for war in Inuktitut. As no praiseworthy value can be semantically attached to mass murder in their language the Inuit must find conceptualizing and discussing warfare difficult.

This semantic obstacle to thoughts of war must necessarily impede efforts to coordinate the social behaviour of war. No such impediment exists in Alaska amongst the Unalit (Nelson, 1899).*

Political organization

In the Central Arctic the political system is the kin system which mirrors, fairly closely, the biological social order (Irwin, 1985a). Amongst the Inuit the matriarchs and patriarchs of the extended families had authority over the younger members of the extended families (Steenhoven, 1959). As there was no institutionalized political authority beyond these respected elders command and control at the tribal level of behaviour would have been difficult to organize. As has already been pointed out the Unalit of the Bering Strait region of Alaska, called the warrior who planned an attack *mu-gokh-ch-ta* (Nelson, 1899) and Weyer tells us that:

> Among the Alaskan Eskimos the office of chieftain is possibly developed to a higher degree than in any other group. This is not strange among the Eskimos of Bering Strait and the adjacent coast of Alaska, where the natives are outstanding for their war-like tendencies.
> Elsewhere among the Eskimo, even in other parts of Alaska, there is less centralization of authority. (Weyer, 1932)

Socialization of non-aggression

In her paper, 'The Origins of Nonviolence; Inuit Management of Aggression', Briggs (1978) theorizes that a combination of affection, playful aggression, teasing and ridicule socialize Inuit children into adults with an unusually well developed capacity for controlling their own potentially aggressive behaviour. Although this kind of psychological thesis is difficult to prove empirically, and although I find myself in disagreement with much of the detail of her thesis (perhaps, in part, because I lived and worked amongst different Inuit tribes), the conclusion may be substantially correct. If Briggs' thesis is sound then specific patterns of reward (affection) and punishment (playful aggression, teasing and ridicule), administered during a child's development may be partly responsible for adult Inuit self control. However, as Briggs points out, this socialization is only one interdependent element in an array of complex processes involved in the management of Inuit aggression. From the theoretical perspective developed here it should be understood as a cultural

*It should also be noted that Briggs (1978) considers the existence of certain Inuit semantic concepts such as *ihuma* (mind), *naklik* (love), and *ilira* (fear) to be important elements in the development of Inuit non-aggression.

proximate mechanism not an evolutionary ultimate cause or 'Origins of Nonviolence'.

Magical conflict

Another element in the culture type that reduces the frequency of murder (both ingroup and outgroup) may involve the 'black arts' of Shamanism. Frequently Inuit, in traditional times, would attempt to kill other Inuit by employing magic of some kind (Irwin, field notes):

> The shamans could curse people in different ways. If a shaman hated or disliked a man or a woman and one of their relatives had just died, he would take some hair from the dead relative and keep it to curse other people. When the shaman hated somebody or was jealous of them because they were a better hunter than himself, or if he wanted to marry someone's wife, then he would put the hair of the dead relative on the ground where that person who he wanted to kill would walk. Or he would put it near or in front of the person he hated and when that person got that hair stuck onto his cloths then that person was cursed. Then all kinds of hair would grow from his face, from his hands, from his legs, from his feet or his back. Hair would grow and every time that person found it he would pull it out. If he kept pulling them all out he would win but if he couldn't catch up with the hair growing on him, then he would die. That is how they used to curse other people if they hated them or disliked them or were jealous of them.
>
> Aupudluck, 1980

My living with the Inuit for many years has not left me with the impression that they are a stupid people. On the contrary their empirical knowledge of their environment with regard to hunting, tools, materials and navigation (Irwin, 1985b) is indispensable to their survival. Why then is magic of this kind believed in and practised? Clearly magic is not so nearly as effective a killer as a knife and perhaps this is precisely the point. A group of individuals involved in murder by magic will be fitter than a group of individuals involved in real acts of physical murder. In other words, by some process of cultural kingroup or group selection, magic may have evolved to allow for the relatively harmless outlet of pent-up hatred or jealousy that might otherwise have precipitated a string of real deaths.

Threat of retribution

As has already been mentioned it was the right and duty of a male relative to avenge the murder of a family member. However, in practice, very few murders were executed in this way (Steenhoven, 1959; Balikci, 1970). It is interesting to note that all the myths concerning murder do end in the execution of the murderer. Why should this discrepancy exist between myth and fact? Perhaps, as with magic, holding a false belief, in this case the belief that a murder would always be revenged, reduced the rate of deadly conflict. From a biological perspective execution can be understood as a genuine cost to the behaviour of murder. This real cost may not be very high. However the apparent cost created by the cultural mythology is high and may act as a false but effective deterrent.

Formal conflict

We are used to institutionalized mechanisms for limiting and controlling conflict within all societies. For example amongst the Inuit two individuals who had a grievance, that could not be settled by a common elder relative, might engage in a 'song duel' in which the disputants would try to ridicule their opponent until they withdrew their complaint out of embarrassment before their peers (Balikci, 1970). If this stratagem failed the disputants might take it in turns to hit each other on the shoulder or temple, with closed fist, until one of them gave up (Balikci, 1970; Irwin, field notes). Of particular interest to the thesis being developed here is the possibility that these efforts to restrict violence may have extended to the worst of intertribal hostility as pointed out by Itimangnerk to Steenhoven in his discussion of the revenge conflict described earlier by Balikci (1970) and Rasmussen (1931).

> All would travel to the enemy's place and, at a short distance from it, they would pitch camp: for the others, probably being unaware of their arrival, should be able to prepare themselves. Then, the revenging party would send forward an old man or old woman to find out whether the others were prepared. For those old ones, according to our customs, were never attacked. Thus the old woman asks them if they are ready to come outside. If so, the other party would advance and, while they were approaching each other, each would choose an opponent whom he considered his match in strength. They would shoot with bow and arrow; and if they were not many, they might fight with snow knives also. Harpoons for bear and muskox hunting were also used. Now, if one party had suffered, say four casualties or wounded, they would concede the victory and the winners would let the defeated go home. (Itimangnerk, in Steenhoven, 1959)

Additionally, in an effort to reduce the possibility of a conflict from ever starting between members of different tribes a formal greeting procedure would be followed when strangers met for the first time (Klutschak, 1881; Boas, 1888; Stefansson, 1914; Jenness, 1922; Mathiassen, 1928; Birket-Smith, 1929; Rasmussen, 1929, 1931; Burwash, 1931; Weyer, 1932), for example a woman might be sent out as a herald of peace:

> . . . as she came nearer and nearer to our group the slower and shorter

> became her steps, and her facial expression betrayed a certain uneasiness. She had a small knife with her as a weapon, and however strange is this custom of sending a woman against the strangers, it is employed constantly by the tribe (Netsilingmiut). (Klutschak, 1881)

Such a greeting might be followed by other ceremonies that would establish formal ties between the strangers. This will be discussed below.

10.7.4 Establishing intergroup ties

From a sociobiological perspective most of the foregoing cultural adaptations can be understood as mechanisms that reduce conflict by manipulating the cost to benefit ratio of social behaviour (either real or apparent). By way of contrast the cultural adaptations to be described below reduce conflict by manipulating the coefficients of relationship (either real or apparent) of potential combatants.

Kinship

Intermarriage alliances can be understood biologically as increases in the coefficients of relationship (r) of the members of two extended families by virtue of their common inclusive fitness interests created when a marriage between their kin is consummated. Although intertribal marriage was rare between the major tribal groups in traditional times marriage between subtribes was common (Boas, 1888; Irwin, field notes). In more recent times rates of intertribal marriage have increased. For example, in the Netsilingmiut extended family, of which I am a member by marriage, the migration rate in my wife's generation is 55%, in her parents' generation 40%, in her grandparents' generation 30%, in her great-grandparents' generation 2% and in her great-great-grandparents' generation 0%. The sample size in the great-great-grandparent generation is too small to conclude that the migration rate is indeed as low as 0% and the sharp increase in the migration rate in the grandparents' generation was due to the general migration of this family to Hudson Bay in the 1920s. Thus although intermarriage may have improved contemporary intertribal relationships, this sample suggests, like the ethnographies, that intertribal marriage probably did not produce extensive networks of alliances between the larger tribal groups in the past.

Pseudokinship

In the absence of the establishment of genuine kinship between the members of the major tribes pseudokinship can be culturally instituted to give the appearance of increasing coefficients of relationship (apparent r) between extended families. Boas describes elaborate formal duels that took place between members of different tribes:

> If a stranger unknown to the inhabitants of a settlement arrives on a visit he is welcomed by the celebration of a great feast. Among the southeastern tribes the natives arrange themselves in a row, one man standing in front of it. The stranger approaches slowly, his arms folded and his head inclined toward the right side. Then the native strikes him, with all his strength on the right cheek and in his turn inclines his head awaiting the stranger's blow. While this is going on the other men are playing at ball and singing. Thus they continue until one of the combatants is vanquished.
>
> Boas, 1988

My informants have provided me with much more detail on these contests that culminated in the establishment of formal pseudokinship ties between the leading members of antagonistic tribal groups. These partnerships were called Itlureet, generally this term translates as cousins, however in this context it translates as ritual cousins. From a number of interviews the following composite description can be given (Irwin, field notes):

> Itlureet were very happy partners but they used to compete with songs. If you and I were Itlureet I would sing about all the things I saw you do wrong and you could do the same to me. Itlureet were not normally from the same place or tribe, they were strangers. They would meet once in a while and then they would compete. In those years long ago the Itlureet were like the law, people had to be in order, get their guilt out. Itlureet did that.
>
> Ipiak, 1982

> My father had an Itluk. His name was Sougriq but later on people called him Marilayuk. After they Itoktutiyuk (became Itlureet by ritual contest) they became friends and gave each other meat or anything they needed. They had fighting contests. Every young man had to find an Itluk in the old days if they thought their strength, size and age matched.
>
> Tavok, 1982

> Itlureet were people who were not from the same place. When they saw each other they wanted to compete, they were happy, they wanted to be friends. I missed my only chance to see two such men. They were Kidlinermiut and Netsilingmiut. I was in an igloo and I didn't know they were having a contest. The Netsilingmiut lost that time, he had a bleeding nose. Maybe the Kidlinermiut was using his shamanism to have more power. The Kidlinermiut had heard the Netsilingmiut was strong so they became Itlureet. They were the strongest of each tribe.
>
> Tungrilik, 1982

Badging

For behaviour to evolve at the tribal level of population structure proximate mechanisms are required to identify the members of such groups. This can be done with cultural badges such as dialect or accent although any variation in culture will do (Irwin, 1985a, 1987). Given the thesis presented earlier, that any difference in coefficients of relationship across a population boundary should produce a polarizing of ingroup/outgroup behaviour then it theoretically follows that any difference in cultural variation between groups (the proximate mechanism of group identity) should have the same effect. This, however, may not always be true in practice as proximate mechanisms are open to manipulation during the development of an organism (Bateson, P., 1979, 1982, 1983a). If human group behaviour evolved in a population structure that on average produced say X variation in culture for every Y variation in coefficient of relationship and if this population structure was, for the most part, marked by considerable variations in both X and Y then a certain variation in X (cultural badging) would evolve to be associated with a certain level of group polarization. If, now, by some process of cultural evolution, variation in cultural badging (X) was kept low then unusually low levels of intergroup hostility, not in keeping with real differences in intergroup coefficients of relationship (Y), could be phenotypically expressed. As with pseudokinship this cultural manipulation of social behaviour can be understood in terms of apparent r.

All Eskimo, from North Alaska to Greenland speak essentially the same language with the result that they can converse with their tribal neighbours. In contrast, this megapopulation is separated from the Indians by a solid linguistic wall. No Eskimo in traditional times could converse with their Indian neighbours who might be only 200 miles distant, while they could have conversed with other Eskimo 2000 miles distant. As has already been noted these two megapopulations did not consider the other human, migration rates were nearly zero and all conflicts between them were particularly bloody. Given the theoretical perspective developed thus far this latter situation is to be expected, however, most ethnographers find it unusual, if not unique, that essentially the same language is spoken by the Eskimo over a 3000 mile stretch of the Earth's surface (Heinbecker and Irvine-Jones, 1928; Weyer, 1932). If this observation is correct then it may be speculated that linguistic variation between Eskimo tribes has been culturally limited so as to reduce conflict between them. As would be expected the relatively peaceful tribes of the central Canadian Arctic have very similar dialects (Birket-Smith, 1928; Irwin, 1985a, 1987) and although I do not have any comparative data on this subject at present I would expect the warring Eskimo tribes of Alaska to show greater linguistic variation. It may be significant to note that Rasmussen had little difficulty understanding the various Eskimo dialects on his travels from Greenland to Alaska, until he reached the Yukon (Rasmussen, 1927), in Western Alaska.

10.8 A SYNTHESIS OF BIOLOGICAL AND CULTURAL MODELS

Mapping the causal chain of any complex human behaviour is probably destined to failure as the chain is in reality a shifting causal network. Nevertheless a model of some sort, within which to systematize the analysis made here, would be welcome in order to understand more clearly what has just been done and provide a framework within which to build similar analysis of other human behaviours.

Tracing the historical origins of phenomena can frequently help us focus on the most relevant causes of phenomena. Cosmology is dominated by inquiries into the history of the universe. Similarly tracing the history of ideology and documenting human sociocultural history provides us with important insights into contemporary cultures and social practices as they are manifest. In biology the historical approach to the analysis of adaptive function has been advocated by Pittendrigh (1958); Williams (1966); Mayr (1974) and Barash (1982). This analytic method may help in developing a better understanding of human behaviours in general (Tinbergen, 1953); Irwin, 1985a, 1987).

The theoretical approach used here is essentially a combination of sociobiology (Wilson, 1975a), neo-ethology (Bateson, P., 1966; Crook, 1970; Hinde, 1970; Tobach, 1978) and cultural evolution (Cavalli-Sforza and Feldman, 1973, 1981; Campbell, D., 1975, 1977, 1983; Alexander, 1979; Boyd and Richerson, 1980, 1983, 1985; Pulliam and Dunford, 1980; Lumsden and Wilson, 1981; Bateson, 1983a,b). A synthesis of these perspectives can be created by tracing the historical events that lead to a behaviour from its biogenetic evolution, through its sociocultural history and individual development, to the manifest expression of the behaviour.

In this case the behaviour under examination is limited intergroup aggression amongst the Inuit of the Central Arctic. Figure 10.1 takes the causes of this behaviour, identified in this chapter, and attempts to arrange them in order of their evolutionary emergence and recent natural history. By way of a thought experiment I will speculatively review this history in an effort to illustrate some of the potential benefits of the synthesis being attempted. First, and perhaps always, we must begin with the imperatives of life, fitness, most commonly understood in terms of reproductive success and survival.* Relevant to the behaviour being examined here the advent of sexual reproduction and associated mammalian demic population structures are important events. In humans this turn of evolution gave rise to the ingroup/

*Fecundity may be a useful index of fitness when hunter/gatherer societies, such as the Inuit, are being examined. However efficiency (MacArthur, 1962; Slobodkin, 1972), biomass (Carson, 1961), competitive ability (Claringbold and Baker, 1961), energy flux (Lotka, 1922) or trophic energy (Van Valen, 1976) may be more useful indexes of fitness for the comparative study of more complex human societies (Irwin, 1985a).

Fitness and the necessities of life, survival and reproduction	Sexual reproduction and genetic/social benefits of limited inbreeding	Demic population structure and ingroup outgroup behaviour	Sexual dimorphism and male specialization as hunter and warrior	Migration to Central Arctic environment	Male hunters become a critical limiting resource	Cultural proximate mechanisms evolved to reduce conflict	Limited Inuit/Inuit conflict within and between populations

Genotype ⟷ Environment ⟷ Phenotype

Genetic evolution ⟷ Cultural evolution ⟷ Individual development ⟷ Phenotype/behaviour

Phylogeny ⟷ Ontogeny ⟷ Phenotype/behaviour

Ultimate cause ⟷ Proximate cause ⟷ Phenotype/behaviour

Sociobiology ⟷ Cultural ecology ⟷ Cognitive, developmental, social, anthro./psyc.

Figure 10.1 'Evolutionary Spectrum' for the behaviour of Inuit/Inuit limited conflict. The model is generated by tracing the natural history of the behaviour under examination back through its recent development then through its cultural evolution and finally through its genetic evolutionary history.

outgroup behaviour we have come to associate with tribalism. At perhaps the same time a number of factors (for example, male competition for females, differences in male and female parental investment, extended nursing) contributed to human sexual dimorphism, that emphasized hunting and the prosecution of war as predominantly male activities. Hunter gatherers extended their range into the Arctic some seven to fifteen thousand years ago. As these migrants moved into the increasingly less hospitable regions of the Central Arctic the techniques of hunting large Arctic mammals were developed and practised almost exclusively by males while females specialized in child care and the manufacture of protective clothing. The rate of male mortality rose and more male hunters were needed to feed the females and their infants. Males were now a critical limiting resource. Female infanticide was introduced to increase the male to female sex ratio but many young males, just approaching their hunting prime, were being lost to the costly behaviour of war. At some point this cost of war, the losing of male protein harvesters, came to exceed the benefits of war, more females and territory. Phylogenetic inertia prevented the genetic evolution of an adjustment to the human biogram that would be phenotypically expressed as an innate capacity for restrained male conflict. By a number of processes of cultural evolution, ranging from individual rational thought to the cultural group selection of traits, numerous cultural proximate mechanisms came into existence to curb male hostility. These traits were enculturated during the development of each Inuk, from birth through puberty, rendering them less prone to aggressive acts and less inclined to be involved in acts of war against their neighbours. When compared to their ancestors or other groups of humans in more hospitable habitats these Inuit might be judged to live in a state of relative peace.

This historical abstraction of the major arguments presented in this chapter, generates the model termed an 'evolutionary spectrum' depicted in Figure 10.1. This description moves from left to right, from genotype to environment to phenotype, from sociobiology through the various social sciences (for example, cultural ecology, sociocultural and cognitive anthropology, developmental and social psychology) to behaviour, from genetic evolution through cultural evolution and individual development to behaviour, from phylogeny through ontogeny to behaviour and from ultimate cause through proximate cause to behaviour. (Further discussion of this theory, and how it relates to other social science and biological theory, is presented in my paper, 'The Sociocultural Biology of Netsilingmiut Female Infanticide', Irwin (1989).)

It should be noted that the evolutionary spectrum only depicts what may be considered the pivotal events in the evolution of limited group conflict amongst the Inuit of the Central Arctic. The spectrum can be greatly expanded by focusing the inquiry on any of the more particular causes of this behaviour. For example the second to last cell on the right hand end of the spectrum, 'Cultural proximate mechanisms evolved to reduce conflict', contains the 12 mechanisms (and there are probably more) listed in Table 10.3. Any one of these mechanisms, if inserted at this point in the spectrum, will trace a slightly different evolutionary path from the general one given. For example, Figure 10.2a depicts some of the extra elements that should be inserted into the evolutionary spectrum for the cultural adaptation of pseudokinship (see section 10.7.4). this particular cultural proximate mechanism requires the evolution of symbolic language and the mapping of biological kin by symbolic language. Figure 10.2b depicts some of the extra elements required for the socialization of non-aggression (see section 10.7.3). This particular cultural manipulation requires the evolution of a parental ability to modify progeny behaviour through the manipulation of rewards and punishments.

10.9 DISCUSSION

The main purpose of this chapter is to uncover knowledge about the nature of Inuit group conflict resolution in the hope that it might be of some practical value to contemporary societies. Although the discussion that is to follow is very speculative I will attempt to review all the cultural proximate mechanisms listed in Table 10.3 with a view to noting what mechanisms we may have some equivalents of, what mechanisms we lack and finally what novel mechanisms could possibly be adapted to meet our own needs in the modern world. These speculations must be considered preliminary and are made to stimulate discussion, not to conclude it.

10.9.1 Sexual and marriage relationships

In most of the western world competition for females is reduced by the legal enforcement of monogamy. However the recent move toward increased polygamy in the form of serial monogamy may increase male competition for optimally reproductively mature females. Enforced monogamy may have both an internal and external prosocial, stabilizing effect. In contrast institutionalized polygamy may increase conflict and may have, for example, contributed to the expansion of the Islamic empire in the middle ages by allowing captured outgroup females to become additional ingroup wives (The Koran).

10.9.2 Property rights

Different political ideologies deal with the problem of the equitable distribution of the resources of a society in different ways. For example the Marxist

Figure 10.2(a) Pseudokinship elements to be included in the 'Evolutionary Spectrum' for the behaviour of limited Inuit/Inuit conflict.

			Symbolic language	Symbolic linguistic mapping of biological kin		Cultural manipulation of symbolic kin map (pseudokinship) to increase intertribal reciprocity and ties	
Fitness and the necessities of life, survival and reproduction	Sexual reproduction and genetic/ social benefits of limited inbreeding	Demic population structure and ingroup outgroup behaviour	Sexual dimorphism and male specialization as hunter and warrior	Migration to Central Arctic environment	Male hunters become a critical limiting resource	Cultural proximate mechanisms evolved to reduce conflict	Limited Inuit/Inuit conflict within and between populations

Genotype ↔ Environment ↔ Phenotype

Genetic evolution ↔ Cultural evolution ↔ Individual development ↔ Phenotype/behaviour

Phylogeny ↔ Ontogeny ↔ Phenotype/behaviour

Ultimate cause ↔ Proximate cause ↔ Phenotype/behaviour

Sociobiology ↔ Cultural ecology ↔ Cognitive, developmental, social, anthro./psyc.

Fitness and the necessities of life, survival and reproduction	Sexual reproduction and genetic/ social benefits of limited inbreeding	Demic population structure and ingroup outgroup behaviour	Sexual dimorphism and male specialization as hunter and warrior	Migration to Central Arctic environment	Male hunters become a critical limiting resource	Cultural proximate mechanisms evolved to reduce conflict	Limited Inuit/Inuit conflict within and between populations
		Parental ability to manipulate progeny behaviour by rewards and punishments	Infants respond to parents' affection and aggression		Cultural manipulation of child rearing practices develops self control of adult aggression		

Figure 10.2(b) Socialization of non-aggression elements to be included in the 'Evolutionary Spectrum' for the behaviour of limited Inuit/Inuit conflict.

solution is Communism and the Capitalist solution is the judicial administration of a Meritocracy. Numerous political systems have been tried out in an effort to create 'the just society' with various degrees of success. To enter into a discussion of the merits of these different systems is beyond the scope of this chapter. However, it is perhaps all too clear that between societies there is little or no attempt to institutionalize equitable access to the resources of the world. So long as this inequality exists then intergroup conflict may be difficult to avoid.

Language

Unfortunately the English language, and so far as I know the languages of all societies that do battle, have cognates that define intergroup hostility in praiseworthy terms. Perhaps it would be more difficult for us to send our young men off to fight an enemy if our semantics only allowed us to tell them to march forth bravely, to confront their fellow man and then, with a warm heart, commit mass murder.

Political organization

The societies that most effectively engage in the behaviour of war possess culture traits that can efficiently organize their society for combat. As with language there may be no simple way to undo these elements in our cultural evolution except to suggest that knowing that these characteristics of our social fabric have come into existence, at least in part, to facilitate war, should put us on guard against the more deadly aspects of these institutions that frequently portray themselves as benign. It should be noted that many of my Inuit informants believed the Royal Canadian Mounted Police did much to reduce intertribal hostility, such as it was. Clearly the imposition of order by an alien force can effectively reduce intergroup conflict. Perhaps it is unfortunate for us that no paternal alien is about to impose order on our war wrecked international world.

Socialization of non-aggression

Even if the early socialization of Inuit children does reduce adult aggression the process is probably not well enough understood for mass export to other more aggressive societies. However, societies that blatantly enculturate their children to be aggressive probably could be socially 'treated' to produce a gentler population. Unfortunately, as with the preceding culture traits, this may not be easy as culture is a most effective behaviour modifier as it seems to attach itself, almost like a virus infection, to the most insidious characteristics of human nature. In many societies playing with death toys and killing games are simply a lot of fun.

Magic

Perhaps we would all be better off if, like the Inuit, we believed we could kill our enemies with magic. Instead of ministries of war with all their deadly weaponry we would have ministries of sorcery with their ineffective potions and spells. Unfortunately our instruments of death are all too effective. However magic, or rather metaphysics, does play an important role in some modern wars. Islamic warriors believe they will go directly to heaven if they die while fighting a holy war. This metaphysical belief may have been adaptive when fighting was done with swords. In medieval times young men would rush forward to fearlessly engage their foe. Today's young men rushed forward only to engage bullets and gas in the marsh lands that divide Iran and Iraq.

Threat of retribution

Just like Inuit mythology, story telling in our culture teaches the costs of murder. Fictitious murderers in books, on the radio and on television invariably do not get away with their crimes. In reality, as with Inuit society, they probably get away with their crimes more often. However, unlike Inuit society, international killing is frequently left unpunished as it is not classed as murder when a state of war exists, and sometimes when it does not exist. Perhaps effective retribution for international killing could reduce international conflict? Perhaps a tradition of story telling, that taught children that acts of international killing would always be revenged, would also discourage many of them from growing up to be soldiers or terrorists.

Formal conflict

Our world is awash with formal and ritual conflict: I am fighting for my hypothesis in this chapter which, no doubt, will come under attack from searching questions from the reader. In the international courts and in the international political forums of the world wars are fought with semantics and rhetoric. We fight economic war with our industrial institutions, we fought a 'Cold War', we fight arms races and our athletes fight on the sports fields of international arenas. Some of these forms of conflict, such as arms races, are more dangerous than others. We would do well to curtail them. Others, such as sports, semantic battles and rhetoric jousts, present little or no risks to the participants. These forms of ritual conflict should be encouraged.

Kinship

It is entirely possible that patterns of international migration would be a fair index of patterns of international alliance. The trick, with respect to kinship and peace making, is to increase rates of intermarriage between worst enemies. Unfortunately, rates of East–West migration are still very low. Opening up borders would help, as it has in Germany since 1989. If the leaders of the Super Power nations held their offices for life then perhaps marriages could be arranged between their sons and daughters. However, in the absence of such a turn of cultural events then, like the Inuit, we will have to make do with pseudokinship.

Pseudokinship

Up till now this discussion of cultural proximate mechanisms for reducing human conflict has contained few or no surprises. However the Inuit cultural invention of ritual cousins, Itlureet, may be unique so we should make a special effort to see if it can't be adapted to our needs. Ritual cousins, it should be remembered, were the leaders of their communities, they became best friends for life through tests of strength, tests of humour and by sharing their material possessions. As with intermarriage, pseudokinship is probably most effective at reducing intergroup conflict when it involves the most prominent members of societies. However, as there may be some resistance, on the part of our political leaders, to initiate ritual cousin ties with their sworn adversaries, it may be up to others to start the ball rolling. Perhaps, with the help of their professional or union organizations individuals could seek out an Itluk in the country of their supposed enemy. As role models, it might prove most effective if athletes, teachers and university professors took the lead in establishing Itlureet ties.

Badging

If cultural variation, 'badging', is a proximate mechanism for the establishment of ingroup/outgroup identity then it follows that cultural heterogeneity should correlate with human conflict. In *Deadly Quarrels* Wilkinson (1980) analyses 780 wars that took place between 1820 and 1952 in order to test numerous empirical generalizations concerning the nature of war. On the question of heterogeneity he concludes:

> The propensity of any two groups to fight increases as the differences between them (in language, religion, race, and culture style) increase. A homogeneous world would probably be a more peaceful one. (Wilkinson, 1980)

The Eskimo, from Siberia through North America to Greenland, could converse with each other. Perhaps this commonality of cultural badging contributed to low levels of intertribal hostility. The nations that make up the East and West blocks appeared to be behaving more like two adjacent mega-populations, like the Indian and Inuit. These behaviours, the proximate mechanisms of deadly competition, can possibly be reversed with cultural exchange, student exchange, intermarriage, worker exchange and the learning and practise of each other's language and customs.

10.10 CONCLUSION

The reason for the evolution of limited conflict amongst the Central Arctic Inuit and limited conflict between modern nuclear Super Powers is probably the same in ultimate cause terms. Specifically the costs of war exceed the benefits. However the proximate causes of peace are very different. For us it is the threat of mutual assured destruction that keeps us at bay. Unfortunately this mechanism for maintaining peace is inherently unstable.

However, one of the cultural mechanisms maintaining Inuit peace was partnership ties between the leading antagonists of any potential conflict. In this case, if there were a shift in the balance of power, it would have been possible to create new partnerships to restore stability. This point can possibly be made clearer by comparing Alaskan partnerships with Central Arctic partnerships. The latter were made between enemies to diffuse conflict while Alaskan partnerships were made between enemies or friends and could be exploited to create peaceful or warring alliances (Burch and Correll, 1971). Central Arctic partnerships were never used for organized conflict. If there is a lesson to be learnt from this piece of ethnography it is that we must actively seek partnerships with our worst enemies. In time our enemies may change, at which point we must shift our institutionalized partnership making activities to our new principal adversary.

The possibility of manipulating the cultural proximate mechanisms associated with the regulation of human hostility, in order to reduce the frequency and intensity of conflict between human groups, may be welcome. However, the 'Evolutionary Spectrum' model of Inuit group conflict resolution suggests that the proximate mechanisms of limited intertribal hostility evolved as a consequence of evolutionary necessities which rendered the costs of warfare higher than the benefits of warfare in the marginal environment of the Central Arctic. If the evolutionary perspective used here can be applied to human group conflict in general, as I believe it substantially can, then we may be forewarned that although manipulating the cultural proximate mechanisms associated with the regulation of human hostility may lead to a reduction in group tensions it will not, in the long term, prevent group conflict, unless the underlying ultimate causes of group conflict are also dealt with. This sad realization threatens to make much of the discussion and conclusions of the chapter redundant. However, in what may be an effort on my part to finish on an optimistic note, I wish to suggest that a reduction in tensions produced by manipulating the cultural proximate mechanisms associated with the regulation of human hostility may help generate a sociopolitical environment in which the more fundamental causes of intergroup hostility can be dealt with by, for example, making the access to the resources of the world, so necessary for survival and reproduction, more equitable.*

In the past it has always been possible to allow social evolution to follow its natural course in the knowledge that given sufficient time selection would edit out the unfit cultural mutations. However selection requires a node of selection (Campbell, 1974); a point about which the unfit adapt or die out; a point about which the fit survive and prosper. In human cultural evolution the units of selection have become less numerous and more deadly. From 10 million hunters and gatherers which would have made up some 20 thousand tribes we find the present world of nearly 5 billion polarized about two or three Super Powers. From prehistoric until recent times some form of group selection could take place without endangering humanity. One tribe or nation could extinguish another. Unfortunately the node of selection has now moved away from the arena of tribes and nations to the arena of worlds. Those worlds of sentient beings that come to terms with the problems of social behaviour will be selected for. Those worlds of sentient beings that fail to come to terms with the problems of social behaviour will be selected against. Human ideologies, at least in so far as they are followed in practice, would seem to have failed to save man from his worst enemy, himself. Indeed, ideology as badges frequently increases human hostility. It may be a slim hope, but perhaps a rigorous scientific understanding of human sociality can succeed where ideology has not. If this can be accomplished, then man and a science of human nature will have come of age by cheating natural selection with her own instruments. In this circumstance, human fitness is dependent on human enlightenment.

*To some extent equitable means different things to different peoples, Communism, Socialism or perhaps the institution of a just Meritocracy. I will not attempt to adjudicate between these ideologies here although efforts to use evolutionary theory for this purpose have been made (Peterson and Somit, 1978). For the foreseeable future equality within societies will probably have to remain a matter of taste, a value judgment. However, between societies, the signing of the Helsinki Agreement, the 'North–South Dialogue' and the 'Law of the Sea' debate hopefully demonstrate a possibility for reaching a degree of consensus on questions of equitability at the international level.

CHAPTER ELEVEN

Human nature and the function of war in social evolution. A critical review of a recent form of the naturalistic fallacy

P. Meyer

The nature of war is one of the recurrent topics in the history of the social sciences. This interest may be partly due to the impact of recent wars, revolutions and other forms of collective violence in this century. Yet another inspiration to the study of these phenomena may have come from the growing knowledge about their impact throughout history. In fact, the science of history has always shown a strong concern for the influence of battles and wars on the course of events (McNeill, 1963; 1982). From the nineteenth century onwards, these accounts were put on an even broader basis: ethnographic studies showed that wars were a social phenomenon very frequent among tribal societies too.

Regarding the development of theories on the nature of war and even on the nature of society, ethnographic accounts on these more primitive societies progressively gained more importance. It seemed that ethnography could in fact provide knowledge about very distant historical phases which all human societies had passed through in their evolution (Mühlmann, 1962, p. 250). Even more important than this addition of quasi-historical information was the corroboration of some anthropological concepts which put social science on a new basis. But why should such a corroboration of social scientific concepts be necessary?

Any science requires a set of regulative ideas, as has been pointed out by Agassi (1977, p. 24). These regulative ideas are necessary because any problem of some relevance can be approached from different scientific perspectives. Human behaviour, for example, may be accounted for in terms of cognitions or ideas, social rules, as well as in terms of the physiological processes underlying overt behaviours, or even in terms of genetic replication patterns. Obviously explanations will offer very different insights into human behaviour if either social rules, physiological processes, or genes will be used as explanatory categories. Undoubtedly every single one of these perspectives can shed some light on the structure of behaviour. An acceptable explanation of war and its potential impact on social evolution will evidently depend on the use of the appropriate categories. The main reason is that the selection of categories will determine what empirical phenomena are to be considered relevant in any given theoretical enterprise.

11.1 IS WAR A UNIVERSAL INSTITUTION?

Turning to war, some general characteristics can be presented in order to obtain criteria for categories suited for an explanation of this social phenomenon. Three propositions concerning the 'social nature' of war can be put forward:

1. war is a very complex behaviour requiring a considerable number of participants and a subtle synchronization of their behaviours,
2. there are various elements of human communication, but language is its major component – it is therefore most unlikely that
3. war can be understood without recourse to some principles of languages.

It will be pointed out later on that some authors obviously consider these propositions as truisms and do in fact suggest explanations of war without any recourse to the impact of language and mind. It will be demonstrated further that this method 'explains away' exactly those aspects of war which are crucial for an understanding of this phenomenon (Agassi, 1977, p. 46).

The first proposition points to the fact that war is not a universal institution. Hunter-gatherer populations very frequently number less than 50 persons, including women, infants and the old (Lenski, 1966, p. 96). Population size is not sufficient here to bring about a social organization of some complexity which could be termed war. There are neither any differentiated social roles in this type of society, nor is there a discernible organization of the violent encounter itself. After suffering some casualties in battle one party will retreat.

This is done not for the purpose of adjusting tactics but in order to give up the entire operation. Obviously the entire scope of actions is very undifferentiated on this level of social evolution. Following Turney-High (1971), it is suggested here that this type of violent action should be termed 'primitive war' or 'sublimitary war'.

Considering the fact that hunter-gatherer societies prevailed throughout the most extended period of human history, primitive war is actually the most frequent type of collective violence. The general understanding of collective violence, therefore, could be furthered if a basic motivational structure of this type of war was to be revealed. Some relevant features of this structure will be pointed out later on. At present, it seems advisable to pay some further attention to the socio-economic conditions of hunter-gatherer societies.

According to Lenski (1966, p. 96), these societies are comparatively small regarding the number of members and primitive with regard to division of labour and social differentiation. 'Primitivity' (Douglas, 1978, p. 77) is an all-pervasive characteristic of various aspects of economic life such as dwellings, weaponry, etc. At the same time, nearly every adult person is able to produce all the tools necessary for his subsistence, not considering the division of labour between the genders here.

These remarks may suffice to become aware of the total absence of economic motives for collective violence between the social groups mentioned. Since these groups lead a nomadic life, following the migrations of game, the possession of territories is much less important than in other stages of social evolution, as for example, horticultural or agrarian societies.

It has become apparent already that war is not a unitary phenomenon throughout social evolution. Quite to the contrary, this institution has changed its form, its underlying causes, and consequently its general function for society as a whole. In other words, the institution of war itself has been invented at some particular time in social evolution (Mead, 1964).

This proposition does not imply, however, that collective violence of minor social complexity did not exist before this invention. Human ethology and sociobiology as well as innumerable studies on the psychophysiology of aggression, proved beyond any doubt that aggression, or agonistic behaviour (Eibl-Eibesfeldt, 1984, p. 474) as the more recent label for this topic, is a common feature of most animal behaviour. In the field of human behaviour this has been corroborated by the findings of the science of history as well as ethnography.

Among the contributions of human ethology and sociobiology to the study of violent behaviour, the basic similarity of physiological processes as well as of some signals, postures and gestures between humans and his closer animal relatives has been emphasized (Eibl-Eibesfeldt, 1984, p. 492); these findings are obviously suited as further corroboration of the Darwinian thesis of a link between man's and other primates' behaviour (Agassi, 1972, p. 131). Undoubtedly the life sciences can account for numerous aspects of human social behaviour, taking into consideration these similarities. It will be disputed, however, that findings about homologies in the physiology of aggression can account for any relevant aspect of war. Whatever the contribution of human biology to the understanding of human action – considering the entire complexity of action – biology can account for some of its aspects only. It will be pointed out later what the contributions of human biology to the understanding of human actions, particularly war, may be. In the following section some general propositions, as suggested by various naturalistic approaches on the relation between human nature and war, will be presented.

11.2 CAUSES FOR WAR: SOCIOBIOLOGY'S CONCEPT OF 'RESOURCES'

In his *On Human Nature* Wilson (1978, p. 34) points out that hunter-gatherer societies should be the primary object of human sociobiology. His major argument for this preference is that their comparatively primitive social organization, and the importance of kin categories for this organization, makes them a close to perfect object for the study of human behaviour. Regarding the causes of warfare among these more primitive tribes, Wilson suggests a link between human aggressiveness and warfare (Wilson, 1978, p. 100).

Wilson's hypotheses on this point include a severe critique of the Freud–Lorenz drive discharge model of aggression. 'Aggression does not resemble a fluid that continuously builds pressure against the walls of its containers' (Wilson, 1978, p. 101).

Quite to the contrary, aggression according to Wilson, should be regarded as a behaviour which arises in the defence of territories, i.e., dependent on certain environmental conditions – 'territoriality is one of the variants of aggressive behavior' (Wilson, 1978, p. 107) – and intruders into these territories will definitely be attacked. The evolutionary logic of territoriality and the aggressive or rather agonistic behaviour, has in fact been brought to light by numerous studies on animal behaviour in recent years (Barash, 1980, p. 249). Control of territories in many bird species, for example, is a precondition of successful mating for males (Barash, 1980, p. 249). In other words, the control over territory functions as a sort of signal of male fitness to female birds: under the assumption of a general striving for reproduction, male competition for these resources seems to be in accordance with evolutionary logic.

In recent years evolutionary biologists have pointed out why explanations in terms of ultimate causality, i.e. explanations presenting the reasons for the evolution of a particular behaviour, add more to the general understanding of behaviour than a compilation of proximate causes, i.e. descriptions of overt behaviours and their elicitors (cf. Barash, 1980, p. 46).

W.H. Durham (1976) has suggested a terse formulation of this evolutionary logic: 'that there is a biologically limited amount of time and energy to each organism. It is believed that natural selection adjusts the genetic influences on behaviour so that this time/energy are spent in ways that tend to maximize the representation of a given individual's genes.' In particular phases of a species' reproductive cycle, it makes sense to invest time/energy into agonistic behaviours between males because this is the way natural selection has organized differential reproduction. It is clear that reproductive chances are the primary resource being strived for in agonistic encounters – territories, dominance hierarchies etc. are of secondary importance only. Control of a territory is not an end in itself, but evidence of an individual's fitness.

An explanation of agonistic behaviour in terms of proximate causality would point to the impact of certain physiological states on this behaviour whereas an account in terms of ultimate causality would stress the advantages in individual fitness which these behaviours entail. Hence 'ultimate causality' is more general and can account for the evolution of these particular behaviours. It will be pointed out somewhat later, however, that explanations in terms of ultimate causality cannot 'explain away' complexities on the level of proximate mechanisms. An example of the method of explaining away is the attribution of war to human nature, as is occasionally suggested by evolutionary biologists.

Returning to the topic of war, Wilson (1978, p. 108) suggests that primitive war adheres to the same type of evolutionary logic as agonistic systems in animal species do. Territories, according to this author, comprise the most important resources such as fruit, game, water etc., and striving for the control of these resources is the most general cause for violent competition among human groups. While it cannot be denied that these and similar resources do indeed play an important role in human conflict, it will be disputed that cultural ideas on the nature and relevance of such resources are negligible in an evolutionary explanation.

Eibl-Eibesfeldt (1984) holds a slightly different view on the function of territoriality and the universality of war. Referring to Chagnon's well known account of Yanomamö warfare (Chagnon, 1968), he comments: 'Whatever reasons people might give for their acts there are always consequences which can be identified: war imposes a pressure upon neighbors and causes a keeping of one's distances. And it is these results which matter' (Eibl-Eibesfeldt, 1984, p. 528). These consequences of war are suited for studies in terms of evolutionary theory: what does war add to the individual or group time/energy budgets? Eibl-Eibesfeldt suggests both a selectionist and an ecological answer to this question.

Casualties in primitive wars range between 20 and 30% of adult males (Eibl-Eibesfeldt, 1984, p. 530). Most reports on primitive war agree on this quantitive effect of warfare (Wright, 1965). Warfare, therefore, is in fact a selective agent and an important cause of differential reproduction.

In terms of ecology, two inferences are suggested: the first one concerns the stabilizing effect of limited population growth. A second consequence of tribal warfare is territorial dispersion, bringing in its wake the widest possible use of natural resources. Vayda has brought to light these ecological consequences in his studies of Maori warfare (Vayda, 1970): the major point of these studies being that the introduction of new weapons tends to destabilize military as well as ecological balances (Melotti, 1986b, p. 75; Meyer, 1977, p. 35).

Summarizing, Wilson, Eibl, Vayda, to name but a few, suggest that territories are a limiting resource and warfare is an important means of achieving control of this resource. Warfare is, in other words, an important mechanism in human evolution bringing about differential reproduction. Whatever reasons tribal mythology and lore might give for war, it is the consequences in terms of resource control which count.

Undoubtedly the study of consequences in terms of reproduction is a legitimate and in many respects fertile research programme. Its fertility has been corroborated in numerous studies of agonistic behaviour in animal species; so why not apply it to the study of human behaviour?

11.3 ON THE SOCIAL CONSTRUCTION OF 'RESOURCES'

To put it briefly: people do not fight for resources, but for ideas of resources. Some of these ideas are closer to the 'real' importance of these resources in terms of metabolic needs, others are more distant from such realistic assessment. It is quite obvious that people may hold very subjective ideas on these properties, but cultural ideas as a set of traditionally shared ideas on the world are a more appropriate category for further discussion. These cultural ideas vary enormously (Meyer, 1981, p. 69) – nevertheless many social scientists agree on some general structural features which permit talk of a cultural evolution. One need not adopt a law-like succession of stages as Comte did with his 'law of three stages' (Comte, 1972) in order to accept the persuasive power of the idea of succession. According to Comte, an animistic-totemistic stage is succeeded by a metaphysical and finally by a scientific stage. Whereas the teleological bias in Comte's theory cannot stand a Darwinian critique, the stages are to be considered valid classificatory tools. Although well understood, the notion of such stages does not explain anything, but does assist in identifying certain structural similarities in a seemingly chaotic variety of cultural systems. The general importance of such systems for warfare will be discussed more extensively later. At first the problem of resources strived for in tribal wars has to be taken up again.

It has been pointed out before that the selectionist-ecological hypothesis stresses the importance of material resources. Contrary to this practice, a cultural notion of resources emphasizes the necessity of mental processing of resources or any other objects of human striving. Mental processing presupposes terms or names for these objects, otherwise they would not exist as objects of social action (Bateson, 1983:81).These processes are in fact a perceptual filter so that some objects are perceived, strived for etc., others not. An example may illustrate this.

In Mesoamerica land was not a scarce resource until about 5 000 and 2000 BC. In this period an 'accidental "kick" . . . made maize cultivation the most profitable single activity' (Phillips, 1987, p. 235). Subsequently the control of land suited for the cultivation of maize became a limiting resource for societies' differential reproduction. The adoption of the new agricultural technology offered evolutionary advantages.

It should be emphasized, however, that this adoption was basically a mental and cultural process. For reasons not to be further discussed here some societies did adopt the new agriculture, others did not. Major consequences were the differential growth of populations, the differential allocation of sources of power and, last but not least, a differential ability of transforming these sources into militarily organized warfare.

These considerations on the mental nature of resources are not trying to explain away the importance of material resources for human behaviour. Man's metabolic needs are a truism and society and its institutions must somehow provide its members with them. However, it is well known that neither all societies nor all biological species have succeeded in this task. More than 99.99% of all species once existent on this globe are now extinct (Wuketits, 1985, p. 332) and the same is true for innumerable human societies.

In most cases of extinction of human societies it can be hypothesized that the gap between mentally constructed resources, the real resources of the physical world, and action patterns to get hold of them, grew too wide. Warfare may well have played a decisive role in this process. It will soon become apparent, however, that warfare could only play this role once it existed in a full-fledged sense. This was not the case, as may be recalled from previous considerations, for most hunter-gatherer societies which prevailed during the most extensive period of human history.

According to the present argument, the nature of resources is Janus faced: only as cultural artifacts do they enter social exchange and only the control of certain material resources will enable society to stand the competition from other societies for these resources once their instrumental value has been realized. It is obvious that a number of social theorists with a life sciences background do not bear this double role of resources in mind. They advocate a one-sided but homogeneous concept of resources which, as in the case of the selectionist-ecological hypothesis, can account for the effects of human behaviour but never for the reasons of this behaviour. This negligence of the socio-cultural construction of reality and hence what relevant resources are, could remain undisputed if it were not for the sake of a better understanding of war (Berger and Luckmann, 1966).

More recent analyses on the impact of material resources on tribal warfare demonstrate that land, in fact, is not a limiting factor in more primitive social settings (Ferguson, R., 1984a, p. 31). Quite the contrary, many reports 'documented an abundance of land' (Ferguson, R., 1984a, p. 34). Hence game and protein were made the limiting resource in materialistic interpretations. 'However', Ferguson suggests in his review of the literature, 'the focus on land and game has created an oversimplified picture of ecological explanations. War is never a simple function of the natural environment' (Ferguson, R., 1984a, p. 32), but if striving for material resources is refuted as a general cause of war, what are the qualities of mentally constructed resources that can motivate primitive people to invest their most precious resource, i.e. life, into 'senseless' wars?

In terms of modern, scientific reasoning, these qualities are to be classified as metaphysical, i.e. these qualities are not observable, hence their existence cannot be falsified. Whatever pejorative connotations scientists might use in the classification of the epistemological status of such concepts – it is these concepts, these ideas which motivated people in pre-scientific phases to invest their lives in wars. These are not necessarily the only motivations, some sort of materialistic ideas might mix. However, non-materialistic motivations must prevail and material gains must be considered a side effect. If this were not so, and materialistic ideas were dominant, the character of the entire sociocultural system would have gone through decisive changes. These considerations are not at all arbitrary, or of a merely 'idealistic' nature. Quite to the contrary, it is the very nature of warfare as a social process which provides the basic categories for these considerations.

Non-materialistic and metaphysical notions of resources prevail in 'primitive and endemic war' (Meyer, 1981). The term primitive here, refers to the absence of any tactical anticipation: a first decisive blow will bring the entire operation to an end. However, this campaign will almost certainly be ensued by another one: a companion's death calls for revenge. This in turn starts a virtually unlimited cycle of blood feuds. Primitive and endemic war usually does not strive for any material gain which could stand for a sort of economic motivation. Its major objective seems to be the restoration of the legal status quo ante.

Huber (1975, p. 659) gives an account of New Guinea warfare, which brings to light these characteristics in a most impressive manner. According to this author, the Anggor, one of the local tribes, consider any companion's death a murder. This calls for revenge and it is the shaman's task to find out how this killing was brought about and who was responsible for it. Subsequently, the murderer must be killed in order to secure one's peace of mind. Anthropological literature abounds with further examples of this type of warfare. It should be emphasized here that booty or 'other material considerations were far from being dominant' (Berndt, 1964, p. 184).

Another important characteristic of cultures pertaining to this stage is the absence of a clear cut separation between the state of war and the state of peace. People asked about their ethnic identity, very frequently answered that they belonged to group A and group B was their enemy. The construction of one's identity requires, in other words, a definition of the sort: A is non-B. The function of B, in this kind of thinking, is serving as a sort of negation of A and thereby identifying A (Mühlmann, 1962, p. 223; Bohannan, 1963, p. 305). It is no surprise then that the notion that war could be employed as an instrument to provide certain material resources is alien to these societies.

In many tribal societies still another typical motivation to recurrent war parties is the concept of power. According to Demallie and Lavenda (1977, p. 155), the Sioux Indians of North America saw themselves exposed to metaphysical powers. 'Power' was essentially a cosmic property. All creatures possessed minor quantities of this power which they called *wakan*. According to Sioux reasoning, it was the major objective of warriors to test their *wakan* in combat. If they were victorious, they could even enlarge their share of *wakan*. Trophies signalled the individual's power, and material booty like horses were of course not disdained. It should be stressed that material gains, for example, horses, were an important resource for Plains Indians. Studies about Cheyenne warfare (Grinnell, 1910, p. 297) indicate, however, that the primary goal of war was the acquisition of honour, and horses and other booty actually were of secondary importance.

Summarizing, these considerations on typical cultural ideas leading to recurrent wars demonstrated that material resources were of minor importance here. This does not matter, a sociobiologist could counter: the effects of warfare in terms of differential reproduction are the only interesting aspect. Why these people fought their wars, in terms of intentions, is negligible on an evolutionary background.

This may well be true, but how could the fact that some cultures pertaining to the same stage of social evolution practise war whereas others do not then be accounted for? It is this point which needs special emphasis here: people do not fight for the resources but for their ideas of resources. And in some cultures like the Pueblo Indians, ideas must have prevailed which prevented them from most types of war for a considerably long time.

11.4 IS WAR A PRIME MOVER OF SOCIAL EVOLUTION?

Regarding the attribute of 'prime mover', for the institution of war, the foregoing considerations on primitive and endemic war have demonstrated that the virtually incessant cycles of primitive war did not set in motion anything relevant for social evolution. However, once the threshold between primitive and full-fledged war was crossed, a major push was exerted on social evolution. In the present context a few remarks on the complex relation of war and social evolution must suffice.

This relationship will be analysed along two lines of closely interrelated characteristics of social evolution: (a) social differentiation, and (b) instrumentalization. Concerning differentiation, primitive war has been characterized as a comparatively undifferentiated social process. The same applies to the social composition of war parties: there is no privileged stratum of warriors, nearly all males within certain age limits may take part. In fact, participation is considered an important *rite de passage*: the acquisition of the power of the slain enemy gives additional power and 'magic fertility' to the victor.

It is impossible to identify a date historically when the dividing line was crossed between primitive war and war in the form of militarily organized armies. The main reason for this is the fact that this happened before the archaic states arose and before writing was invented. However, ethnographic reports about certain regions of America and Africa can give some idea on general characterisics of this process.

Perhaps the single most important social change was brought about by the evolution of a specialized stratum of warriors. In some American Indian tribes privileged warrior societies existed (Ellis, 1951, p. 180) – these societies developed a specific code of honour, very similar in this respect to aristocratic codes of medieval Europe, Japan and some other regions. These codes had a strong impact on cultural ideas and values. In general they became more and more warlike. At the same time, warfare progressively turned into an instrument of specialists who strived for the acquisition of honour to legitimize their various privileges.

These tendencies were further intensified by various social and technological inventions. The notion of integrating defeated foes and their kin was unfamiliar to tribal societies. In fact, there was no economic need or possibility of doing so: there was no way of using additional labour in primitive economy. Another decisive barrier against such integration was the absence of political institutions providing loyalty to a social structure above the tribal level.

Once the possibility of integrating other clans and tribes into a more embracing structure had been made, however, social evolution had made a great leap forward. Ethnohistory reports about this social invention in various cultural environments, for example, the evolution of the Zulu proto-state and the League of the Six Nations in North America. Especially in the case of the Zulu (Otterbein, 1967) this social invention was a side effect of certain military needs. The Zulu king Shaka built up a huge army and conquered an empire of some 20000 miles, which embraced about 300 tribes.

Shaka's and many other historical personalities' important invention was that aliens could be recruited for an army and this army could in turn be used as a means of acquiring material resources like land, or the furs of animals in these territories, as in the case of the Iroquois (Turney-High, 1971, p. 170). In this type of situation war has become an instrument in achieving certain goals

– metaphysical ideas about resources may still have existed, but war had been changed irrevocably into an instrument. No longer was war endemic, incessant as in more primitive settings.

The evolution of the state turned these 'inventions' into multi-faceted and stable structures, such as bureaucracies, political institutions, armies, etc. The combinatorial effects of these institutions (Corning (1983) suggests the term 'synergism' for such effects) produced a higher amount of power than any other social structure before. Regarding the relation between states and adjacent pre-state societies, these latter were usually subjugated or integrated into the state. The evolution of the state was accompanied by most decisive changes in all important sectors of society, such as politics, economy, religion and also warfare. The most important feature common to all these sectors was 'differentiation', i.e. economic, political or religious activities turned into specialists' tasks.

It is impossible to review even the most important aspects of warfare in this era. However, one general proposition regarding social values and typical motivations to war can be made: in the relation of states, power seems to be the most important limiting resource. Power is not a thing but an aspect of social relations (Friedrich, 1950, p. 22). If power may be defined, following Weber's suggestion (Weber, 1964, p.38), as one actor's ability to influence another actor against his will so as to do certain things, then this ability obviously depends on a differential access to sources of power. Whatever source of power might prevail in a particular situation, violence directed against a person or a group, and endangering their life, undoubtedly is the most effective form of power (Meyer, 1981). This is the reason why the state, as Weber has put it, tends to monopolize any legitimized use of this form of power (Weber, 1964, p. 1042). At the same time the state tends to control all relevant sources of power which can be put to use in military organizations, in particular socio-economic stages of development; hence they turn into limiting resources for states' competition.

Social motivation to war is very dependent upon a particular state's relation to other states, once state formation has taken place. These motivations are a dependent variable, in other words. At times states, or rather their specialized institutions, use certain sets of stereotypes such as ethnocentrism in order to boost enthusiasm for war. These measures may be more or less successful in terms of these institutions' goals. Ethnocentrism, individual aggressivity or whatever other feature of a directly biological nature will never suffice in starting a war unless one is attributing these characteristics to political leaders.

Is war a 'prime mover' of social evolution as Carneiro, Alexander and a few others have suggested? An answer to this question is in fact dependent on what is meant by the concept 'prime mover'. In a broad sense this concept suggests that the competition between human groups became ever fiercer with the quantitative growth of these groups. Access to resources was limited by this development and warfare became an important means for differential reproduction. In various forms this idea has been put forward by Spencer, Carneiro and Alexander, to name but a few.

While war undoubtedly played a major role in social evolution, at times accelerating change, this was not the case throughout the hunter-gatherer stage.

Certainly hunter-gatherer societies are far from being 'generally peaceable'. Their 'primitive and endemic' war, however, did not exert any major impact on social change for a long period of human history. Only after some 'social inventions' had occurred, war evolved in its military form and turned into an instrument of securing access to mostly material resources. In more primitive social contexts, material resources are not in themselves the most important ends being strived for: 'Australian aborigines, for example, would ordinarily not conceive of taking land of others. Each band's land is sacred and spirit infested' (Kennedy, 1971, p. 44).

11.5 CONCLUSIONS

According to the main lines of thought presented here, war cannot be considered a 'prime mover' since it came into existence as a distinct phenomenon in a comparatively recent stage of social evolution. War played its role in social evolution generally dependent upon developments in other social sectors such as religion, economy and technology (Meyer, 1981, p. 71). So, in a strict sense, war is not as universal and homogeneous a phenomenon as some biologists seem to believe. Eibl-Eibesfeldt's handling leaves this question open: whereas he suggests in one passage (Eibl-Eibesfeldt, 1984, p. 531) that war should be considered a cultural invention, a notion being stressed here, he emphasizes in another context that war 'is as old as humanity' (Eibl-Eibesfeldt, 1984, p. 519).

This vagueness may seem a merely semantic problem. Considering the initial question on the interrelation between human nature and the function of war in social evolution, however, any suggestion concerning the origin of war is of the greatest importance.

Whereas Eibl-Eibesfeldt does not attribute a prime mover function to war as Alexander does (Alexander, 1979), he relates war to human aggressiveness. This is of course a camouflaged way of suggesting the universality of war and linking war up with agonistic behaviour in animals. Contrary to such notions, it is suggested here to clearly separate animal from human behaviour. Agonistic behaviour in animals and aggressive impulses in man may well have a homologous physiological basis. Man's actual aggressive behaviour cannot be understood, however, without recourse to the operation of mental and cultural processes.

Conclusions

Any human behaviour has a genetical and physiological basis, as well as a mental and a cultural basis. The operation of the mind, though extremely difficult to be accounted for, has a natural basis as well. Whatever epistemological problems may arise when proper explanations of mental and cultural processes are being strived for, any methodology explaining away these processes or their relevance is doomed to failure.

Regarding the origin of war, it is suggested that war is a phenomenon started in the mind, i.e. it is neither merely a reaction to the distribution of external resources nor an expression of some inbuilt impulses to mutual destruction. These impulses do obviously exist in man, but their very existence does not provide a satisfactory account for the origin of war.

Societies being very similar in general evolutionary characteristics may well differ in their willingness for war. Hence a basic similarity in terms of sociocultural evolution must not be considered a sufficient motivation for war. However, the study of the more specific ecological conditions as well as the economic and political nature of relations between societies may well provide additional insights into typical motivations for war. Explanations in terms of ultimate causality can account only for a very basic layer of aggressive dispositions. According to the present argument such dispositions are not suited at all for an explanation of war (Meyer, 1981, p. 42; van der Dennen, 1986): while aggression is an individual attribute, war is a special form of collective violence.

Moreover, it has been pointed out that war in its full fledged form is a comparatively recent social phenomenon. Once this phenomenon had come into existence, however, it played a major role in human history, at times favouring the evolution of more complex political structures, sometimes fostering less complex solutions. The evolution of archaic empires and their succession by the less complex feudal societies demonstrate these alternating states of complexity (Meyer, 1977, p. 53).

Regarding the fertility of the concept of ultimate causality for the evolution of warfare it should be emphasized once again that this concept cannot account for the causes which brought about a multitude of cultural attitudes towards warfare. If an explanation in terms of ultimate causality can show why aggressive dispositions exist at all, which it can really account for, its grasp of the actual variety of cultural identities is much less convincing. Why should some desert people like the Bushmen be more peaceable whereas others like some Bedouin tribes should be more warlike? How could an account of such cultural differences do without a recourse to explanations in terms of proximate mechanisms?

As a social phenomenon war has very different characteristics in various stages of social evolution. Consequently, the 'grand functional approach', 'the idea of a complex, closed system readjusting resources to people or people to resources' (Ferguson, R., 1984a, p. 30), as suggested by Vayda, is rejected.

Resources do in fact exist in a physical sense quite independently from human ideas about them. However, their transformation into resources of human social behaviour presupposes a process of cultural definitions. So, if war is a struggle for resources, any account must attempt an understanding of what these resources mean in terms of economy, religion and power relations, for the respective societies.

An explanatory strategy taking these anthropological requirements into consideration becomes aware of the fact that war is not a unitary phenomenon, consequently there can be no single 'function' of war and hence there is no homogeneous human nature which could be blamed for the institution of war.

Such biological approaches studying human nature, as human ethology and human sociobiology, are certainly suited to enlarge knowledge about man's social behaviour. They would be ill advised, however, if they would not pay tribute to the unique importance of the human mind.

The understanding of war as a cultural institution requires the combination of contributions from many fields. Unfortunately it seems to be a common aim of the various social sciences to explain away the complexities of adjacent disciplines. Hence a satisfactory account of war may well have to be postponed to a time when these shortcomings of the social sciences will be overcome by a sound 'philosophical anthropology'. It is very dubious if mankind's need for the banishment of modern war could wait for this event, but perhaps the practical importance of science is minor in this respect to a change of moral attitudes.

CHAPTER TWELVE

War and peace in primitive human societies: a sociobiological view

U. Melotti

12.1 INTRODUCTION

The existence of significant intergroup conflicts in primitive human societies – in particular, war and warlike activities – has been long discussed by scholars of different disciplines: anthropologists; sociologists; psychologists; ethologists. However, the conclusions of these scholars are far from unanimous.

Some of them (for example, Montagu, 1976) affirm the basically peaceful character of primitive human groups, in particular food gatherers and hunters, and suggest that our prehistoric ancestors, although capable of aggressive behaviour, generally must have led peaceful lives and must have been highly cooperative and deeply interested in the welfare of their fellow men. In contrast, other scholars have denounced 'the myth of the aggression-free societies of food gatherers and hunters' (Eibl-Eibesfeldt, 1974a; see also 1974b and 1986) and have presented good evidence of intra-group and inter-group conflicts in both contemporary and prehistoric primitive human populations, food gatherers and hunters not by any means excluded. Therefore, they have strongly refuted 'the neo-Rousseauist view' that Palaeolithic hunter-gatherers were peaceful creatures, in accordance with the thesis of man's original good nature, and the consequent inference that armed conflicts first appeared in Neolithic times, with the developments of agriculture and stock breeding (Eibl-Eibesfeldt, 1979, p. 170).

Nonetheless, this view as well seems to be affected by a strong ideological bias. The idea that war has been an eternal feature of human societies implies (or at least suggests) the existence of an innate instinct to kill one's own kind – an instinct which was previously advocated by Sigmund Freud (1932), Raymond Dart (1953) and Robert Ardrey (1961) – though Eibl-Eibesfeldt, who expressed this view with particular cogency, did not invoke such an instinct, but only social and cultural factors, among which the development of culture itself and 'pseudospeciation', i.e. the process of adaptive radiation of culture, so-called for its analogy to that of the formation of species (Erikson, 1966). Moreover, to show the alleged warlike character of all human populations, these scholars overlook the profound differences in intergroup conflicts at diverse levels of social organization and among diverse types of groups. In effect, there is good evidence that, between some groups, aggression is present only in highly ritualized forms, while between others it is cruel, bloody and destructive. Furthermore, war and warlike activities, though not absent in gathering-hunting societies (as some authors have erroneously assumed), have increased both in frequency and in destructive power with the growth of productive forces, the development of the division of labour, the emergence of hierarchical stratification, the formation of social ranks and social classes and the establishment of some forms of centralized authority.

Biological concepts are not, therefore, sufficient to analyse war even in its most primitive forms. An interdisciplinary approach is needed. Nevertheless, sociobiological theory can contribute to a reassessment of this issue by providing an important conceptual framework for the available ethological, palaeoanthropological and ethnological data. With this conviction, I would try to put forward some suggestions in the light of my previous research on the sociobiological determinants of primitive social organizations (see Melotti, references cited).

12.2 WAR IN PRIMITIVE HUMAN SOCIETIES: SOCIOBIOLOGICAL CONSIDERATIONS

First of all I would like to emphasize that intergroup competition seems to have been a counterpart to intra-group cooperation ever since our most remote origins (Melotti, 1985a and 1986a). However, intergroup competition need not degenerate into warfare.

The peaceful or warlike character of intergroup competition primarily depends upon the ecological conditions and economic factors, as well as upon the type of social structures in the groups involved. Quincy Wright, in his classic *A Study of War* (1965), pointed this out clearly: the higher hunters and lower agriculturalists are more warlike than the food gatherers, lower hunters and horticulturalists, and, in turn, the pastoralists and higher agriculturalists are more warlike than the higher hunters and lower agriculturalists. Similarly, Fried (1967, p. 214) observed that 'the frequency of war increases in relation to the complexity of the society': it is more intense in chiefdoms than in tribal societies, more intense in primitive state-like societies than in chiefdoms, and more intense in complex societies of modern times than in primitive states. In effect, bellicosity seems to reach its maximum in contemporary complex societies.

Archaeological findings do not support the assumption of large-scale war activities among Palaeolithic bands and tribes. Moreover, very low population density, the small size of the human groups, their loose social organization and dispersal over very large areas must have made such conflicts unlikely or at least infrequent.

Nevertheless, as some scholars have suggested, intergroup conflicts could have been instigated even in Palaeolithic times by competition for some critical resources; for instance, drinking water, big game and vegetable food (Bigelow, 1969; Eibl-Eibesfeldt, 1979), or even women, the highly strategic 'goods' that only could convert the other resources controlled by the groups into offspring (Borgia, 1980; Melotti, 1984b and 1986b).

The available data show that intergroup conflicts tend to assume a destructive character when they become battles to conquer or defend a territory. On the other hand, sociobiological theory suggests that such conflicts may only take place regularly when their benefits outweigh their costs – both costs and benefits being measured in terms of inclusive fitness. This occurs when the resources present within an area are rare, defensible, predictable and concentrated (Dyson-Hudson and Smith, 1978). Such situations are unknown to food gatherers and hunters, but become much more frequent with the impressive social and economic changes due to the Neolithic revolution – the development of agriculture and stock breeding, the passage to sedentary life and the emergence of villages and towns.

We should also remember that, at least from Upper Palaeolithic times, human bands probably practised exogamy, the custom forbidding marriage within one's own group. Indeed, this custom seems to be very ancient (see Melotti, 1979; Lopreato, 1984), and some scholars (for example, Reynolds, 1976; Pusey, 1980) provide some data even suggesting its possible foundation in the behaviour of some non-human primates. Exogamy entails important gene exchange between neighbouring groups, and this seems to rule out the hypothesis of continual destructive wars between them. A population could hardly survive if its member groups strived to exterminate each other despite the fact that they to a large extent shared identical genes by common descent (Melotti, cited references except 1985a, b and 1987). Moreover, under exogamous conditions, the migration rate between the groups (up to 50% per generation) is far too high for group selection to be effective (Borgia, 1980; Melotti, 1984b).

This situation changed with the Neolithic revolution. The development of agriculture and stock breeding allowed human populations to grow and expand. Territorial aggression became common, both against the remnants of ancient gathering and hunting populations (which were displaced into marginal areas) and between genetically unrelated groups of herders and cultivators. Later on populations enlarged even more, mixed peoples moved to towns and migrations to better lands and farther fields diluted or demolished the old social units consisting mostly of closely related individuals. This type of intergroup aggression, therefore, could spread out almost without any biological limit and control.

12.3 THE EFFECTS OF EXOGAMOUS AND ENDOGAMOUS RULES

Finally, I would like to expand upon the effects of exogamous and endogamous rules on the intergroup conflict; a theme that has been almost completely overlooked by recent research on the biological bases of peace and war.

This is a rather strange fact, since the relevance of this custom for peace and war in primitive peoples had already been pointed out by one of the founding fathers of cultural anthropology, Edward B. Tylor. Exogamy, according to Tylor, was an extraordinary factor of peace, for it developed a bond of solidarity between the groups by making them dependent on each other for wives and children. For primitive men, who generally had no ties of other sorts outside of their groups, the choice was, as Tylor emphasized, 'between marrying out and being killed out' (Tylor, 1889, p. 277).

The peace-preserving effects of exogamy have been overlooked by both ethologists and sociobiologists, probably because Tylor himself, as well as almost all the subsequent social and cultural anthropologists, had treated it as a merely cultural rule, without considering its latent biological function. This error – to point out a rather surprising fact – was repeated by Robin Fox, one of the founders of so-called 'bio-social anthropology'. Indeed, in his valuable book on *Kinship and Marriage* (1967), Fox analysed the effects of exogamy on intergroup relationships only in psychological terms. Making a transparent allusion to a passage of the Bible (Genesis, xxxvi, 16), which had already been used by Tylor himself, he wrote: 'You would not try to exterminate a band whose wives were your daughters and whose daughters were your potential wives; you would become, in one sense at least, one people; you would be dependent on each other for your continuity and survival'.

Exogamy, however, has not only psychological effects; it also has important biological effects. Far from being only an 'exchange of women', as it is usually defined by social and cultural anthropologists, exogamy is also, and basically, an exchange of genes. It does not serve a merely economic function, as some well-known anthropologists have suggested (for example, Frazer, 1918; Lévi-Strauss, 1949 and 1956; Fox, 1967: who maintain that in exogamy women are treated only as a kind of goods). It also serves a primary biological function.

As I have already mentioned, a population could hardly survive if its groups (usually bands, among hunter-gatherers, or clans or moieties, or parts of them, among horticulturalists) tried to exterminate each other regardless of the fact that they, to a considerable extent, shared the same gene pool.

The effects of exogamous and endogamous rules

Exogamy is practised mainly between bands of the same nexus or between clans or moieties of the same tribe, while it is hardly practised between nexuses or tribes. In effect, nexuses and tribes usually observe the opposite rule of endogamy (which obliges one to marry within one's own unit). This fact is consistent with ethnological data on war and warfare in simple societies, for primitive peoples make a clear distinction, as far as these activities are concerned, between related and non-related groups. Usually it is members of the latter groups that they try to kill, while with members of the former they fight only in highly ritualized ways, being careful to avoid battles that could result in excessive bloodshed.

This is not surprising. In segmentary groups, i.e. groups that are not linked with other groups by the effects of the division of labour, exogamy is the hidden structure of social solidarity (Melotti, 1979, 1980 and 1981). Therefore, to explain the existence of friendship-relationships between these groups, it is not necessary to invoke either the 'original goodness of man', following Rousseau (1755), or the mythical 'mechanical solidarity' derived from their uniformity, following Durkheim (1893). In effect – to use more literally the expression that the latter reserved for solidarity arising from the division of labour – solidarity among these groups is quite 'organic'. This is also most interesting from a theoretical point of view. As it implies an exchange of genes (the basic units of biological reproduction), exogamy appears to be the true archetype of reciprocity, which Lévi-Strauss (1949) and other anthropologists have striven to find only at the social and cultural level.

CHAPTER THIRTEEN

Primitive war and the Ethnological Inventory Project

J.M.G. van der Dennen

13.1 INTRODUCTION

The term 'primitive' as used in this chapter may give rise to some misunderstandings. It seems only appropriate, therefore, to make myself perfectly clear on this subject. I do not use the term primitive in any negative or derogatory sense, nor in a non-complex sense (primitive societies may in fact be more complicated than modern ones), but rather in the original Latin meaning: 'of or belonging to the first age, period or stage' and as such has no derogatory implications whatsoever (cf. Hallpike, 1979). I prefer this term to substitutes such as simple, non-literate or preliterate, tribal, band-level, non-state, acephalous, or, as one may find in older literature, 'savage'.

With the grinding monotony of a magical formula it is contended by ethnologists and other students of primitive war – and sociobiologists have been only too eager to reiterate it – that all primitive peoples, with a few notable exceptions such as some Inuit (Eskimo) societies, do have war in their behavioural repertoire. War is considered to be ubiquitous among primitive peoples. Man is considered to be universally belligerent, and he is supposed to have been so from the very first beginning of hominid evolution. I will call this the Central Axiom, or, more properly, the Central Dogma. What is more, most theories of the genesis and evolution of war, including sociobiological ones, are predicated upon such an assumption. Some recent formulations are the following:

> Throughout history, warfare, representing only the most organized technique of aggression, has been endemic to every form of society, from hunter-gatherer bands to industrial states. (Wilson, 1978)
>
> Warfare may be carried out differently for many reasons by more and by

> less complex societies, but it is carried out everywhere. (Cohen, 1984).
> Die Ethnologie ist generell sehr dezidiert in ihren Aussagen, 'dass alle bekannten Gesellschaften irgendwann einmal in bewaffnete Auseinderset-zungen mit einem oder mehreren ihrer Nachbarn verwickelt waren. Ob als Angreifer oder Verteidiger: sie haben gekämpft'. (Krippendorff, 1985 citing Arnd, undated)
> Der Krieg hat die Menschen seit den frühesten Zeiten begleitet. (Eibl-Eibesfeldt, 1986)
> Primitive clashes of force first occurred when groups of Paleolithic men, armed with crude stone implements, fought with other groups for food, women, or land. Somewhere along the prehistoric road, other drives – such as sport, the urge for dominance, or the desire for independence – became further causes for armed conflicts. (Dupuy and Dupuy, 1986)

Apodictic statements to the same or similar effect may be found in Alexander (1971, 1979), Bigelow (1969, 1975), Eibl-Eibesfeldt (1979), Shaw (1985), Shaw and Wong (1987, 1988), and many others (see also Chapter 10).

The eagerness – one would almost say the voracity – with which sociobiologists have parroted the Central Dogma raises some vexing questions. Why has nobody ever taken the trouble to 'check, double-check and check again' and find out the truthfulness or falsehood of the dogma? Does a belligerent world or humanity fit the sociobiological social cosmology better than a peaceful world? And if so, is this an autonomous recent development or does it constitute a regression to the worst 'Nature red in tooth and claw' doctrines of Social Darwinism? Does the sociobiological discipline as such necessitate such a – basically Hobbesian (*vide infra*) – bias (i.e. to see violent conflict, strife, agonism, war everywhere) or, formulated otherwise, is it inherent in the science? I will come back to these questions later on. All one can say at the moment is that students of primitive war have uncritically repeated one another, without even feeling the necessity to check their most pivotal notions.

Where does the Central Dogma come from in the first place? It must have some ultimate source. As far as I am able to trace the origin of the notion, one such ultimate source (although the omni-presence of struggle had already been a favourite theme in nineteenth century Social Darwinism) seems to be Davie's *The Evolution of War: A study of its role in early societies* (1929), and, but to a lesser extent, Steinmetz' *Soziologie des Krieges* (1929). These two authors may be considered to be the last true-blooded representatives of the Social Darwinist school of thought on primitive war, its origin and evolution. Steinmetz's main conclusion (or rather presupposition) was that 'original man' ('der Urmensch') must have been aggressive and cruel right from the start; that these were even the necessary preconditions of man's humanization process because war was the main and most potent instrument of group selection, eliminating the weaker or otherwise inferior races or tribes. Without war no progress, evolutionary or culturally (these notions were not new, of course, they came from Spencer and Bagehot and numerous other Social Darwinist thinkers; but Steinmetz managed to amplify them into a full-blown Apology of War). Steinmetz wrote in German and his influence was mainly limited to this language area.

Curiously, in the same year, van der Bij (1929) refuted Steinmetz's most central hypothesis by showing rather convincingly in his dissertation that the 'lowest peoples', as he calls them, had neither stomach nor appetite for war: '*tot groepsstrijd gaan de laagst bekende volken niet of zeer node over; offensieve strijd komt op den laagsten cultuurtrap niet voor. Waarschijnlijk ontbreekt oorlog aanvankelijk geheel*'. [The lowest peoples known to us do not, or very reluctantly, resort to group-fights; offensive combat does not occur on the lowest cultural level. Very probably war is totally absent in the beginning.] But van der Bij's dissertation was written in Dutch, to my knowledge never translated, and his careful and conscientious work never had the impact it rightly deserved.

More influential in Anglo Saxon circles appears to be Davie's (1929) work. Violence, Davie asserted, was omnipresent, even if war is occasionally absent among preliterate peoples. Compiling the then available ethnographical literature, mainly dating from the nineteenth century, he concluded that warfare in tribal societies is universal and sanguine (with a few notable exceptions like some Eskimo groups). Moreover, Davie's theorizing on the subject was thoroughly materialist (as we would say now), though he was not blind for the religious, supranatural dimension of primitive war, his main explanation being – in the language of modern sociobiology – competition for scarce and vital material resources.

Davie conveniently forgot that much of his material came from tribal societies on continents, like conflict-ridden Africa, where white intrusion and belligerent European colonialism, not to forget the slave trade, had caused wholesale displacement and migratory waves of peoples, eviction, massacres, ethnocide, cultural degeneration and intertribal warfare with fire-weapons (which hardly can be called primitive any more).

It is quite possible that the notion of the ubiquity of primitive war is even older than Social Darwinism. In his *Essay on the History of Civil Society*, Adam Ferguson (1767) had already concluded: 'We had occasion to observe that in every rude state the great business is war; and that in barbarous times, mankind, being generally divided into small parties, are engaged in almost perpetual hostilities.' But I am unable to ascertain whether Ferguson's essay had any impact on later theorists. His work seems largely unknown to nineteenth century writers. Ferguson's work may, however, be placed in a still broader historical context: the so-called Hobbes–Rousseau controversy.

13.2 THE HOBBES–ROUSSEAU CONTROVERSY

Following the Renaissance, two well-defined theories of the origin of war became widely distributed among social thinkers. One of these was sponsored by the social contract theorists and by Hugo Grotius (1625) and Thomas Hobbes (1651) in particular. This theory, which was very widely accepted in the seventeenth and eighteenth centuries, held that the scarcity of resources in relation to the numbers and needs of early man, the lack of adequate technological development among primitive peoples and their inherent suspicion and dislike of strangers made it inevitable that primitive society should be characterized by constant warfare. This theory has many points of resemblance to the modern teaching that pressure of population on the means of subsistence is a chief indirect cause of war (Bernard, 1944). However, the social-contract theorists of the seventeenth century were less concerned with explaining the causes of war than with describing what they conceived to be the war-ridden condition of primitive society (though in his *Leviathan* Hobbes (1651) made an interesting attempt to list the causes as follows: 'So that in the nature of man we find three principal causes of quarrel. First, competition; secondly, diffidence; thirdly, glory'). Nor did they attempt to determine when warfare first began. The lack of an evolutionary theory of society and the general acceptance of the doctrine of human origin set forth in Genesis made a theory of the gradual evolution of war seem unnecessary, if it did not render it inconceivable (Bernard, 1944). Hobbes may be taken as fairly typical of the general advanced opinion of the time regarding the warlike attitudes and proclivities of early man. His aim was to describe human behaviour in terms of a kind of social physics. Thus the tendency of physical objects to pursue their own trajectory – when left to themselves – could be translated into an egoistic principle for human beings, that they pursue their own interests in the line of least resistance. This, of course, is the source of social conflict, of 'Warre, as is of every man against every man' (*bellum omnium contra omnes*). Hobbes describes the '*status naturalis*', the state of nature, as the '*status hostilis*'. This is explained by man's behaviour, which, in its natural state, is governed not by reason, but by passion and desire. These drive him to act in a way in which reason plays a subordinate part. It orientates man's desires. 'For the Thoughts, are to the Desires, as Scouts, and Spies, to range abroad, and find the way to the things desired'. It appears from this that Hobbes, contrary to later philosophers, has little confidence in reason. It is part of human nature. On the other hand it teaches man to be peaceable, loyal and helpful. In Hobbes' philosophy the '*homo homini Deus*' and the '*homo homini lupus*' must be set side by side. But, on the other hand, reason teaches us that desire predominates in man's nature and that his life is constantly in danger (Spits, 1977). The most dangerous of all desires is that for prestige, the craving for honour. Everything that gives joy to the mind relates to honour and fame. It results from vanity. Vanity – and everything is vanity – is the root of all evil, the source of all vice. It makes man crave for power. This brings him to his often quoted statement: 'I put for a generall inclination of all mankind, a perpetuall and restlesse desire for Power after power, that ceaseth onely in Death'.

On this account, states as well as tribal societies are led to war because of competition for material possessions, mistrust, fear and the pursuit of glory, 'with fear being the prime motive in that it supposedly leads to a concern to secure what we already have' (Walker, 1987).

Diametrically opposed views were espoused by Rousseau (1762) in his *Le Contrat Social*, building on Montesquieu's ideas exposed in *L'Esprit des Lois* (1748). Rousseau introduced the concept of the 'Noble Savage', who did not wage war because he simply had no reason for doing so; not only because he lacked the material incentive (there were no benefits to gain), but also because he lacked the necessary infrastructure which was expressed thus by Glover and Ginsberg:

> The antithesis between war and peace is really inapplicable to the simple conditions in which these ['primitive'] peoples live. Anything like the organized and aggressive warfare which we find in early history and among the more advanced of the simpler societies can have no place in the life of the simplest societies, for this implies organization, discipline and differentiation between leaders and led which the people of the lowest culture do not possess. But if these do not have war, neither have they peace. We must think of war not as a genus uniquely opposed to peace but as a species of violence opposed to social order and security. (Glover and Ginsberg, 1934)

Property which Rousseau singles out in the second discourse as a crucial factor in inequality and consequently in violence, should not be seen as a cause of war but as a consequence of the 'cupidity' and insecurity that dominate men once their original isolation comes to an end (Hoffmann, 1965).

This so-called Hobbes–Rousseau controversy, a persistent and irreconcilable one, has dominated the anthropological, sociological and psychological literature until today. The first to tackle the problem empirically was Adam Ferguson. In 1767 Ferguson published *An Essay on the History of Civil Society*, which he based on Montesquieu while extending the latter's method by drawing more upon the observations of primitive peoples made by contemporaneous missionaries and travellers such as Charlevoix, Lafitau, Dampier, and others, as well as the classical sources. His attempts to describe human nature, or 'man in a state of nature', was in terms of a careful evaluation of the evidence from primitive life. He had at once more data to work with than Montesquieu and also a more severely questioning attitude toward its validity

– although it should be remarked that evaluation of data had a much more central place in Montesquieu's work than in that of the other 'philosophes' such as Rousseau (Service, 1975). Ferguson concluded that war – as armed group antagonism – 'has been the great business of mankind since time immemorial'.

Ferguson more than agreed with Montesquieu about the falseness of the common idea that 'man in a state of nature' was free to be his natural self. Man is governed by society, and never was outside it – he has always wandered or settled 'in troops and companies'. Man's nature as seen in a study of man in society is highly composite, in the primitive world as well as in civilization – which is but a continuation and accumulation of devices from earlier times. Ferguson saw human nature as being composed of many opposite propensities – sociability and egoism, love and hostility, cooperation and conflict – an amalgam that is necessary to allow for the different kinds of characteristics demanded by society in different times and places. Ferguson felt – contrary to Rousseau's belief – that conflict had a positive function in cultural evolution, and for that matter even in individual psychology: the stronger the hostility to outsiders, the closer the internal bonds of the collectivity; the very meaning of friendship is acquired from a knowledge of enmity (Service, 1975).

The condition of 'almost perpetual hostilities', or relative peacelessness among primitive peoples and early hominids was confirmed, after Ferguson, by Waitz (1859–62), Lyell (1863), Lubbock (1870), Tylor (1874), Jaehns (1880, 1893), Gumplowicz (1883 *et seq.*), Hellwald (1883), Bagehot (1884), Spencer (1885 *et seq.*), Vaccaro (1886), Maine (1888), James (1890, 1910), Ratzel (1894–95), Novikow (1896), Vierkandt (1896), de Molinari (1898), Schultze (1900), Schaeffle (1900), Topinard (1900), Frobenius (1903), Lagorgette (1906), Steinmetz (1907, 1929), Sumner (1911), Boas (1912), McDougall (1915), Hartmann (1915), Jerusalem (1915), Weule (1916), Knabenhans (1917), Keller (1918), Mueller-Lyer (1921), Hobhouse (1924), Sumner and Keller (1927), van Bemmelen (1928), Davie (1929), Andreski (1954), and most contemporary evolutionary anthropologists and sociobiologists.

Relative peace as the primeval condition of mankind was advocated by Montesquieu (1748), Rousseau (1755, 1762), Letourneau (1895 – although Letourneau is ambiguous on this point), Westermarck (1889, 1907), Kropotkin (1902), Holsti (1912, 1913), Anthony (1917), Perry (1917, 1923), de Lavessan (1918), Dickinson (1920), Dewey (1922), Rivers (1922), Smith, G.E. (1924), Wheeler, W.M. (1928), Cleland (1928), van der Bij (1929), Schmitthenner (1930), MacLeod (1931), Benedict (1934), Malinowski (1936, 1941), Mead (1940 *et seq.*), and most contemporary cultural anthropologists.

These visions are not, however, as diametrically opposed as might *prima facie* appear. They are differences of emphasis rather than essential distinctions. Although 'belligerence is a concomitant of increasing civilization' (Broch and Galtung, 1966), that does not mean that most primitive cultures lived in a paradisiac condition of perpetual peace and blissful harmony. On the contrary, the terms 'pseudo-peace' (Garlan, 1975) or '*Friedlosigkeit*' (Hartmann, 1915) are more appropriate. 'It is interesting', Service (1975) notes, 'that the actual nature of primitive prestate society as we now know it ethnologically can support both Hobbes and Rousseau, each in part. War, as Hobbes meant it – as threat or imminence as much as action – certainly is an omnipresent feature of primitive life, as is, in part, an appearance of the Rousseauian peace and generosity. As we shall see, these two aspects of social life coexist; the threats of violence caused by the ego-demands of individuals are countered by social demands of generosity, kindness, and courtesy'. Long ago, Sumner (1911) answered the question whether man began in a state of peace or a state of war as follows:

> They began with both together. Which preponderated is a question of the intensity of the competition of life at the time. When that competition was intense, war was frequent and fierce, the weaker were exterminated or absorbed by the stronger, the internal discipline of the conquerors became stronger, chiefs got more absolute power, laws became more stringent, religious observances won greater authority, and so the whole social system was more firmly integrated. On the other hand, when there were no close or powerful neighbors, there was little or no war, the internal organization remained lax and feeble, chiefs had little power, and a societal system scarcely existed.

13.3 A STUDY OF WAR

A further impetus to the notion of the universality of primitive war probably came from Quincy Wright's (1965) *opus magnum: A Study of War* (actually a collective enterprise, and numbering some 1600 pages), in which a small section is devoted to primitive war: 'Appendix IX. Relation between warlikeness and other characteristics of primitive peoples', comprising a cross-cultural sample of some 650 distinctive primitive peoples, arranged alphabetically by continent and categorized with respect to warlikeness and other characteristics, and based on the list of peoples used by Hobhouse, Wheeler and Ginsberg (1915). This is followed by statistical tabulations indicating the relationship between warlikeness and a series of other variables such as habitat, political and social organization etc. Furthermore, Wright introduced the important distinction between four types of primitive war: defensive; social; economic; political.

Defensive war occurs if war is never embarked upon except for immediate defence of the group against attack, with the inclusion of a few tribes who do not even defend themselves from attack. This category comprised 5% of the total sample.

If war is embarked upon for purposes of revenge, religious expiation, sport, or personal prestige, this was coded social war. This category comprised 59% of the total sample.

If war in addition to utilization for defensive and social purposes, is an important method for acquiring slaves, women, cattle, pastures, agricultural lands, or other economic assets (including the provision of victims for human sacrifice), this was coded economic war. This category comprised 29% of the total sample. Finally, if war is fought not only for defensive, social and economic purposes but also to maintain a ruling class in power and to expand the area of empire or political control, this was coded political war. This category comprised 7% of the total sample.

Please note that:

1. *In toto* this means 95% warlike peoples versus 5% unwarlike, a result which seems to substantiate the notion of universal belligerence.

2. In terms of cultural level, 'it seems clear that the collectors, lower hunters and lower agriculturists are the least warlike, the higher hunters and higher agriculturists are more warlike, while the highest agriculturalists and the pastorals are the most warlike of all'. A conclusion which seems to support van der Bij's (1929) general conclusion, and which seems to indicate that belligerence is a concomitant of increasing civilization (fully confirmed by Broch and Galtung's (1966) reanalysis of Wright's data; cf. also: Holsti, 1913; Hobhouse, Wheeler and Ginsberg, 1915; Toynbee, 1950; Sorokin, 1957; Sahlins, 1960; Russell, 1972; Harrison, 1973; Eckhardt, 1975, 1981, 1982; Leavitt, 1977; Harris, 1978; Beer, 1981; van der Dennen, 1981, 1986; Lenski and Lenski, 1987). Furthermore, these studies indicate that wanton cruelty, human sacrifice and a general low valuation of life do not seem to be a function of 'primitivity'; they are rather phenomena manifesting themselves on the level of chiefdoms, preliterate kingdoms and hierocracies.

3. Regarding the motives of primitive war, the modern emphasis on 'economism' (or 'materialism') is emphatically not endorsed by Wright (1942). At several places in the text he states this quite clearly: primitive wars 'seldom have the object of territorial aggression or defense until the pastoral or agricultural stage of culture are reached, when they become a major cause of war' (p. 76). And 'primitive peoples only rarely conduct formal hostilities with the object of achieving a tangible economic or political result' (p. 58). 'If acquisitive motives play a part in primitive war, the commodity sought is likely to be an object of magic, ritual or prestige value rather than of food value' (p. 75). The majority of primitive wars belong to Wright's social category.

4. Wright offers a number of cautionary remarks, which still ought to give us food for thought:

(a) The terms 'warlike' and 'unwarlike' are used in very different senses by different writers. Some refer to the frequency of war, some to the functional importance of war in the culture, some to the cruelties of war methods as judged by civilized standards, some to the proportion of casualties in battle, and some to the degree of divergence of war objectives from those of civilized people.
(b) The writers frequently do not adequately identify the groups which they characterize as warlike or unwarlike.
(c) Descriptions of war practices are frequently omitted or inadequate in materials dealing with a particular people.
(d) Changes in war practices occur within the same group especially after it comes into contact with a more civilized people, and it is not always easy to determine the time to which an author refers in considering war practices of a primitive people.

All in all, Wright's conclusions seem to confirm the notion of universal belligerence, and so it has entered the cocktail-party wisdom of the average western intellectual. There are, however, apart from some minor flaws and shortcomings in Wright's sample (for example, Wurrunjerri and Yara Yara refer to the same Australian tribe, and not to two different ones), reasons to doubt the representativeness of his list.

Interestingly, Hobhouse, Wheeler and Ginsberg (1915), who compiled the original list which Wright used for his statistical analysis, were much more reserved in their conclusions. They write:

> The question has been raised whether the traditional view of early society as one of constant warfare is really justified by the facts. There is, in fact, no doubt that to speak of a state of war as normal is in general a gross exaggeration. Relations between neighbouring communities are in general friendly, but they are apt to be interrupted by charges of murder owing to the belief in witchcraft, and feuds result which may take more or less organized form. In the lower stages it is in fact not very easy to distinguish between private retaliation when exercised by a kinfolk or a body of friends, and a war which is perhaps organized by a leader chosen for the occasion, followed by a party of volunteers. Strictly we take it that external retaliation means a quarrel exercised by a part of a community only upon members of another community as a whole. Feuds would thus also be the appropriate name for reprisals exercised by one branch of a community upon another, e.g., as between two clans or two local groups within a tribe. As distinguished from a feud, war implies a certain development of social organisation, and is probably not so common at the lowest stages as it becomes higher up.
>
> We have sought to distinguish war and feuds, but must confess that the evidence is often very inadequate for the purpose. We have, however, set

> down the cases where one or other is reported, together with those in which war is said to be nonexistent, and we have in the aggregate 298 cases of war or feuds distributed through all the grades, and nine certain and four doubtful cases of 'no war'. These are mainly confined to the lowest grades, there being 4½ among the lower, 3½ among the higher hunters, and 2 in the lowest agriculture. Thus there are a few very primitive societies, mostly of the jungle folk, that seem quite peaceful, but there is no indication of any association of peaceful propensities with the lowest stages of culture as such. At most it may be said that organised war develops with the advance of industry and of social organisation in general. (Hobhouse, Wheeler and Ginsberg, 1915)

There seems to be a persistent misunderstanding in the literature about Hobhouse *et al.*'s list. For example, Malmberg (1983), in his *Human Territoriality*, writes:

> In their examination of *311* 'primitive' societies Hobhouse, Wheeler and Ginsberg (1930: 228–233) found that 298 showed war and feud behaviour distributed through all the grades while in 9 certain and 4 doubtful cases war was not known. (Malmberg, 1983, p. 107; emphasis added)

This author misquotes the figures, obviously because he did not consult the original source thoroughly, thus suggesting that virtually all peoples had war in their repertoire. Such misquotations have a nasty habit of assuming a life of their own. For a similar uncritically parotting of fanciful figures on contemporary wars, see Jongman and van der Dennen (1988). The question is: why do we so eagerly want to believe the myth of universal belligerence, even to the point of manipulating the evidence?

Such a vast discrepancy between the figures provided by Hobhouse *et al.* (298 cases of war or feuds) and those of Wright (95% of the total sample: almost twice as much) requires explanation. Wright states that he used additional literature published after 1915 to code his sample. It is quite possible that he found information on the presence of war and/or feuding which was not available to Hobhouse *et al.* However, this is unlikely to explain all the difference.

Unfortunately, Wright lumped together in his category 'social warfare' all the subtle but crucial distinctions made by Hobhouse *et al.* regarding the maintenance of public justice (for example, retaliation and self-help, regulated fight, expiatory fight etc.). This is, for instance, what they have to say about the Australian tribes in their list:

> The expiatory combats and the regulated fights of the Australians are also all of them palpably means of ending a quarrel, or marking a point beyond which it is not to go. They do not seek to punish a wrong but to arrest

> vengeance for wrong at a point which will save the breaking-out of a devastating fight.

I will leave it to the reader to decide whether such conflict-limiting procedures merit to be labelled 'social warfare'. I suspect that many such cases of private or public redress, retaliation etc., have been classified as 'social warfare' by Wright. The remaining numerical difference, however, is still considerable. The puzzle is, as yet, unsolved.

I have been growing more and more suspicious and skeptical about both the Central Dogma and the representativeness of Wright's sample over time, and some five years ago I decided to make an inventory of my own. I had already made a list of all the peoples mentioned in Holsti (1913), Davie (1929), van der Bij (1929), Steinmetz (1929), Wright (1965), Turney-High (1949), Numelin (1950), and many other authorities on primitive war. Gradually it dawned upon me that I had been searching in the wrong direction. What I should do was place the body of ethnographical findings on the presence of primitive war within the context of the *total* number of peoples ever described in the ethnographical, anthropological and historical literature. The inventory should also contain the Minimal Conflict Units at subtribal level (such as moieties, phratries, clans, bands, hordes, castes etc.) if feuding was to be included.

This search necessitated a careful reexamination of more than a century of ethnological, ethnographical and anthropological periodical volumes, hundreds of anthropological handbooks and numerous monographs and other sources. My 'Ethnological Inventory', as I named it, at the moment comprises more than 50000 entries, containing the band-level and tribal societies, peoples and ethno-linguistic groups of the world and in the world's history, alphabetically classified according to continent and cross-referenced with regard to synonyms, variant spellings etc., together with all the references to the pertinent literature. The next edition, in preparation, will contain some 100000 entries.

13.4 SOME PRELIMINARY OBSERVATIONS AND CONCLUSIONS

It makes little sense to make the distinction belligerent/peaceful, warlike/unwarlike. It is not a neat, static and historically fixed dichotomy. Peace and war represent the two extremes of a whole array of collective survival strategies, ranging from collective retreat and cultural insulation to imperialist war. Many adjacent peoples lived or live in what may be termed a state of permanent peacelessness: not exactly a state of perpetual war but neither a state of perpetual peace. Sometimes originally 'peaceful' peoples are forced by circumstances to wage defensive wars, which, in turn, generates its own dynamics toward an optimal adaptation to a potentially hostile environment.

War has high opportunity costs, while peace carries with it high existential costs in the form of loss of life, territory, vital resources, cultural integrity etc. Thus most peoples may be seen manœuvering, 'cybernating' between Scylla and Charibdis, in a continual effort to reach an optimal balance. To say that man is belligerent by nature is a phrase devoid of any meaning (the savage, one might paraphrase Rousseau, is neither noble nor ignoble: he is just utterly human). But for the sake of argument let us for the moment regard the distinction belligerent versus peaceful as a simple dichotomy.

The whole history of European colonialism is neatly reflected in the literature: Spanish and Portuguese references for South America; French and English for North America; Portuguese, French, English, Dutch, German, Italian, Belgian references for Africa etc. I mention this only for its curiosity value. It should remind us of the fact that it always was white 'outsiders' with their own Western-centred and ethnocentric biases who described post-contact phenomena in peoples (savages), who were often first conquered, routed and evicted from their territory, decimated, or who otherwise received the blessings of civilization in the form of Bible, epidemics and alcohol. Only one century after Columbus, hundreds of Indian peoples were tottering on the brink of extinction if not extinct already. Africa remained largely *terra incognita* until far into the nineteenth century. By that time, reports from explorers, missionaries, ethnographers etc. had been accumulating to such an extent that the African Negro was considered to be the most 'warlike race' by Davie (1929). One cannot escape the impression, however, that much of the havoc that befell this poor continent (and tribal war in Africa could be devastating in its ferocity, lethality and sequelae) has been caused, directly or indirectly, by European colonial intrusion, the White Man's Burden, and the abominations of the slave trade. We should be aware, in other words, that much of the ethnographical literature does not describe original primitive war at all. Emphatically I am not saying that war was a post-contact phenomenon almost *in toto*, but intertribal warfare must have been greatly exacerbated by it (see Appendix A, p. 264).

Of the overwhelming majority of tribal societies and primitive peoples in the Ethnological Inventory, we either do not know much more than their name and presumable habitat (or their former presumable habitat in the case of those extinct), or no war activities have been reported. This is not to say that they therefore were 'peaceful' only that it is not documented (at least I could find no evidence). The latter may have two reasons: (a) it may be due to sampling error, the likelihood of which increases inversely proportional to the number of independent observations made at specific points in time (in other words: the less sources, the greater the chances of sampling error), or (b) it may simply reflect true absence of war.

Now some elementary calculations: if we subtract some 2000 historical peoples as not primitive, and some 2000 primitive peoples of whom we have positive evidence of war and/or feuding (van der Dennen, 1984), and even if we eliminate half the remainder of the list as due to sampling error, we are still left with more peaceful than warlike peoples, if, and this is crucial, we are prepared to give them the benefit of the doubt, as long as there are no logically compelling reasons to assume otherwise. So the Central Dogma is refuted and has to be rejected as unsubstantiated.

In a strict sense, of course, I have not proved anything: one cannot prove the absence or non-existence of a phenomenon. Yet, I believe these results reverse the burden of proof back to the advocates of man's universal belligerence.

In order to countercheck my own results, I used Markham's (1910) enlarged and extended list of Amazonian tribes, the Amazonian region traditionally being considered a very warlike one. Now, Markham's list contains 1047 entries, including sub-tribes, cross-references, and synonyms: a total of 485 distinct tribes. Altogether 106 of these tribes were already extinct in Markham's time, and about 15 more were practically extinct (and we know virtually nothing more about them). Markham's descriptions of these tribes contain 23 explicit qualifications such as 'warlike' and 'given to feuding', and 22 (partly overlapping) implicit ones (such as 'fierce', 'ferocious', 'pugnacious', 'hostile', 'savage', 'barbarous', 'cruel', 'sanguinary'), while 20 tribes are explicitly described as 'peaceful', 'gentle', 'friendly', 'docile' or 'inoffensive'. Some 15 peoples are alleged to be cannibals, but it is often unclear whether exo- or ceremonial endocannibalism is at stake, and how far exocannibalism figures as a motive for warfare. Thus, even in such an allegedly belligerent region as the Amazonian, of the overwhelming majority of peoples we do not know much more than their names and presumable habitats. A result which accords well with my own list.

Similarly, of the nearly 1000 tribelets and bands living in northern Mexico and southern Texas at the time of the Conquista, and which I listed as Coahuiltec and Tamaulipec, virtually nothing else is known but their names and presumable location. Also data about many thousands of Australian tribes, bands and hordes listed by Tindale (1974) are well in accordance with my main finding.

There is more circumstantial evidence that does not point to universal human belligerence. In many cases of extinction – in 1895 when Markham compiled his first list of Amazonian tribes, a good deal of them were already 'legendary' – these peoples were massacred, famished, forcibly evicted, or decimated by imported diseases, without putting up much resistance on their part, which does not typically indicate a warlike spirit (see Appendix B, p. 264).

One of the most zealous and ardent Social Darwinists, Gumplowicz, author of *Der Rassenkampf*, who envisaged war as an evolutionary Agent of

Progress, wrote in 1892 on the subject of peaceful peoples: 'Die Völkerkunde bietet uns unzählige Beispiele solcher 'friedlichen' Völker: sie bleiben auf der Stufe der Affen, sie kennen keinen Krieg, keine Führung, keinen Befehl, keinen Zusammenstosz mit Fremden, sie 'beuten nicht aus' und werden nicht 'ausgebeutet' . . . sie sind die vollkommenste Affen' [Ethnology offers us numerous examples of such "peaceful" peoples: they remain on the level of monkeys, they have no war, no leadership, no command, no collision with strangers, they do not "exploit" and are not "exploited" . . . they are the most perfect monkeys.] (Notice the arrogance and derogation in his tone.)

So Gumplowicz, without giving any explanation of why and how so many peaceful peoples have been able to survive in so warlike a world, so much as admits that ethnology offers numerous examples of them.

Now, to return to the questions posed in the first paragraph, why did so many writers in the Social Darwinist tradition, and contemporary human sociobiologists, depict the relations between primitive communities in the darkest colours and the most gloomy terms? This question was already raised, and partially answered, by Holsti (1913). I shall try to enumerate a number of reasons. For the Social Darwinists the answer may be quite simple: they envisaged each war-mitigation as a phase in a process of linear progress, which logically implied that original war (presumably reflected in primitive warfare) must have been unmitigating, all-out and sanguinary. For the contemporary writers on the subject the answer may be equally simple: regrettably, most human sociobiologists and ethnologists writing about human warfare, primitive or contemporaneous, do not seem to be bothered by their stupendous ignorance of the subject, which does not prevent them from espousing vulgar materialist notions of war causation, or linking war in an astoundingly simplistic fashion to human 'aggression' (van der Dennen, 1986). But there may be other, more subtle, reasons:

A deficient classification

Considerable difference exists between the various modes of primitive war. In Holsti's (1913) words:

> When therefore certain authors use statements concerning primitive warfare without properly classifying them it follows that actual wars by comparatively advanced primitives in order to procure slaves, victims for sacrifices, or cattle, or to make conquests or to exterminate the enemy, are put side by side with small open combats fought between savages of the lowest type, while these statements again are mixed up with instances of indiscriminate slaughter after sudden attacks. When to this is added the tendency more or less systematically to overlook facts which imply mitigation of the cruel and destructive character of primitive warfare, as

> well as the prevalence of other important rules bearing on the relation between primitive communities, it becomes still clearer that such a treatment of the subject cannot throw any light either on the relations between primitive communities in general, or on the character of primitive warfare in particular.

Most contemporary theorists of primitive war still treat it as if it were a unitary and homogeneous concept, and try to devise unitary theories to explain what is in reality a highly heterogeneous and kaleidoscopic set of phenomena.

Selective perception

This possibility was already acknowledged by Steinmetz himself. In his *Ethnologische Studien zur ersten Entwicklung der Strafe*, Steinmetz (1892) refers to 50 instances in which primitive groups are stated to be revengeful and bloodthirsty, and to 20 other cases among whom injuries received are not always avenged but are soon forgotten. He then continues:

> We have been able to collect 20 other cases where certainly no long-lasting, intense search for revenge was to be seen; these when compared with the 50 opposite cases make up no mean minority, which has all the more weight in that the tendency of the observers probably was for the most part to assume the pitiless seeking of revenge as a natural feature of the savage character; and this must have prevented their finding and duly valuing phenomena pointing in an opposite direction.

The same selective perception mechanism operates, of course, in the ardent believers of the Central Dogma, who manage to construe the slightest agonistic episode between individuals as somehow evidence for universal human belligerence.

Exaggerated accounts of historical episodes

If one remembers the accounts of civilized peoples of their own warlike achievements and exploits, we can hardly expect the imagination of primitives to be less inventive in this respect. War is certainly the last thing that primitive tradition forgets. Now, owing to the tendency to exaggerate their warlike events, primitives are easily, gradually led to make huge battles out of even comparatively small strifes. For example, the Eskimo are one of the most peaceful group of peoples that ever existed, and yet, if we are to believe the Eskimo themselves, they have had great battles. But these can undoubtedly be reduced to cases of petty fighting at wide intervals of time, which a vivid imagination has gradually magnified, as Ratzel (1885) already noted. The primitive warrior is generally more concerned with the show of ferocity, his reputation as a fierce warrior, than with any real brave or heroic act. Thus, the Abipones boast of martial souls but are at the same time far from being heroes (Dobrizhoffer, 1822). Similarly, Williams and Calvert (1858) wrote about the Fijian warriors: 'the deeds of which they boast most proudly, are such as the truly brave would scorn'. Spencer and Gillen (1912) point out that the Australian aborigines 'have a marvellous capacity for exaggerating the reports of such fights that really take place.' One last anecdote to illustrate the point: speaking of the Papuans of New Guinea, Rawling (1913) states that he once asked a furious war party whether the warriors would eat their enemies. 'Yes, yes' was the unhesitating reply. 'It seemed to us, however, that speaking thus, they were actuated more by bravado than by any real intention'. After furious excitement and preparations the party started. Soon, however, warrior after warrior turned home, until the whole expedition was postponed until tomorrow and when the next day came the very enemies happened to come to visit. Giving a minute description of the great festivities which were arranged to celebrate the visit, Rawling observes: 'Had we not witnessed both events, it would have been hard to believe that such violent anger and thirst for revenge could have evaporated in so short a time'. (Margaret Mead was not the first anthropologist to be framed by informers, nor will she be the last, one might conclude.)

'When thus the stories of primitives themselves so greatly harmonize with the inclination of white travellers and other authors to look upon them as, before all, restless warriors, it follows more or less as a matter of course that we are often provided with accounts of primitive warfare which can by no means always serve as a basis on which to build scientific theories' (Holsti, 1913). And, one might add, let us not forget the vanity of ethnographers who rather cater to the public taste with spectacular stories of hideous wars, than carefully describing the rather dull and monotonous day-to-day life of primitive peoples in times of peace.

13.5 POSTSCRIPT

Wilson (1975a) justly criticized Dart's (1953) slaughterhouse vision of human evolution as obviously bad biology and bad anthropology, but up till now sociobiologists have not been too eager to remedy the situation and correct the view. On the contrary, many seem to revel in reconstructing mankind's evolution and history in the most gory terms. Shaw (1985) even goes so far as to give sociobiologists undue credit by asserting: 'A second contribution of sociobiologists to the study of human aggression has been to demolish the belief that truly peaceful cultures exist'.

Yet, sociobiology *as such* does not necessarily generate the presupposition of universal belligerence, nor does it necessitate a view of human evolution as an abattoir or a perpetual concatenation of carnage. But why, then, the attraction, the power of enchantment of the dogma? Any real answer to this question is beyond the scope of this humble contribution, but I cannot refrain from suggesting that (a) the sanguinary 'nature-red-in-tooth-and-claw' view of nineteenth century Social Darwinism may not really have been transcended (in other words, it never left us completely), and (b) the realistic-conflict paradigm of old may have degenerated into a vulgar materialist, truncatedly 'realist', and thoroughly 'dementalized' caricature of its former self (see Meyer's contribution in this volume for an elaboration of this point). Finally, sociobiologists should be extremely cautious in their statements on these matters lest they be stigmatized as just another emerging new brand of War Apologists.

13.6 A NOTE ON METHODOLOGY

The Ethnological Inventory Project has been criticized for its lack of methodology. It is not so much the lack of methodology, however, but its sheer simplicity. In collecting information and literature on the many thousands of peoples, societies, ethnies etc. of the world and in the world's history, I am not hampered in any way by (a) fixation of a particular historical point in time for reasons of comparison, and (b) Galton's problem (cultural diffusion); problems which vex all attempts to establish a Standard Cross-Cultural Sample (see Naroll, 1964, 1970; Murdock and White, 1969), such as the Human Relations Area File (HRAF).

At the moment I am interested in whether war, feuding or other collective conflict behaviour has ever been documented for a particular people, regardless where this information comes from, and irrespective of whether that society has been adequately described or not by academic anthropologists.

I am not trying to establish another cross-cultural standard sample, which eliminates the need to carefully identify culture areas and select particular cases in order to avoid Galton's problem, i.e. overrepresentedness in a sample because of cultural diffusion of a particular trait. I am at the moment not so much interested whether the warring and/or feuding behaviour resulted from cultural borrowing or not. I intend to use my list for more elaborate and sophisticated codings later on.

I believe that at the moment my list allows the following general conclusion: all statements and theories regarding primitive warfare, which are based on current cross-cultural samples, should be interpreted with extreme caution, because of the biases inherent in their methodology (especially the elimination of 'inadequately described' societies, which are likely candidates for the benefit of the doubt).

APPENDIX A

We have evidence for the post-contact exacerbation of warfare for the following peoples (based on Holsti, 1913):

Maori	Gudgeon, 1902; Manning, 1836; Smith, 1899;
Fijians	Thomson, J. 1908; Williams and Calvert, 1858;
Tonga Islanders	Mariner, 1817;
South Sea General	Ellis, 1829;
Dyaks	Bayle, 1865; Low, 1848; Ratzel, 1894;
Masai	Kallenberg, 1892; Merker, 1904; Thomson, B.H. 1885;
Galla	Ratzel,1895;
Kafir	Kidd, 1904;
Blackfeet	Dunn, 1844;
Iroquois	Mooney, 1894; Morgan, 1851;
Omaha	Fletcher and LaFlesche, 1911;
Indians General	Dellenbaugh, 1901; Powell, 1891;

APPENDIX B

Peaceful peoples as listed by

Hobhouse *et al.* (1915)	van der Bij (1929)	Smith (1930)
Bahima	African Pygmies	African Pygmies
Curetus	Andamanese	Algonquin
Chepewyans	Australians	Andamanese
Ghiliaks	Botocudo	Aru Islanders
Greenland Eskimo	Kubus	Australians
Kubus	Negritos	Beothuc
Paumari	Punan	Californians
Point Barrow Eskimo	Semang	Dene Indians
Punan	Senoi	Eskimo
Sakai	Tasmanians	Kubus
Semang	Tierra del Fuegians	Negritos
Tjumba	Veddahs	Paiutes
Wintun		Punan
		Sakai
		Salish
		Semang
		Siberians
		Tierra del Fuegians
		Veddahs

Appendix B

A list of peaceful peoples from miscellaneous sources (Davie, 1929; Holsti, 1913; Markham, 1910; Numelin, 1950; among others); with references:

Ainu	Batchelor, 1895, 1901; Murdock, 1934;
Alacaluf	Bird, 1946; Oyarzun, 1922;
Amazonian Tribelets	Markham, 1895; von Martius, 1832;
Andamanese	Radcliffe-Brown, 1933;
Aquitequedichagas	de Azara, 1809;
Arafuras	Farrer, 1880; Kolff,1840;
Arapesh	Fortune, 1939; Mead, 1935;
Arawak of Bahamas	Rouse, 1948;
Arawak of Colombia	Nicholas, 1901; Rouse, 1948;
Arawak of Cuba	Rouse, 1948;
Arawak of Jamaica	Rouse, 1948;
Aros	Partridge, 1905;
Australians General	Frazer, 1910; Howitt, 1904; Ratzel, 1896; Smyth, 1878; Sumner, 1911; Wood, 1868–70;
Badagas	Harkness, 1832; Murdock, 1934;
Bahima	Roscoe, 1907;
Bakonjo	Johnstone, 1902;
Bambuti (Mbuti)	Fabbro, 1978; Schebesta, 1941; Turnbull, 1965, 1966;
Bannock (Diggers)	Ross, 1824; Service, 1968;
Bara	Koch-Gruenberg, 1906, 1921;
Barea	Munzinger, 1864;
Batti	Ratzel, 1895;
Bodo	Hodgson, 1850;
Bushmen of Kalahari	Fabbro, 1978; Fourie, 1960; Lee, 1968; Marshall, 1965; Schapera, 1930; Thomas, 1969;
Cayapa	Murra, 1948;
Cayua	Coudreau, 1893; Fransen Herderschee, 1905; de Goeje, 1908; von Koenigswald, 1908; Schomburgk, 1847;
Central Eskimo	Boas, 1888;
Chichas Orejones	Markham, 1910;
Chickasaw	Adair, 1775;
Chiquitos	Markham, 1895, 1910;
Chocktaw	Adair, 1775;
Chunchos	Moss, 1909;
Chunipies	Markham, 1910;
Churumatas	Markham, 1910;
Cipos	Markham, 1910;
Conibos (Manoas)	Markham, 1910;

Copper Eskimo	Fabbro, 1978; Jenness, 1922, 1946; Palmer, 1965; Rasmussen, 1932;
Cree	Ross, 1855;
Cupeño	Bean & Smith, 1978;
Curetus	Markham, 1910;
Desana	Durham, 1976; Reichel-Dolmatoff, 1971;
Dhimal	Hodgson, 1850;
Dorobo (Okiek)	Blackburn, 1974, 1982;
Early Chinese	Douglas, 1899;
Early Esthonians	Weinberg, 1903;
Early Finns	Ailio, 1911; Appelgren-Kivalo, 1911; Koskinen, 1881; Yrjo-Koskinen, 1890;
Early India General	Maine, 1876;
Early Lapps	Bosi, 1960; Keane, 1886;
Early Slavs	Wilser, 1904;
Early Toltecs	Clavigero, 1787;
Ellice Islanders	Turner, 1884;
Eskimo General	Bancroft, 1875; Jenness, 1922; Rasmussen, 1932; Rink, 1887;
Esselen	Hester, 1978;
Fida	Farrer, 1880;
Flatheads and Neighbors	Bancroft, 1875; Dunn, 1844; McKenney & Hall, 1836; Scouler, 1848;
Fox Islanders	Coxe, 1787;
Garos	Godwin-Austen, 1873;
Greenland Eskimo	Nansen, 1893; Sumner, 1911;
Guana	de Azara, 1809;
Guarayos	Markham, 1910;
Guato	de Azara, 1809; Metraux, 1946;
Guayaki	Dobrizhoffer, 1822; Ehrenreich, 1898; de la Hitte & ten Kate, 1897; Vogt, 1902;
Hadza	Bagshawe, 1924; Kohl-Larsen, 1958; Woodburn, 1968;
Hierro	Cook, 1899;
Hopi	Eggan, 1943; Murdock, 1934;
Hudson Bay Eskimo	Turner, 1894;
Humboldt Bay Papuans	Krieger, 1899;
Humphrey Islanders	Turner, 1884;
Ifaluk	Bates, 1953; Bates & Abbott, 1958;
Inland Columbian Tribes	Bancroft, 1875;
Irulas	Murdock, 1934;
Kalapuya	Gatschet, 1900;
Karamojo	Johnstone, 1902;

Kavirondo	Northcote, 1907;
Khasis	Godwin-Austen, 1873;
Kotas	Harkness, 1832; Murdock, 1934;
Krepi	Ratzel, 1895;
Kubus of Sumatra	van der Bij, 1929; Forbes, 1884, 1885;
Kukatas	Schuermann, 1879;
Kurumbas	Murdock, 1934;
Ladaki	Ratzel, 1895;
Land Dyaks of Sarawak	Geddes, 1961;
Lepchas of Sikkim	Gorer, 1938; Hooker, 1854;
Makalakas	Davie, 1929;
Maku	Koch-Gruenberg, 1906, 1921;
Malays General	Ratzel, 1894; Skeat, 1902; Skeat & Blagden, 1906; Winstedt, 1950;
Manansas	Holub, 1882;
Manganja	Moggridge, 1902;
Manties	Orton, 1876;
Manua	Mead, 1938;
Manyaris	Markham, 1910;
Maya	Brinton, 1882;
Mishmis	Dalton, 1872;
Montagnais-Naskapi	Lips, 1947; Speck, 1933; Thwaites, 1896;
Moriori	Mair, 1905; Tregear, 1904;
Motilones (Lamistas)	Markham, 1910;
Moxos (Musus)	Markham, 1895, 1910;
Mrabri (Yumbri)	Bernatzik, 1938, 1941;
Nago Group	Ratzel, 1895;
Napo	Simson, 1883;
Navaho	Davis, 1857;
Ninquiguilas	de Azara, 1809;
North-East African Agricultural Tribes General	Munzinger, 1864;
Northern Ojibwa	Driver, 1961;
Ona	Bridges, 1893; Cooper, 1946;
Ovambo	Ratzel, 1895;
Pacaguaras	Markham, 1910;
Pageh Islanders	Frazer, 1910;
Papago	Browne, 1869;
Parexis	Markham, 1910;
Passes	Markham, 1895, 1910;
Paumari (Pammarys)	Markham, 1895, 1910;
Pima	Browne, 1869;
Pueblo Indians	Benedict, 1934; Davis, 1857; Driver, 1961;

Punans of Borneo	Furness, 1902; Hose, 1926;
Puri	von Koenigswald, 1908; zu Wied-Neuwied, 1820;
Quissama (Kisama)	Monteiro, 1876;
Rio Elvira Tribelets	Stegelmann, 1903;
Riukiu Archipelago	Letourneau, 1881;
Samoans	Kraemer, 1902; Mead, 1939;
Saulteaux	Hallowell, 1940; Kane, 1859;
Selangor Pahang Tribes	Knocker, 1907;
Semai (Semai Senoi)	Dentan, 1968, 1978; Fabbro, 1978;
Semang	Murdock, 1934; Schebesta, 1923, 1927;
Setebos	Markham, 1910;
Shoshone	Driver, 1961; Hoffmann, 1896; Steward & Voegelin, 1974;
Siberian Peoples General	Levin & Potapov, 1964;
Similkameen	Allison, 1892;
Singhalese of Ceylon	Meaden, 1866;
Siriono	Fabbro, 1978; Holmberg, 1948, 1966;
Soones	Fremont & Emory, 1849;
Suk	Johnstone, 1902;
Sungei Ujong Tribes	Knocker, 1907;
Tacullies	Bancroft, 1875;
Tamils of Ceylon	Meaden, 1866;
Tarahumara	Bennett & Zingg, 1935; Cassell, 1969; Montagu, 1974; West *et al.*, 1969;
Tasmanians	Montagu, 1974; Murdock, 1934; Roth, 1890, 1899; Thirkell, 1874; Turnbull, 1948;
Tenae	Dalton, 1872;
Tibetan Peoples General	Bell, 1928;
Tikopians	Firth, 1957, 1963; Otterbein, 1970;
Timorlaut	van der Bij, 1929; Forbes, 1884, 1885;
Toala of Celebes	Sarasin & Sarasin, 1905;
Todas	Farrer, 1880; Harkness, 1832; Murdock, 1934; Rivers, 1906;
Tokelau Islanders	Turner, 1884;
Tonga	Ratzel, 1895;
Ticunas (Jumanas)	Markham, 1910;
Tucuna	Nimuendaju, 1948;
Tukano	Reichel-Dolmatoff, 1971, 1975;
Tupi	Markham, 1910;
Uaupes	Markham, 1910; Orton, 1876;
Veddahs of Ceylon	Sarasin & Sarasin, 1893; Seligman, 1911;
Wafipa	Ratzel, 1895;

Yahgan	Bridges, 1893; Cooper, 1946;
Zuñi	Benedict, 1934;

Note that this (rather incomplete) list does not include the paragons of primitive peacefulness, the Tasaday of the Philippines, as these people were recently reported to be perpetrators (or victims?) of a clever hoax (*International Herald Tribune*, April 14, 1986, p. 4).

Because there is contradictory evidence on the alleged peacefulness of some of the above-mentioned ethnies, it is probably better to call them Highly Unwarlike Peoples, or some equivalent term. At least the material presented above casts some severe doubts on the dogma of universal belligerence.

A simple bipolar continuum, incorporating Wright's categories, and ranging from Highly Unwarlike, through Endemic, to Instrumental War, might be the following:

1. Highly Unwarlike, no cognates of war present (for example, some Inuit);
2. Highly Unwarlike, defence not developed (for example Andaman Islanders);
3. Highly Unwarlike, defensive warfare sophisticated (for example, Pueblo Indians);
4. Unwarlike, feuding present but infrequent;
5. Unwarlike, feuding frequent;
6. Warlike, low-scale random skirmishes with few casualties;
7. Warlike, low-scale raiding private initiative;
8. Warlike, middle-scale raiding tribal initiative;
9. Highly warlike, retaliatory warfare institutionalized;
10. Highly warlike, economic warfare institutionalized;
11. Highly warlike, political warfare institutionalized.

These tentative categories are roughly and globally correlated with social (Turney-High, 1949) and military (Andreski, 1954) organization (cf. also Malinowski, 1941; Fried, 1961, 1967; Feest, 1980; van der Dennen, 1989), and perception and evaluation of the enemy (Speier, 1941).

PART FOUR

The Conflict about Sociobiology

CHAPTER FOURTEEN

The sociobiology of conflict and the conflict about sociobiology: science and morals in the larger debate

U. Segerstråle

14.1 INTRODUCTION

The sociobiology debate has now gone on for over a decade, ever since the publication of Edward O. Wilson's book *Sociobiology: The New Synthesis* in 1975 and the violent reaction to it which followed. Time has come to take stock of the situation. What has really been going on during all these years? In the following I will argue that although the sociobiology controversy is often unproblematically seen as a typically politically motivated 'nature–nurture' controversy where conservative hereditarians clash with progressive environmentalists, the characterization of the whole controversy in this way is largely a smoke-screen, which is held up in the interests of many parties involved.

It is also a mistake to believe that the reason why this controversy is still going on is that we here have a traditionally unresolvable debate about things that are basically unknowable, as Thomas (1981) would have it. We are not basically dealing with a case of miscommunication because of incommensurable paradigms or worldviews, where the opponents talk through each other. On the contrary, as shown later in this chapter, the opponents actively seek to miscommunicate. At the same time, very few scientists have tried to mediate in this conflict.

In the following, I will attempt to demonstrate that much of the behaviour of the scientists involved in the sociobiology controversy can in fact be explained with the help of some insights from current sociology of science. I will further argue that the real issues under dispute in this controversy cannot be easily characterized as scientific or political in a direct sense. Rather, they lie largely at a meta-scientific level. This means that they basically deal with such things as the meaning and social function of good science and reliable knowledge, the moral role of the scientist etc. What is happening in the sociobiology controversy now is that these larger problems of contemporary science get expressed as scientific and moral errors of individual scientists.

In the following I will first deal with some examples of general misunderstandings and/or misconstruals in the sociobiology debate, particularly concerning issues relevant to the theme of social conflict. I will then attempt an analysis of the controversy as a cognitive clash between two different views of science and the role of the scientist. I will, however, proceed to demonstrate that a purely cognitive analysis is insufficient, and that for a proper understanding of this debate strategical factors have to be taken into account. Finally, I will discuss my own role as an analyst of this debate and the reactions of others to my analysis. My overall conclusions are based on my own empirical studies, especially on extensive interviews with participants, observers and 'arbiters' in this conflict (Segerstråle, 1983).

14.2 THE POLITICIZATION OF THE DISCUSSION ABOUT SOCIOBIOLOGY

By now, 'sociobiology' has unfortunately become a very loaded word among academics and laymen alike. Because of Wilson's 1975 book and the attack on it, it is often ignored that sociobiology (or behavioural ecology, as the English will have it) is really a new discipline attempting to explain the evolutionary basis of animal social behaviour with the help of new insights from evolutionary biology. Instead many have come to believe that the primary aim of sociobiology is the biological explanation of human social behaviour both in the United States and in Europe. Because of Wilson's special interpretation of the sociobiological project, many do not know that other 'sociobiologists' such as, for example, Richard Dawkins, author of *The Selfish Gene* (1976), are quite happy to leave humans out and grant them special status. (Incidentally, Wilson's take-over of the field was a source of considerable irritation among his British colleagues.) Neither is it well-known that Wilson does not even represent the mainstream of current thinking within this new field in regard to animal behaviour (for example, he is one of the few current biologists still believing in the reality of group selection).

Instead, the field of sociobiology in its entirety has been connected to various unsavoury earlier attempts to justify social inequalities, racial supremacy etc. It has even been equated with Nazi science. In this conjunction, it has been assumed that the very wish to investigate into the biological basis of human nature or human social behaviour, or into genetic differences between human groups, *cannot* be scientifically motivated but *must* have political reasons of the most sinister kind.

In an earlier article (Segerstråle, 1986) I have examined the initial reaction to Wilson's book by the Sociobiology Study Group, which included his colleagues Richard Lewontin and Stephen J. Gould. In that article I have concentrated especially on the protagonists in this conflict – Wilson and Lewontin – and tried to show how their long-range moral-cum-scientific agendas were, in fact, on countercourse already before the start of the sociobiology controversy. There I follow the development of the sociobiological and critical programmes through the publication of *Genes, Mind, and Culture* (Lumsden and Wilson, 1981), and demonstrate how that book invited another head-on collision between Wilson and Lewontin. In the present chapter I extend the discussion of the sociobiology conflict to the opposition within the larger camps that were formed within the academic community as a result of this controversy. (The basic lines, however, were drawn already in conjunction with the earlier IQ controversy, from which the sociobiology one inherited much of its criticism and many of its critics.)

An important point is that because of the very strong initial political charge of the issues involved, a real scientific debate about the merits and dismerits of sociobiology as a new scientific discipline was prevented from taking place. At the very outset of the sociobiology controversy, scientific issues got tangled in with moral/political ones. One result of this was that some scientific reviews critical of Wilson's book were simply withdrawn because the reviewers did not want their own negative views to be associated with those of Wilson's original critics, the Sociobiology Study Group (Ernst Mayr, interview).

The initial politicization of the controversy irritated and puzzled those biologists who were looking forward to a good scientific battle about Wilson's claims. One among these was Wilson's Harvard colleague Ernst Mayr. According to him, the critics had a lot of 'good' scientific criticism, and there was no need for additional political criticism. Actually, the political criticism was simply counterproductive (*ibid.*).

In this conjunction, it is interesting to note that not all left-wing activists in the Boston area saw it necessary or even useful to get involved in the sociobiology controversy. For example, Salvador Luria, molecular biologist at M.I.T., wondered why the critics were wasting so much time and energy on a detailed criticism of Wilson. According to him, sociobiology was not even a good political issue, in the sense that it could be used for political mobilization of people in society at large (Luria, interview). Others, like George Wald, a Harvard biologist, thought that there existed many much more important things for academics to deal with, such as the nuclear threat, and that sociobiology was a waste of time (Wald, interview). Finally, Noam Chomsky, the M.I.T. linguist known for his radical views, could not be persuaded to join the anti-Wilson campaign, largely because he could not see anything wrong in Wilson's search for a general human nature (Chomsky, interview). However, as I will show below, for various reasons the political criticism was really quite crucial to the critics.

But a perhaps even more important result of the initial polarization and politicization of the issues is that any discussion about any messages that the new discipline, sociobiology, might have for humans and human society, was made absolutely impossible. In the zeal to expose the negative social implications of biological research on human behaviour, some of the potentially positive uses of recent biological knowledge in social argumentation got lost. Now, arguments from biology could have been especially relevant for the discussion about social conflict, both concerning such things as the biological basis of aggression, or biological justifications for social inequalities, racism, sexism etc. Let me here review some arguments, both those that were made and those that were not.

14.3 THE SOCIAL RELEVANCE OF THE SOCIOBIOLOGY OF CONFLICT AND COOPERATION

The critics of sociobiology have consistently been emphasizing the interest of sociobiologists to treat aggression as a genetically determined trait and have pointed to the immorality and unscientificness of such a statement. But in doing this they have played down the message that for many human sociobiologists is an equally, or more, important one: that there may be a genetic basis for altruism in humans. That this is a paramount concern of especially Wilson's can be seen quite clearly already from his claim in his 1975 book that altruism is the central problem for sociobiology, and the title of a more popular article: 'Human Decency is Animal' (Wilson, 1975b). Also, his above mentioned interest in group selection models (where animals pay attention to the benefit of the group instead of their own individual interests), and his great fascination with cooperative behaviour in social insects, are testimony to this. As I have argued elswhere (Segerstråle, 1986), the moral aim of Wilson's scientific-cum-moral agenda, is to put ethics and religion on a solid materialist basis. With the help of biology (whose present capabilities he overrates), he wants mankind to be able to rationally plan its own future as against divine prophecy, history etc.

If we now turn to the issue of aggression, the most central point of the sociobiological message has been missed in the general discussion. The novelty of sociobiology in respect to aggressive behaviour is that, unlike ethology, sociobiology treats aggression (like altruism) as a genetic trait, not as an unchangeable instinct. According to modern interactionist theory, any genetic predisposition (aggression, altruism) is expressed differently depending on the environment (including the social one). Ethology, on the other hand, at least the Konrad Lorenz school, sees aggression as innate in man and therefore unavoidable. Thus the Lorenz school believes in biologically determined instincts, while sociobiology is a much more environmentalist creed. In fact, its social message (if we insist on such a thing), could be that if we only adjust the environment accordingly, we would be able to avoid aggressive behaviour altogether.

Of course, one of the obvious criticisms as regards the very idea that aggression could be treated as a genetic trait is the same as regards all other claims about animal traits: these are all hypothetical. No one has yet seen an aggressive (or altruistic) gene. Furthermore, to talk about genes 'for' traits is highly misleading biological jargon. (As biologists point out when they are not using this shorthand, there are no genes for traits; there are genes that make the difference between one phenotypical expression and another, for example, blue eyes or brown eyes.) The most basic objection voiced by some critics is that aggression is an entirely human construct and that it is therefore meaningless to operationalize aggressive behaviour (for example, Lewontin, 1976). But the same is true of any other traits: they are all human constructs. Still, a counterargument to this is that breeding experiments based on such postulations of an underlying genetic basis for phenotypical traits, including quite complex behavioural ones, have succeeded. This would indicate that there is no inherent reason why aggression (however operationalized) could not also be treated this way (see for example, Maynard Smith, 1976, 1986). As regards aggressive behaviour in humans, the greatest objection is of course the relationship between individual human aggressive acts and organized, rational group aggression, such as warfare. The critics of Wilson's sociobiology have claimed that he believes in simple reduction of group aggression to individual aggression, thus making him into an arch-reductionist (Allen *et al.*, 1975; Lewontin, 1976). The situation is more complicated than that. In *On Human Nature* (1978), Wilson's second book entirely devoted to human sociobiology, he deals with exactly the issue of human warfare. Here it is clear that the biological model he follows as regards human evolution is based on a view of an interaction between two selection processes: group selection on the one hand, and individual selection on the other. According to Wilson, we can explain our readiness to group level solidarity, expressed in such phenomena as religious ceremony, willingness to sacrifice oneself in war, etc., through mankind's long developmental process. This type of altruistic behaviour was important in tribal warfare, while in times of peace normal selfish behaviour was prevalent. Interestingly, in his review of *sociobiology*, John Maynard Smith, who is respected by Wilson's critics, pointed out that he basically shared Wilson's above mentioned views on human evolution (Maynard Smith, 1976). This view, then, appears to be at the very least a biologically feasible one.

Thus, according to Wilson, we have a genetic basis both for making an in-group/out-group distinction and for solidarity with the in-group. The circumstances will determine which one is salient at any particular time. But, as I have shown elsewhere (Segerstråle, 1983, 1986), Wilson has additional reasons for his interest in human willingness to sacrifice for a higher cause, and in human responsiveness to propaganda in general.

When he states that 'Human beings are incredibly easy to indoctrinate. They seek it' – one of the most attacked expressions in his infamous last chapter – what he is really trying to do, at least in part, is trying to explain to himself his own earlier deep religious involvement as a Southern Baptist (cf. Segerstråle, 1986).

Wilson's message is often really quite radical. In *On Human Nature* (1978) he explains among others how the crusaders lent themselves to manipulation by power interests, and how the military and the Church throughout history have been systematically manipulating the human capability for collective response and sacrifice. It is the critics' unwillingness to grant Wilson any other than political motives that make them ignore the many radically liberal attempts included in the latter's sociobiological programme.

In the following, I will now try to explain the larger debate about sociobiology first as a cognitive misunderstanding between two main camps within the academic community, involving both epistemological and moral/political factors. Then I will proceed to demonstrate that a cognitive view is in fact insufficient and that strategical factors inevitably come into play in controversies of this kind, paradoxically resulting in a short-circuit of some of the proclaimed aims of the critics of sociobiology.

14.4 COMMUNICATIVE NATURALISTS AND CRITICAL UNIVERSALISTS, PLANTERS AND WEEDERS IN THE GARDEN OF SCIENCE

A cognitive explanation of the conflict would now be the following. Different fields in science have their own standards for tolerable plausibility arguments, reasonable assumptions etc. Those who have been socialized to different fields are also trained in how to read a scientific text in the 'right spirit' (cf. Bazerman, 1981). Now, in a situation where scientists from different fields enter the same scientific discourse – which is what happened in the IQ and sociobiology controversies – these scientists will have widely different conceptions of 'good science' and 'normal' scientific belief, and will want an explanation for any deviance.

In both the IQ and the sociobiology controversy in their American setting one can perceive a polarization of the academic community into two main camps, across disciplines. Typical of the first camp is a prediction-oriented, 'operational' or model-building approach, combined with a positivist belief in the inherent morality of science and a 'communicative interest' in reporting scientific progress to the laymen. At the same time, this coincides with the traditional approach of naturalistically oriented evolutionary biologists. The typical features of the other camp is a cognitive orientation to a search for an 'underlying reality', a belief in scientific experiments as representing the best science, a critical attitude to scientific epistemology and, particularly, a critical attitude to statements of members of the first camp, which are searched for hidden agendas. At the same time, this is typical of a newer breed of reductionistically trained biologists (and other scientists). We thus have an opposition between 'communicative naturalists' and 'critical universalists' within the biological community itself. In the larger current debate, one could label these camps 'planters' and 'weeders'. These two general epistemological orientations now seem to be connected to two different scientific and moral/political socialization processes: traditional 'positivist' scientists, believing in the democratic process, and the 1960s radical generation who believe in manipulation by power holders. The idea for the latter is to prevent the power holders from abusing science for political purposes. Thus, we have the planters who want to produce useful knowledge, and the weeders who want to weed out bad knowledge from the very beginning, by attacking bad scientists. (As to the characteristics of the political approach of the 1960s generation in the US, see Bouchier, 1977 and Jacoby, 1973.)

But there is an additional cognitive process operating in both cases: an unproblematic coupling of scientific and moral interests. For the first community of scientists, scientific knowledge is objective and thus unconnected to its social origin or social use. For the second community, bad science is ideologically motivated and therefore lends itself to social abuse, while good science is/does not. The logic here is that if dangerous consequences can be inferred from a theory, then it *must* be ideologically motivated as well as scientifically bad or oldfashioned (for a closer analysis, see my 1986 article).

This is why in the sociobiology controversy, even such arguments that Wilson (and other biologists) found scientifically reasonable, or scientific beliefs that Wilson shares with many members of the academic community, had to be both scientifically wrong and ideologically traceable. (Such thinking can be found especially in the Sociobiology Study Group's first letter to *The New York Review of Books*, in many of Lewontin's writings, and in interviews with members of the group. For a closer analysis, see my 1986 article.) In addition, of course, Wilson's personal moral reasons for emphasizing the power of sociobiology as against religious dogma were taken to express political goals of his. We can now start understanding why the sociobiologists' scientific work was subjected to a severe moral reading by the critics. The plausibility arguments presented by sociobiologists as a matter of course could not from a critical hard data point of view be intended as serious science and were therefore instead analysed for social implications, which now became equated with the personal moral error of particular scientists. In turn, the idea of attributing extrascientific reasons of different kinds to those scientists with whose views one disagrees in science appears to be quite common in scientific reasoning, if we are to believe the sociologists Mulkay and Gilbert (1982). They argue that scientists routinely regard one another's divergent views on the same issue as errors. I myself have seen the word 'error' used rather often in biological writings, quite outside controversies. Thus, one can rather far-reachingly try to analyse recent nature–nurture controversy as due to a profound clash in the cognitive realm, in respect to both scientific and moral concerns of the participants.

14.5 OPTIMIZATION STRATEGIES AMONG SOCIOBIOLOGISTS AND THEIR CRITICS

But something more than a cognitive misunderstanding is involved. In the sociobiology and IQ controversies, these basic cognitive differences are now exploited for strategical reasons. According to the French sociologist Pierre Bourdieu (1975), the scientists' routine quest for peer recognition should be regarded as nothing but a competitive struggle for symbolic capital. What I believe is taking place in these controversies is that this struggle has now been expanded to include the moral realm as well. I would thus like to claim that an important motive for the participants in controversies dealing with 'genes vs. environment' is the acquisition of moral capital, in addition to any ordinary scientific capital that may be gained. When it comes to moral capital, the critics have a certain advantage, because of the current socio-historical feasibility in the US to obtain moral recognition for unmasking racist or biological determinist scientists. In fact, the critics often undertake a moral reading of the target of their criticism, the purpose of which is to demonstrate how scientific error leads to conclusions with maximally undesirable social implications. This way maximal moral recognition can accrue to the revealer of this fact. In this endeavour, the critics can be shown to pursue optimization strategies, by which both scientific and moral recognition can simultaneously be obtained from scientific criticism of carefully selected targets. One of the major strategies for criticism used by the opponents of sociobiology is the radicalization of plausibility arguments into strong, easily dismissible claims. (This does not prevent the critics themselves from employing plausibility arguments when it suits them.)

Interestingly, the critics of sociobiology stick to the negative implications of sociobiological knowledge. They do not, like Kropotkin, bring up the positive message of altruism as a possible evolutionary trait in humans. Neither do they make too much of the profoundly anti-racist message of new developments in biological thinking.

Sociobiologists and their critics agree that it is biologically incorrect today to think in terms of 'average types', because modern biology shows that there is a tremendous variation among individuals within any population group (race). Old 'typological' thinking has now been supplanted by modern 'populational' thinking. Most significantly, the critics do not discuss in a more abstract way (as one would expect of analytical Marxists or Critical Theorists), the different ways in which ideology may (and does) creep into scientific discourse, the unavoidable anthropomorphism of biological language etc. They simply do not seem to be able to disentangle themselves from their moral/political condemnation of individuals.

But the defenders of sociobiology and IQ research can reap moral benefits too. Their audience is now those who have a traditional view of science. The defenders, in fact, have a rather easy match: they can simply uphold a traditional image of unproblematic objective science against the threat by left-wing ideologues. Bernard Davis, the most vehement spokesman for Wilson, academic freedom and objectivity etc., has even invented a word: 'neo-Lysenkoist', to describe the critics' approach, which he believes is exclusively politically motivated (Davis, 1985; for a critical analysis, see Segerstråle, 1987). In this way the traditionalist scientists can also dismiss serious scientific criticism by the critics as ideologically motivated, and sweep clear errors under the rug.

This is a pity, because part of the critics' attack has been focused on real errors and sloppy research among sociobiologists and IQ-researchers. The problem here, again, has been that the critics persist in attacking individuals for moral errors, instead of addressing larger problems, such as the consequences of the 'publish-or-perish' syndrome in contemporary science.

14.6 SOCIAL AND SCIENTIFIC CONSEQUENCES OF THE SOCIOBIOLOGY CONTROVERSY

The continuous possibility of accumulating new moral capital can now explain why it is in both parties' interest to keep the controversies going. The limiting factor for the critics is the supply of new 'biological determinists', who can be subjected to their critical routine (who, indeed, will be the next candidate after Wilson?) The potential targets, again, will be able to draw support from traditional elements within the scientific community and the larger society. Because of the public nature of these recent gene–environment debates and the conduct of much of the controversy in popular books and book reviews, interestingly the general public too in this way gets involved in granting moral recognition to different co-existing scientific claims.

At the most abstract level, however, I believe that recent controversies represent a larger process that is going on in an invisible way. What is taking place in these debates is an indirect negotiation about what is in fact to be counted these days as 'good science'. This involves both the scientific and the lay community. It is ironical that while the critics of sociobiology and IQ research are in principle concerned with such problems as epistemological and ethical problems of science, as well as inherent problems of today's scientific practice, in practice their strong socialization to the competitive struggle and to profit maximization in the realm of symbolic capital induces them to attack individual scientists, thus morally short-circuiting their meta-scientific ambitions.

The direct meta-scientific debate that would today be needed in science would among others deal with the degree of 'product control' necessary as regards scientific knowledge products in general, and especially potentially dangerous knowledge. The basic question here would be just how well-established a scientific claim would have to be in order to be passed out to laymen or politicians as scientific knowledge. (Science journalists would of course have a great responsibility too in this respect.) Another problem is the current reward system of science, which favours quick and dirty research. A third is the above mentioned publish-or-perish syndrome which easily leads to sloppy research, sometimes even fraud, and insufficient peer criticism (no one has time to check anybody's primary sources). In other contexts, the critics of sociobiology have voiced real concern about these things. However, in conjunction with the sociobiology debate, these aims have been obscured because of the collapse of the general argument into attacks on individuals. The American left-wing radicals, even those who claim to be Marxists (see for example, Levins and Lewontin, 1985), have not, despite all, been able to disassociate themselves from the individualistic/moralistic programming typical of their society (see among other Bouchier, 1977; Jacoby, 1973).

But when one closely examines the arguments in these debates, one sees that we are really dealing with plausibility statements. Different camps are arguing about the reasonability of their own claims and the impossibility of the opponent's claims. Because of the unclarity of results of many studies, especially those dealing with humans, the controversy many times boils down to what are really scientific and social *beliefs* at a particular time. However, in scientific controversy these plausibility statements are formulated as strong claims based on solid evidence.

What are the implications of all this for the general public? From the writings of the critics of sociobiology and IQ-research one can see that one of their self-imposed missions is to enlighten the innocent layman. My guess is that the innocent layman remains rather puzzled. This is due to the fact that because of moral capital gain, the critics of biological determinist scientific theories have tended to keep the discussion within the realm of their own scientific expertise, often turning it into quite technical matters, to which they have usually added a moral/political dimension. What the critics have *not* said, for example, even though they might have, is that social and ethical decision making ought to be independent of any particular scientific findings about the biological differences between individuals, and based on totally other criteria.

By making plausibility claims sound like hard data, and accusing each other of ideological bias in their scientific endeavours, both sides in these controversies have probably only reinforced the laymen in their view that (good) scientific knowledge is objective and absolute, experts are to be trusted, and that the only problem is to select between good and bad experts.

Thus, there is a reinforcement of the prevailing view of scientific knowledge as reliable. The view among the innocent laymen then easily becomes that there is no basic problem with science as such, it is only that it needs weeding out of bad science. The public can rest assured, however, because there are some very alert gardeners around taking care of the weeding. In a way, one could say that the targets of moral outrage, like Jensen and Wilson, are sacrifices of the scientific priesthood in order to guarantee to the lay community the continuous reliability of scientific expertise.

In this way, while they are indirectly negotiating the nature of science, both sides in these recent controversies are really reinforcing the popular view that (good) science provides firm knowledge of vital importance to social and ethical decisions. It is not in the interest of either side to play down the role of science in society. Thus, the opposing camps in recent controversy are in fact engaged in a strange symbiosis: while they are indirectly negotiating the game rules of science, they are together meanwhile upholding the social authority of science!

14.7 THE SOCIOLOGIST IN THE SOCIOBIOLOGY CONTROVERSY

What is then my own role as analyst of scientific controversy? One could say that in a sense I have appointed myself detective: I have tried to find out what really goes on in the recent American debate. I call this approach to the study of scientific debate *in vivo* analysis, since I am interviewing living persons and their colleagues, not sticking to historical case studies (cf. Segerstråle, 1986). There is also an element of danger involved, because part of my *in vivo* approach is to serve back my results to the people I have interviewed and their colleagues, and they may or may not like them. I believe, namely, that one should, if possible, try to validate one's analysis in different ways. This is, one should not only rely on the judgement of one's own colleagues in the general field of the social study of science, but also try to see whether one's objects of study recognize the result as true in some sense. Otherwise, it may just remain a 'just-so story', however convincing in the abstract.

Biologists like to talk, and I have been swamped with a lot of useful gossip. Of course, everyone told me that they knew the real truth of the story. I am not claiming that I was checking alibis, but as a sociologist I reserved the right to sort out people and explain them from a certain distance. This, again, I regard as another duty for the sociologist of science: not to leave the study at the level of one's study objects' description of the situation (which may, sometimes, amount to mutually contradictory accounts). Expectedly, the analysis elated some and disappointed others.

For instance, some Harvard colleagues of mine who initially followed my research with great interest, later told me that I have 'betrayed the cause'. I now understand that they evidently expected me to further document the criminality of Wilson's sociobiology, instead of trying to find out new things about the controversy. As an answer to such critics, I want to add a personal note. I myself started out very critical of sociobiology, and it took considerable emotional work for me as a sociologist to bring myself even to read sociobiological texts. I had to work hard to be able merely to understand the biological or sociobiological approach to the world. Has this made me into a convert to sociobiology? Not at all, I still agree with much of the philosophical criticism of sociobiology, I think a non-reductionist biology is a thrilling prospect, and I believe that, unfortunately, human sociobiology may easily be politically abused – as may many other theories. But I refuse to condone in moral guilt-tripping of individuals, which I think is a disgrace to academic discourse.

What is then the status of my explanation? Am I telling you yet another 'just-so' story? The answer depends on who one regards as the correct judge of this matter. Should one ask biologists in general what they think? In that case, I have indeed got a lot of reprint requests and enthusiastic letters from biologists and biologically oriented anthropologists. Very few biologists have told me I am wrong. On the contrary, several 'grand old men' having read my 1986 article (having a much narrower scope than this chapter) have told me that they think my analysis is convincing. Or should one ask philosophers, historians or sociologists for their views? Also here, I have had almost only positive comments. (The most encouraging one is from a sociologist who told me that I had succeeded in making him feel some sympathy for Wilson, something that he would never have believed possible.) Or is perhaps the correct standard for assessment of the truth of my conclusions the protagonists themselves?

In that case we get into a more tricky situation. I refer to an article by Colin Campbell in *The New York Times* in November 1986. Here he interviewed Wilson and Lewontin about their reactions to my 1986 article. I will quote from there. Wilson says about me that I have 'bent over backwards to be fair to everybody'. Lewontin, after first having refused to be interviewed, explained that he had not read the article. He also said that he had 'lost or thrown away' my thesis. His bottom line was that he does not 'want to worry about what some fairly dumb person says or doesn't'. So where does that leave my analysis?

My agents at Harvard and among the biological community recently told me that Lewontin had been very bothered by the fact that Maynard Smith apparently does not regard his scientific efforts as inspired by Marxism. (That this is indeed the latter's view is evident both from a quotation in my article and in Maynard Smith's own review of Levins and Lewontin's *The Dialectical Biologist* (1985), in *The London Review of Books* (February, 1986). Meanwhile, the debate continues, which means that we analysts are constantly provided with more avenues to increase our own scientific and moral capital, as long as there are people who will find our accounts convincing . . .

Bibliography

Abrams, S. and Neubauer, P.B. (1976) Object orientedness: The person or the thing. *Psychoanalytic Quart.*, **45**(1), 73–99.

Agassi, J. (1977) *Towards a Rational Philosophical Anthropology.* Martinus Nijhoff, The Hague.

Alcock, N.Z. (1972) *The War Disease.* CPRI Press, Ontario.

Alexander, R.D. (1971) The search for an evolutionary philosophy of man. *Proc. Roy. Soc. Victoria*, **84**(1), 99–120.

Alexander, R.D. (1974) The evolution of social behavior. *Ann. Rev. Ecol. Systematics*, **5**, 325–83.

Alexander, R.D. (1979) *Darwinism and Human Affairs.* University Washington Press, Seattle.

Alexander, R.D. and Tinkle, D.W. (1968) A comparative review. *Bioscience*, **18**, 245–8.

Alexander, R.D. and Tinkle, D.W. (eds) (1981) *Natural Selection and Social Behavior.* University of Michigan Press, Ann Arbor.

Allen *et al.* (1975) Letter. *The New York Review of Books*, November 13, 182, 184–6.

Altman, I. (1970) Territorial behavior in humans: An analysis of the concept, in *Spatial Behaviour of Older People*, (eds L.A. Pastalan and D.A. Carson), University of Michigan Press, Anne Arbor.

Altman, I. and Haythorn, W. (1967) The ecology of isolated groups. *Behavioral Science*, **12**, 169–82.

Altum, J.B.T. (1868) *Der Vogel und sein Leben.* Niemann, Münster.

Ammon, O. (1893) *Die natürliche Auslese beim Menschen.* G. Fischer, Jena.

Ammon, O. (1900) *Die Gesellschaftsordnung und ihre natürlichen Grundlagen.* G. Fischer, Jena.

Amnesty International (1973) *Report on Torture*, Amnesty International, London.

Andreski, S. (1954) *Military Organization and Society.* Routledge and Kegan Paul, London.

Andreski, S. (1964) Origins of war, in *The Natural History of Aggression*, (eds J.D. Carthy and F.J. Ebling), Academic Press, New York, 129–36.

Andreski, S. (1971) Evolution and war. *Science J.*, **7**(1), 89–92.

Angell, R.C. (1965) The sociology of human conflict, in *The Nature of Human Conflict*, (ed. E.B. McNeil), Prentice Hall, Englewood Cliffs, 91–115.

Anthony, R. (1917) *La Force et le Droit*, Alcan, Paris.

Apter, M.J. and Smith, K.C.P. (1976) Negativism in Adolescence, *The Counsellor*, **23/24**, 25–30.

Apter, M.J. and Smith, K.C.P. (1985) Experiencing personal relationships, in *Reversal Theory: Applications and Developments*, (eds M.J. Apter, D. Fontana and S.

Murgatroyd), University College Cardiff Press, Cardiff, UK, 161–78.
Apter, M.J., Fontana, D. and Murgatroyd, S. (eds) *Reversal Theory: Applications and Developments*, University College Cardiff Press, Cardiff, UK.
Archer, J.E. (1988) *The Behavioural Biology of Aggression*, Cambridge University Press, Cambridge.
Ardrey, R. (1961) *African Genesis*, Atheneum, New York.
Ardrey, R. (1966) *The Territorial Imperative*, Atheneum, New York.
Ardrey, R. (1976) *The Hunting Hypothesis*, Atheneum, New York.
Argyle, M. (1976a) *Bodily Communication*, London, Methuen.
Argyle, M. (1976b) Personality and social behaviour, in *Personality*, (ed. R. Harre), Oxford, Blackwell, 145–88.
Arndt, N. (n.d.) Kriegführung, in Staatliche Museen preusischer Kulturbesitz Berlin, Museum für Völkerkunde, Abt. Südsee, Nr. 079.
Ausubel, D.P. (1952) *Ego Maturation and the Personality Disorders*, Grune & Stratton, New York.
Axelrod, R. (1984) *The Evolution of Cooperation*, Basic Books, New York.
Axelrod, R. and Hamilton, W.D. (1981) The evolution of cooperation. *Science*, **211**, 1390–96.
Baer, D. and McEachron, D.L. (1982) A review of selected sociobiological principles: application to hominid evolution: the development of group social structure. *J. Soc. Biol. Struc.*, **5**(1), 69–90.
Baerends, G.P. (1979) The functional organization of behaviours. *Animal Behav.* **27**, 726–38.
Bagehot, W. (1873) *Physics and Politics: Thoughts on the application of the principles of natural selection and inheritance to political society.* Appleton, New York, (2nd edn 1884).
Bakke, E.W. (1965) Concept of the social organisation, in *Modern Organization Theory* (ed. M.Haire), Wiley, New York.
Bales, R.F. (1953) *Interaction Process Analysis*, Addison-Wesley, Reading, Mass.
Balikci, A. (1967) Female infanticide on the arctic coast, *Man*, **2**, 615–25.
Balikci, A. (1968) The Netsilik Eskimos: adaptive processes, in *Man the Hunter* (eds R.B.Lee and I.DeVore) Aldine Atherton, Chicago, 78–82.
Balikci, A. (1970) *The Netsilik Eskimo*. The Natural History Press, Garden City, N.Y.
Balkind, J. (1978) A critique of military sociology: lessons from Vietnam. *Jour. Strategic Studies*, **1**, 235–54.
Barash, D.P. (1977) *Sociobiology and Behavior*, Elsevier, New York.
Barash, D.P. (1980) *Soziobiologie und Verhalten*, Paul Parey, Hamburg.
Barash, D.P. (1980) *Sociobiology: The whisperings within.* Souvenir, London.
Barash, D.P. (1982) *Sociobiology and Behaviour*, 2nd edn, Elsevier, Amsterdam.
Barash, D.P. and Lipton, J.E. (1985) *The Caveman and the Bomb; Human nature, evolution, and nuclear war*, McGraw-Hill, New York.
Bastock, M., Morris, D. and Moynihan, M. (1953) Some comments on conflict and thwarting in animals. *Behaviour*, **6**, 66–84.
Bates, D.G. and Lees, S.H. (1979) The myth of population regulation, in *Evolutionary Biology and Human Social Behaviour: An Anthropological Perspective* (eds N.A. Chagnon and W. Irons) Duxbury Press, Massachusetts, 273–89.
Bateson, G. (1983) *Ökologie des Geistes*, Suhrkamp, Frankfurt a.M.
Bateson, P.P.G. (1966) The characteristics and context of imprinting. *Biol. Rev.*, **41**, 177–220.
Bateson, P.P.G. (1979) How do sensitive periods arise and what are they for? *Animal Behav.*, **27**, 470–481.

Bibliography

Bateson, P.P.G. (1982) Behavioral development and evolutionary processes, in *Current Problems in Sociobiology*, (ed. King's College Sociobiology Group) Cambridge University Press, Cambridge.

Bateson, P.P.G. (1983a) Genes, environment and the development of behaviour, *Animal Behaviour*, Vol. 3. *Genes, Development and Learning* (eds T.R. Halliday and P.J.B. Slater), Blackwell, Oxford, 52–81.

Bateson, P.P.G. (1983b) Rules of changing the rules, in *Evolution from Molecules to Men* (ed. D.S. Bendall) Cambridge University Press, Cambridge.

Bauer, H.R. (1980) Chimpanzee society and social dominance in evolutionary perspective, in *Dominance Relations: An Ethological View of Human Conflict and Social Interaction*, (eds D.R. Omark, F.F. Strayer, and D.G. Freedman) Garland Press, New York, 49–119.

Bayle, F. (1865) *Adventures among the Dyaks of Borneo*, Murray, London.

Bazerman, C. (1981) What written knowledge does. *Phil. Soc. Sci.*, **11**, 361–387.

Beals, A.R. and Siegel, B.J. (1966) *Divisiveness and Social Conflict: An anthropological approach*, Stanford University Press, Stanford.

Beer, F.A. (1981) *Peace against War: The ecology of international violence*, Freeman, San Francisco.

van Bemmelen, J.F. (1928) *Kriegsdrang ein sexueller Instinkt*. Verh. 1. Internat. Kongr. f. Sexualforschung.

Benedict, R. (1934) *Patterns of Culture*, Mentor Books, New York.

Benjamin, L. Smith- (1974) Structural analysis of social behavior, *Psych. Rev.*, **81**, 392–425.

Bercovitch, F.B. (1986) Male rank and reproductive activity in savanna baboons. *International Journal of Primatology*, **7**, 533–50.

Berger, P. and Luckmann, T.H. (1966) *The Social Construction of Reality*, Doubleday, New York.

Berkowitz, L. (1962) *Aggression: A Social Psychological Analysis*, McGraw-Hill, New York.

Berkowitz, L. (1981) The concept of aggression, in *Multidisciplinary Approaches to Aggression Research* (eds P. Brain and D. Benton), Elsevier/North Holland, Amsterdam, 3–15.

Bernard, J. (1950) Where is the modern sociology of conflict? *American Journal of Sociology*, **56**, 11–16.

Bernard, J. (1951) The conceptualization of intergroup relations with special reference to conflict. *Soc. Forces*, **29**, 243–51.

Bernard, J. (1957) Parties and issues in conflict. *Journal of Conflict Resolution*, **1**, 111–21.

Bernard, J. (1957) The sociological study of conflict, in *The Nature of Conflict*, (eds J. Bernard, T. Pear, R. Aron *et al.*) UNESCO, Belgium.

Bernard, L.L. (1924) *Instinct: A study in social psychology*, Henry Holt, New York.

Bernard, L.L. (1944) *War and its Causes*, Henry Holt, New York.

Berndt, R.M. (1964) Warfare in the New Guinea Highlands. *Am. Anthropol.*, **66**(4ii), 183–203.

Bernstein, I.S. (1976) Dominance, aggression and reproduction in primate societies. *J. of Theor. Biol.*, **60**, 459–72.

Bernstein, I.S. and Gordon, T.P. (1980) The social component of dominance relationships in rhesus monkeys (*Macaca mulatta*). *Animal Behav.*, **28**, 1033–39.

Bernstein, I.S., Gordon, T.P. and Rose, R.M. (1974) Factors influencing the expression of aggression during introductions to rhesus monkey groups, in R.L. Holloway (ed.) *Primate Aggression, Territoriality, and Xenophobia*, Academic Press, New York, 221–40.

Bertalanffy, L. Von (1968) *General System Theory*, Braziller, New York.

Bertram, B.C.R. (1975) The social system of lions, *Sci. Amer.*, **232**(5), 54–65.

Bertram, B.C.R. (1976) Kin selection in lions and evolution, in *Growing Points in Ethology*, (eds P.P.G.Bateson and R.A.Hinde) Cambridge University Press, Cambridge, 281–301.

Bertrand, M. (1969) *Behavioural Repertoire of the Stumptail Macaque, Macaca speciosa*, S. Karger AG, Basel.

Betzig, L.L. (1986) *Despotism and Differential Reproduction: A Darwinian View of History*, Aldine, New York.

Bigelow, R. (1969) *The Dawn Warriors; Man's evolution towards peace*, Little Brown and Co., Boston.

Bigelow, R. (1972) The evolution of cooperation, aggression and self-control, *Neb. Symp. Mot.*, **20**, 1–57.

Bigelow, R. (1975) The role of competition and cooperation in human evolution, in *War, Its Causes and Correlates*, (eds M.A. Nettleship, R.D. Givens and A. Nettleship) Elsevier, Amsterdam, 235–67.

van der Bij, T.S. (1929) *Ontstaan en eerste ontwikkeling van den oorlog*, Wolters, Groningen.

Birdsell, J.B. (1972) *Human Evolution: An Introduction to the New Physical Anthropology*, Houghton Mifflin Co., Boston.

Birket-Smith, K. (1928) Five hundred eskimo words: a comparative vocabulary from Greenland and central Eskimo dialects. *Report of the Fifth Thule Expedition*, 1921–1924, **3** (3).

Birket-Smith, K. (1929) The Caribou Eskimos. *Report of the Fifth Thule Expedition*, 1921–1924.

Birket-Smith, K. (1959) *The Eskimos*. Methuen, London.

Blackey, R. (1976) *Modern Revolution and Revolutionists*, Olio Books, Santa Barbara, Ca.

Boas, F. (1888) The central Eskimo. *Bur. Amer. Ethnol.*, **6** (1884–1885), 399–699.

Boas, F. (1912) An anthropologist's view of war, in *International Conciliation*, March, New York, 5–14.

Boesch, C. and Boesch, H. (1989) Hunting behavior of wild chimpanzees in the Tai National Park. *Amer. J. Phys. Anthropol.*, **78**, 547–73.

Boggess, J. (1979) Troop male membership changes and infant killing in langurs (*Presbytis entellus*). *Folia primatol.*, **32**, 65–107.

Boggess, J. (1984) Infant killing and male reproductive strategies in langurs (*Presbytis entellus*) in *Infanticide: Comparative and Evolutionary Perspectives.* (eds G. Hausfater and S.B. Hrdy), Aldine, Hawthorne, 283–310.

Bohannan, P. (1963) *Social Anthropology*. Holt, Rinehart and Winston, New York.

Borgia, G. (1980) Human aggression as a biological adaptation. In *The Evolution of Human Social Behavior*, (ed. J.S. Lockard) Elsevier, New York, 165–185.

Bouchier, D. (1977) Radical ideologies and the sociology of knowledge. *Sociology*, **11**, 29–46.

Boulding, K.E. (1957) Organization and conflict. *J. Conflict Res.*, **1**(2), 122–34.

Boulding, K.E. (1962) *Conflict and Defence; A general theory*, Harper, New York.

Bourdieu, P. (1975) The specificity of the scientific field and the social conditions of the progress of reason. *Int. Soc. Sci. Infor.*, **14**(5–6), 19–47.

Boyd, R. and Richerson, J. (1980) *Cultural Transmission and the Evolution of Cooperative Behavior*, Presented at the annual meeting of the American Anthropological Association, held in Washington, D.C., December 5.

Boyd, R. and Richerson, J. (1983) *Culture and the Evolutionary Process*. University of Chicago Press.

Bibliography

Boyd, R. and Richerson, J. (1985) Why is culture adaptive? *Quart. Rev. Biol.*, **58**, 209–214.

Brian, M.V. (1983) *Social Insects: Ecology and Behavioural Biology*, Chapman and Hall, London.

Brickman, P. (1974) *Social Conflict: Readings in rule structures and conflict relationships*, Heath, Lexington.

Briggs, J.L. (1978) The origins of nonviolence: Inuit management of aggression, in *Learning Non-Aggression* (ed. A Montagu), Oxford University Press, 54–93.

Bright, J.R. (1964) *Research, Development and Technological Innovation*, Irwen, Homewood, Ill.

Brinton, C. (1965) *The Anatomy of Revolution*. Vintage Books, New York.

Broch, T. and Galtung, J. (1966) Belligerence among the Primitives; A re-analysis of Quincy Wright's data. *J. Peace Res.*, **3**(1), 33–45.

Brodie, B. (1974) *War and Politics*, Cassell, London.

Brodie, F.M. (1974) *Thomas Jefferson: An Intimate History*. W.W. Norton, New York.

Brown, J.L. (1964) The evolution of diversity in avian territorial systems. *Wil. Bull.*, **76**, 160–69.

Brown, J.L. (1975) *The Evolution of Behavior*. Norton, New York.

Buechner, H.K. (1961) Territorial behavior in Uganda kob. *Science*, **133**, 698–9.

Bullock, A. (1962) *Hitler: A Study in Tyranny*, Harper and Row, New York.

Burch, E.S. (1975) *Eskimo Kinsmen, Changing Family Relationships in Northwest Alaska*, West Publishing Co., Boston.

Burch, E.S. and Correll, T. (1971) Alliance and Conflict: Inter-Regional Relations in North Alaska in Alliance in Eskimo Society. *Proceedings of the American Ethological Society*, 1971, *Supplement*, University of Washington Press.

Burt, W.H. (1943) Territoriality and home range concepts as applied to mammals. *J. Mammalogy*, **24**(3), 346–52.

Burwash, L.T. (1931) *Canada's Western Arctic. Report on Investigations in 1925–29, and 1930*. Acland, Ottowa.

Buss, A.H. and Plomin, R.A. (1975) *A Temperament Theory of Personality Development*, Wiley, New York.

Buss, A.H., Plomin, R. and Willerman, L. (1973) The inheritance of temperaments. *J. Pers.*, **41**, 513–552.

Bygott, J.D. (1972) Cannibalism among wild chimpanzees ('Pan troglodytus'). *Nature*, **238**, 410–11.

Cadzow, D.A. (1929) Clean and Honest Eskimos. *Sci. Amer.*, **140**, Feb, 105–7.

Calhoun, J.B. (1974) *The Universal City of Ideas*, presentation at a conference on The Exploding Cities, held at Worcester College and the Taylor Institution, Oxford, 1–6 April 1974, Conference Transcripts, 301–306.

Campbell, C. (1986) Anatomy of a fierce academic feud. *The New York Times*, 9 November, Section 12.

Campbell, D.T. (1965) Variation and selective retention in sociocultural evolution, in *Social Change in Underdeveloped Areas*, (eds R.W. Mack *et al.*), Schenkman, Cambridge, Mass.

Campbell, D.T. (1972) On the genetics of altruism and the counter-hedonic components in human culture. *Journal of Social Issues*, **28**(3), 21–37.

Campbell, D.T. (1974) Downward causation in hierarchically organized biological systems, in *Studies in the Philosophy of Biology*, (eds F.J. Ayala and T. Dobzhansky), Macmillan, London, 179–86.

Campbell, D.T. (1975) On the conflicts between biological and social evolution and between psychology and moral tradition. *American Psychologist*, **10**, 1111–12.

Bibliography

Campbell, D.T. (1977) *Descriptive Epistemology: Psychological, Sociological, and Evolutionary*. William James Lectures, Harvard University, Cambridge, Mass.

Campbell, D.T. (1983) The two distinct routs beyond kin selection to ultrasociality: Implications for the humanities and social sciences, in *The Nature of Pro-social Development: Interdisciplinary Theories and Strategies*. (ed. D.L. Bridgeman), Academic Press, New York, 11–41.

Carneiro, R.L. (1970) A theory of the origin of the state. *Science*, **1969**, 733–8.

Carpenter, C.R. (1940) A field study in Siam of the behavior and social relations of the gibbon. *Comparative Psychology Monographs*, **16**(5), 1–212.

Carr-Saunders, A.M. (1922) *The Population Problem: A Study in Human Evolution*. Clarendon Press, Oxford.

Carson, H.L. (1961) Heterosis and fitness in experimental populations of Drosophila melanogaster. *Evolution*, **15**, 496–509.

Carver, T.N. (1908) The basis of social conflict. *Amer. J. Sociology*, **13**, 628–37.

Carver, T.N. (1915) The forms of human conflict, in *Essays in Social Justice*, Harvard University Press, Cambridge.

Cattell, R.B., Eber, H.W. and Tatsuoka, M.M. (1970) *Handbook for the Sixteen Personality Factor Questionnaire*, Institute for Personality and Ability Testing, Champaign.

Cavalli-Sforza, L. and Feldman, M. (1973) Models for Cultural Inheritance I. Group Mean and Within Group Variation. *Theoretical Population Biology*, **4**(69), 42–55.

Cavalli-Sforza, L. and Feldman, M. (1981) *Cultural Transmission and Evolution: A Quantitative Approach*. Princeton University Press, Princeton.

Chagnon, N.A. (1967) Yanomamö social organization and warfare, in *War: The Anthropology of Armed Conflict and Aggression*, (eds M. Fried, M. Harris and R. Murphy), Natural History Press, New York, 109–59.

Chagnon, N.A. (1968) *Yanomamö – The Fierce People*. Holt, Rinehart & Winston, New York.

Chagnon, N.A. (1974) *Studying the Yanomamö*. Holt, Rinehart & Winston, New York.

Chagnon, N.A. (1988) Life histories, blood revenge, and warfare in a tribal population. *Science*, **239**, 985–92.

Chagnon, N.A. and Irons, W.G. (eds)(1979) *Evolutionary Biology and Human Social Behavior: An Anthropological Perspective*. Duxbury Press, Scituate.

Chamberlain, H.S. (1899; trans.1911) *The Foundations of the Nineteenth Century*. Lane, London.

Chance, M.R.A. and Jolly, C. (1970) *Social Groups of Monkeys, Apes and Men*. Cape, London.

Chase, S. (1951) *Roads to Agreement*. Harper, New York.

Cheney, D.L. (1977) The acquisition of rank and the development of reciprocal alliances among free-ranging immature baboons. *Behavioral Ecology and Sociobiology*, **2**, 303–18.

Cheney, D.L. (1987) Interactions and relationships between groups, in *Primate Societies* (eds B.B. Smuts, D.L. Cheney, R.M. Seyfarth, R.W. Wrangham and Th.T. Struhsaker). University of Chicago Press, Chicago, 267–81.

Chou, E. (1980) *Mao Tse-Tung: The Man and the Myth*. Stein and Day, New York.

Christian, J.J. (1970) Social subordination, population density, and mammalian evolution, *Science*, **168**, 84–90.

Claridge, G.A., Canter, S. and Hume, W.J. (1973) *Personality Differences and Biological Variations: A Study of Twins*, Pergamon Press, New York.

Claringbold, P.J. and Barker, J.S.F. (1961) The estimation of relative fitness of Drosophila populations. *Jour. Theor. Biol.*, **1**, 190–203.

Clausewitz, K. von (1832) *Vom Kriege*. Ferdinand Dümmler, Berlin.
Cleland, H.F. (1928) *Our Prehistoric Ancestors*, Coward-McCann, New York.
Clements, F.E. and Shelford, V.E. (1939) *Bio-Ecology*. Wiley, New York.
Cloninger, C.R. (1986) A unified biosocial theory of personality and its role in the development of anxiety states, *Psychiatric Development*, **3**, 167–226.
Cloninger, C.R. (1987) A systematic method for clinical description and classification of personality variants, *Archives of General Psychiatry*, **44**, 573–88.
Clutton-Brock, T.H. and Albon, S.D. (1979) The roaring of red deer and the evolution of honest advertisement. *Behaviour*, **69**(3–4), 145–70.
Clutton-Brock, T.H. and Harvey, P.H. (1976) Evolutionary rules and primate societies, in *Growing Points in Ethology*, (eds P. Bateson and R. Hinde), Cambridge University Press, Cambridge, 195–237.
Cohen, R. (1984) Warfare and state-formation: wars make states and states make wars, in *Warfare, Culture and Environment*, (ed. R. Ferguson), Academic Press, New York, 329–58.
Collins, R. (1975) *Conflict Sociology: Toward an explanatory science*. Academic Press, New York.
Comte, A. (1972) Das Drei-Stadien-Gesetz, in *Sozialer Wandel. Zivilisation und Fortschritt als Kategorien der soziologischen Theorie*, 2nd printing, (ed. H.P. Dreitzel), Luchterhand, Neuwied, 111–20.
Cooley, C.H. (1918) *Social Progress*. Scribner's, New York.
Corbey, R. (1988) Aapmensen, Mensapen, Primitieven. *De Gids*, **151**(7/8), 504–13.
Corning, P.A. (1975) An evolutionary paradigm for the study of human aggression, in *War, Its Causes and Correlates*, (eds M.A. Nettleship, R.D. Givens and A. Nettleship) Elsevier, Amsterdam, 359–87.
Corning, P.A. (1983) *The Synergism Hypothesis. A Theory of Progressive Evolution*. McGraw-Hill, New York.
Coser, L.A. (1956) *The Functions of Social Conflict*, Free Press, New York.
Coser, L.A. (1967) *Continuities in the Study of Social Conflict*, Free Press, New York.
Coser, L.A. (1978) *Gulzige Instituties: Patronen van Absolute Toewijding*, Deventer: van Loghum Slaterus, translation of "Greedy Institutions", 1974.
Crawford, M.H. and Enciso, V.B. (1982) Population structure of circumpolar groups of Siberia, Alaska, Canada and Greenland, in *Current Developments in Anthropological Genetics*, Vol. 2 (eds M.H. Crawford and J.H. Mielke), Plenum Press, New York.
Cronin, V. (1972) *Napoleon Bonaparte: An Intimate Biography*. William Morrow, New York.
Crook, J.H. (1970) *Social Behavior in Birds and Mammals*. Academic Press, New York.
Crow, J.F. (1979) Genes that violate Mendel's rules. *Scientific American*, **240**(2), 104–13.
Curio, E. (1963) Probleme des Feinderkennens bei Vögeln. Proceedings 13th International Ornithology Congress, 206–39.
Dahrendorf, R. (1958) Toward a theory of social conflict. *Journal of Conflict Resolution*, **2**(2), 170–83.
Dahrendorf, R. (1959) *Class and Class Conflict in Industrial Society*, Stanford University Press, Stanford.
Dahrendorf, R. (1961) Über einige Probleme der soziologischen Theorie der Revolution. *Archives Européennes de Sociologie*, **1**, 153–63.
Dahrendorf, R. (1967) *Conflict after Class: New perspectives on the theory of social and political conflicts*. Longmans, London.
Darlington, C.D. (1969) *The Evolution of Man and Society*, Allen & Unwin, London.

Darlington, Jr., P.J. (1957) *Zoogeography: The Geographical Distribution of Animals*, Wiley, New York.
Dart, R.A. (1953) The predatory transition from ape to man. *International Anthropological Linguistic Review*, **1**, 201–19.
Darwin, C.R. (1859) *The Origin of Species by Means of Natural Selection, or the Preservation of Favoured Races in the Struggle for Life.* Murray, London.
Darwin, C.R. (1871) *The Descent of Man, and Selection in Relation to Sex.* Murray, London.
Darwin, C.R. (1873) *The Expression of the Emotions in Man and Animals.* Murray, London.
Datta, S. (1988) The acquisition of dominance among free-ranging rhesus monkey siblings, *Animal Behaviour*, **36**, 754–72.
Davie, M.R. (1929) *The Evolution of War; A study of its role in early societies.* Yale University Press, New Haven.
Davies, J.C. (1962) Toward a theory of revolution. *American Sociological Review*, **27**, 5–19.
Davis, B.D. (1985) *Storm over Biology*, Prometheus Books, Buffalo, New York.
Davis, K. (1974) The migrations of human populations, *Scientific American*, **231**(3), 93–105.
Dawkins, R. (1976) *The Selfish Gene.* Oxford University Press, Oxford.
Dawkins, R. (1982) *The Extended Phenotype: The gene as the unit of selection.* Oxford University Press, Oxford.
De Ciantis, S.M. (1987) Management style and thinking style, in *Effective Management*, (ed. W.J. Reddin), McGraw-Hill, New Delhi.
DeKadt, E.J. (1965) Conflict and power in society. *International Social Science Journal*, **17**(3), 454–71.
de Lapouge, G. Vacher (1896) *Les Selections Sociales.* Librairie Thorin & Fils, Paris.
de Lapouge, G. Vacher (1899) *L'Aryen, son rôle social.* Librairie Thorin & Fils, Paris.
Dellenbaugh, F.S. (1901) *The North-Americans of Yesterday*, Putnam, New York.
Demallie, R.J. and Lavenda, R.H. (1977) Plains Siouan concepts of power, in *The Anthropology of Power; Ethnographic Studies from Asia, Oceania and the New World*, (eds R.D. Fogelson and R.N. Adams), Academic Press, London.
de Molinari, G. (1898) *Grandeur et Décadence de la Guerre.* Alcan, Paris.
de Mortillet, G. (1883) *Le Préhistorique – Antiquité de l'Homme.* Reinwald, Paris.
de Savorgnan, F. (1914) *Les Antagonismes Sociaux.* Scientia, Roma.
de Savorgnan, F. (1926) *La Guerra e l'Eugenica.* Scientia, Roma.
Deutsch, K.W. and Senghaas, D. (1971) Die brüchige Vernunft von Staaten, in *Kritische Friedensforschung*, (ed. D. Senghaas), Frankfurt am Main, Suhrkamp, 105–63.
Deutsch, M. (1969) Conflicts: productive and destructive. *Journal of Social Issues*, **25**, 5–41.
de Waal, F.B.M. (1975) The wounded leader; A spontaneous temporary change in the structure of agonistic relations among captive Java-Monkeys (*Macaca fascicularis*). *Netherlands Journal of Zoology*, **25**(4), 529–49.
de Waal, F.B.M. (1976) Straight-aggression and appeal-aggression in *Macaca fascicularis. Experientia*, **32**(10), 1268–70.
de Waal, F.B.M. (1977) Agonistisch gedrag binnen een groep Java-apen, in *Agressief Gedrag; Oorzaken en Functies* (eds P.R. Wiepkema and J.A.R.A.M. van Hooff). Bohn, Scheltema & Holkema, Utrecht, 165–82.
de Waal, F.B.M. (1978) Exploitative and familiarity-dependent support strategies in a colony of semi-free living chimpanzees. *Behaviour*, **66**, 268–312.

Bibliography

de Waal, F.B.M. (1982) *Chimpanzee Politics: Power and Sex Among Apes*. Harper & Row, New York.

de Waal, F.B.M. (1986) The brutal elimination of a rival among captive male chimpanzees. *Ethology and Sociobiology*, **7**, 237–51.

de Waal, F.B.M. and van Hooff, J.A.R.A.M. (1981) Side-directed communication and agonistic interactions in chimpanzees. *Behaviour*, **77**, 164–98.

Dewey, J. (1922) *Human Nature and Conduct*. Henry Holt, New York.

Diamond, J.M. and Terborgh, J.W. (1968) Dual singing in New Guinea birds. *Auk*, **85**, 62–82.

Dickemann, M. (1979) Female infanticide, reproductive strategies, and social stratification: a preliminary model, in *Evolutionary Biology and Human Social Behavior*, (eds N.A. Chagnon and W. Irons), Duxbury Press, Scituate, 321–67.

Dickinson, G.L. (1920) *The Causes of International War*. Swarthmore Press, London.

Dickinson, G.L. (1923) *War: Its nature, cause and cure*. Macmillan, New York.

Divale, W.T. (1970a) An explanation for primitive warfare: population control and the significance of primitive sex ratios. *The New Scholar*, **2**, 173–92.

Divale, W.T. (1970b) An explanation for tribal warfare. Paper 69th Annual Meeting American Anthropological Association, San Diego.

Divale, W.T. (1973) *Warfare in Primitive Societies: A bibliography*. Clio Press, Santa Barbara.

Divale, W.T. (1974) Migration, external warfare, and matrilocal residence: an explanation for matrilineal residence systems. *Behavioural Science Research*, **9**, 75–133.

Divale, W.T. and Harris, M. (1976) Population, warfare, and the male supremacist complex. *American Anthropologist*, **78**, 521–38.

Divale, W.T., Chamberis, F. and Gangloff, D. (1976) War, peace and marital residence in pre-industrial societies. *Journal Conflict Resol.*, **20**(1), 57–78.

Djilas, M. (1957) *The New Class: An Analysis of the Communist System*. Frederick A. Praeger, New York.

Dobrizhoffer, M. (1822) *An Account of the Abipones, an Equestrian People of Paraguay* (3 Vols) Murray, London.

Dollard, J., Doob, L.W., Miller, N.E., Mowrer, O.H. and Sears, R.R. (1939) *Frustration and Aggression*. Yale University Press, New Haven.

Dolhinow, P. (1977) Normal monkeys? *American Scientist*, **65**, 266.

Douglas, M. (1978) *Purity and Danger; An Analysis of Concepts of Pollution and Taboo*. Routledge & Kegan, London.

Drucker, P.F. (1969) Management's New Role, *Harvard Business Review*, **47**, 49–54.

Duke, J.T. (1976) *Conflict and Power in Social Life*. Brigham Young University Press, Provo, Utah.

Dunbar, R.I.M. (1988) *Primate Social Systems*. Croom Helm, London.

Dunn, J.P. (1844) *History of the Oregon Territory and the British North-American fur trade; with an account of the habits and customs of the principal native tribes on the northern continent*, Edward and Hughes, London.

Dupuy, R.E. and Dupuy, T.N. (1986) *The Encyclopedia of Military History, from 3500 BC to the present*. Jane's Publishers, London.

Durbin, E.F.M. and Bowlby, J. (1939) Personal aggressiveness and war, in *War and Democracy*, (eds E. Durbin and G. Catlin), Columbia University Press, New York.

Durham, W.H. (1976) Resource competition and human aggression. Part I: a review of primitive war. *Quarterly Review Biology*, **51**, 385–415.

Durham, W.H. (1976) The adaptive significance of cultural behavior. *Human Ecology*, **4**, 89–121.

Durham, W.H. (1979) Towards a coevolutionary theory of human biology & culture, in *Evolutionary Biology and Human Social Behavior*, (eds N.A. Chagnon and W. Irons), Duxbury, North Scituate, 39–59.
Durkheim, E. (1951) *Suicide.* Free Press, New York.
Durkheim, E. (1893) *De la division du travail social.* Alcan, Paris.
Dyer, G. (1985) *War.* Stoddart Publishing Co. Ltd, Toronto.
Dyson-Hudson, R. and Smith, E.A. (1978) Human territoriality: An ecological reassessment. *American Anthropologist*, **80**(1), 21–24.
Eaton, R.L. (1978) The evolution of trophy hunting. *Carnivore*, **1**, 110–21.
Eaves, L. and Eysenck, H.J. (1975) The Nature of Extraversion: A Genetical Analysis, *Journal of Personality and Social Psychology*, **32**, 102–12.
Eckhardt, W. (1982) Atrocities, civilizations, and savages: ways to avoid a nuclear holocaust. *Bulletin of Peace Proposals*, **13**(4), 343–50.
Eckhardt, W. (1981) Quincy Wright's study of war: an interpretative essay. *Peace Research*, **31**(1), 1–8.
Eckhardt, W. (1975) Primitive militarism. *Journal of Peace Research*, **12**(1), 55–62.
Edwards, L.P. (1970) *The Natural History of Revolution.* University of Chicago Press, Chicago.
Ehardt, C.L. and Bernstein, I.S. (1986) Matrilineal overthrows in rhesus monkey groups. *International Journal of Primatology*, **7**, 157–81.
Eibl-Eibesfeldt, I. (1963) Aggressive behavior and ritualized fighting in animals, in *Violence and War*, (ed. J. Masserman), Grune and Stratton, New York, 8–17.
Eibl-Eibesfeldt, I. (1967) *Grundriss der vergleichenden Verhaltensforschung.* Piper, München.
Eibl-Eibesfeldt, I. (1970) *Ethology, The Biology of Behavior.* Holt Reinhart, New York.
Eibl-Eibesfeldt, I. (1974a) The myth of the aggression-free hunter and gatherer society, in *Primate Aggression, Territoriality and Xenophobia*, (ed. R.L. Holloway) Academic Press, New York, 435–7.
Eibl-Eibesfeldt, I. (1974b) *Grundriss der vergleichenden Verhaltensforschung.* (4th edn), Piper, München.
Eibl-Eibesfeldt, I. (1975) *Krieg und Frieden aus der Sicht der Verhaltensforschung.* Piper, München.
Eibl-Eibesfeldt, I. (1977) Evolution of destructive behavior. *Aggressive Behaviour*, **3**(2), 127–44.
Eibl-Eibesfeldt, I. (1978) Die Evolution der destruktiven Aggression aus der Sicht der Verhaltensforschung. *Beiträge zue Konfliktforschung*, **8**(2), 87–111.
Eibl-Eibesfeldt, I. (1979) *The Biology of Peace and War.* Viking Press, New York.
Eibl-Eibesfeldt, I. (1982) Warfare, Man's indoctrinability and group selection. *Z.f. Tierpsychologie*, **60**(3), 177–98.
Eibl-Eibesfeldt, I. (1984) *Die Biologie des menschlichen Verhaltens. Grundriss der Humanethologie.* Piper, München.
Eibl-Eibesfeldt, I. (1986) *Die Biologie des menschlichen Verhaltens: Grundriss der Humanethologie* (2nd edn.), Piper, München.
Eisenberg, J.F., Muckenhirn, N.A. and Rudran, R. (1972) The relation between ecology and social structure in primates, *Science*, **176**, 863–74.
Eldridge, A.F. (1979) *Images of Conflict.* St. Martin's Press, New York.
Ellis, F.H. (1951) Patterns of aggression and the war cult in Southwestern Pueblos. *Southwestern Journal of Anthropology*, **7**, 177–201.
Ellis, W. (1829) *Polynesian Researches, during a residence of nearly eight years in the Society and Sandwich Islands* (2 Vols). Fisher, Son & Jackson, London.
Ellison, L.N. (1971) Territoriality in Alaska spruce grouse. *Auk*, **88**, 652–4.

Ember, C. R. (1978) Myths about hunter-gatherers. *Ethology*, **17**, 439–48.

Erikson, E.H. (1966) Ontogeny of ritualization in man. *Transactions of the Royal Society of London. Ser. B*, **251**, 337–49.

Erikson, K. (1966) *Wayward Puritans: A Study in the Sociology of Deviance*, Wiley, New York.

Eshel, I. and Motro, U. (1988) The three brothers' problem: kin selection with more than one potential helper. I. The case of immediate help. *American Naturalist*, **132**, 550–66.

Estrada, A. and Estrada, R. (1978) Changes in the social structure and interactions after the introduction of a second group in a free-ranging troop of stumptail macaques (*Macaca arctoides*): social relations II. *Primates*, **19**, 665–80.

Ettlie, J.E. and O'Keefe, R.D. (1982) Innovative attitudes, values and intentions in organizations, *Journal of Management Studies*, **19**, 163–82.

Etzioni, A. (1964) *Modern Organizations*. Prentice-Hall, New York.

Ewer, R.F. (1971) The biology and behaviour of a free-living population of black rats (*Rattus rattus*), *Animal Behaviour Monographs*, Vol.4, Part 3, 127–74.

Eysenck, H.J. (1953) *The Structure of Human Personality*. Methuen, London.

Eysenck, H.J. (1967) *The Biological Basis of Personality*. Thomas, Springfield.

Falger, V.S.E. (1984) Sociobiology and political ideology: comments on the radical point of view, *Journal of Human Evolution*, **13**, 129–35.

Falls, C. (1961) *The Art of War from the Age of Napoleon to the Present Day*. Oxford University Press, London.

Fast, J. (1970) *Body Language*. Pocket Books, New York.

Feest, C.F. (1980) *The Art of War*. Thomas & Hudson, London.

Feij, J.A. (1978) *Temperament: Onderzoek naar de Betekenis van Extraversie, Emotionaliteit, Impulsiviteit en Spannings-be hoefte*. Academic Press, Amsterdam.

Feij, J.A., Orlebeke, J.F., Gazendam, A. and Van Zuilen, R. (1979, 1981) Sensation Seeking: Measurement and Psychophysiological Correlates, paper presented at the International Conference on Temperament, Need for Stimulation and Activity, Grzegor zewice, Poland, Sept. 1979; and in *Biological Foundation of Personality and Behaviour* (eds J. Strelau, F.H. Farley and A. Gale), Hemisphere Press, New York.

Ferguson, A. (1767) *An Essay on the History of Civil Society*. Aldine, Chicago.

Ferguson, R.B. (1984a) Introduction: studying war, in *Warfare, Culture and Environment* (ed. R.B. Ferguson), Academic Press, New York, 1–81.

Ferguson, R.B. (1984b) A re-examination of the causes of Northwest Coast warfare, in *Warfare, Culture and Environment* (ed. R.B. Ferguson), Academic Press, New York, 267–328.

Ferrero, G. (1898) *Il Militarismo*. Treves, Milano.

Ferri, E. (1895) *L'Omicidio*. Fratelli Bocca, Torino.

Fink, C.F. (1968) Some conceptual difficulties in the theory of social conflict. *Journal of Conflict Resolution*, **12**(4), 412–60.

Fischer, H. (1965) Das Triumphgeschrei der Graugans (*Anser anser*). *Zeitschrift für Tierpsychologie*, **22**, 247–304.

Fletcher, A.C. and LaFlesche, F. (1911) The Omaha tribe. *Bureau American Ethnology. Annual Report* 27, Washington.

Flohr, H. (1987) Biological bases of social prejudices, in *The Sociobiology of Ethnocentrism* (eds V. Reynolds, V. Falger and I. Vine). Croom-Helm, London/ Sydney, 190–207.

Flügel, J.C. (1955) *Man, Morals, and Society; a psychoanalytic study*. Penguin, Harmondsworth.

Folk, G.E. Jr. (1966) *Introduction to Environmental Physiology*. Lea and Febiger, Pa.

Fossey, D. (1984) Infanticide in mountain gorillas (*Gorilla gorilla berengei*) with a comparative note on chimpanzees, in *Infanticide, Comparative and Evolutionary Perspectives* (eds G. Hausfater and S.B. Hrdy). Aldine, Hawthorne, 217–36.
Fox, R. (1967) *Kinship and Marriage. An anthropological perspective.* Penguin Books, Harmondsworth.
Fox, R. (1987) Letter, *The Harry Frank Guggenheim Newsletter*, **4**(1).
Fraser, A. (1973) *Cromwell: The Lord Protector.* Knopf, New York.
Frazer, J.G. (1890) *The Golden Bough: A Study in Magic and Religion.* Macmillan, London.
Frazer, J.G. (1918) *Folk-lore in the Old Testament.* Macmillan, London.
Freedman, L.Z. and Roe, A. (1958) Evolution and human behavior, in *Behavior and Evolution*, (eds A. Roe and G.G. Simpson) Yale Univ. Press, New Haven.
Freeman, D. (1973) Human aggression in anthropological perspective. In *Aggression and Evolution* (ed. C.M. Otten). Xerox College Publications, Lexington.
Freeman, D.M. (1972) Social conflict, violence, and planned change: some research hypotheses. *Co-Existence*, **9**, 89–100.
Freeman, M.M.R. (1971) A social and ecologic analysis of systematic female infanticide among the netsilik eskimo. *American Anthropologist*, **73**, 1011–18.
Freud, S. (1913) *Totem und Tabu.* Heller, Wien.
Freud, S. (1932) *Warum Krieg*? International Institute of Intellectual Co-operation, Paris.
Fried, M.H. (1961) Warfare, military organization, and the evolution of society. *Anthropologica*, **3**(2), 134–47.
Fried, M.H. (1967) *The Evolution of Political Society; An essay in political anthropology.* Random House, New York.
Friedrich, C.J. (1950) *Constitutional Government and Democracy.* Xerox College Publishing, Lexington, Mass.
Frobenius, L. (1914) *Menschenjagde und Zweikämpfe.* Thüringer Verlagsanstalt, Jena.
Frobenius, L. (1903) *Weltgeschichte des Krieges.* Thüringer Verlagsanstalt, Jena.
Fromm, E. (1973) *The Anatomy of Human Destructiveness.* Holt, Rinehart & Winston, New York.
Galtung, J. (1965) Institutionalized conflict resolution: A theoretical paradigm. *Journal of Peace Research*, **2**, 348–97.
Galtung, J. (1973) Is Peace Possible? *PRIO Publ.* 25–10, whole issue.
Galtung, J. (1974) *Theories of development, conflict and peace.* Inter-University Centre of Post-Graduate Studies, Dubrovnik (mimeo).
Gardner, R. and Heider, K.G. (1968) *Gardens of War; Life and Death in the New Guinea Stone Age.* Random House, New York.
Garlan, Y. (1975) *War in the Ancient World: A social history.* Chatto & Windus, London.
Geist, V. (1978) On weapons, combat, and ecology, in *Aggression, Dominance, and Individual Spacing* (eds L. Krames, P. Pliner and T. Alloway). Plenum Press, New York, 1–30.
Ghiglieri, M. (1988) *East of the Mountains of the Moon: The Chimpanzees of Kibale Forest.* Free Press, New York.
Gibb, C.A. (1969) *Leadership, Handbook of Social Psychology*, Vol. 4, (eds G. Lindzey and E. Aranson). Addison-Wesley, Reading, Mass., 205–82.
Gilder, W.H. (1881) *Schwatka's Search. Sledging in the Arctic in Quest of the Franklin Records, 1878–80.* Charles Scribner's Sons, New York.
Gilliard, E.T. (1962) On the breeding behavior of the cock-of-the-rock. *Bulletin of American Museum of Natural History*, **124**(2), 31–68.

Gini, C. (1921) Gli effetti eugenici o disgenici della guerra. *Genus*, **1**, 29–42.

Girard, R. (1982) *Le Bouc Emissaire*. Bernard Grasset, Paris.

Glover, E. and Ginsberg, M. (1934) A symposium on the psychology of peace and war. *British Journal of Medical Psychology*, **14**, 274–93.

Gobineau, A. de (1853) *Essay sur l'Inégalité de Races Humaines*. Libraire de Fermin-Didot, Paris.

Goffman, E. (1969) *Behavior in Public Places*. Free Press, New York.

Goffman, E. (1972) *Relations in Public*. Harper & Row, New York.

Goldsmith, R.E. (1986) Convergent validity of four innovativeness scales, *Educational and Psychological Measurement*, **46**, 81–7.

Goldsmith, R.E. (1984) Personality characteristics associated with adaption-innovation, *Journal of Psychology*, **117**, 159–65.

Goldsmith, R.E. (1989) Creative style and personality theory, in *Adaptors and Innovators: styles of creativity and problem solving* (ed. M.J. Kirton), Routledge, London, 37–55.

Goodall, J. (1971) *In the Shadow of Man*. Houghton Mifflin, Boston, Mass.

Goodall, J. (1977) Infant killing and cannibalism in free-living chimpanzees. *Folia Primatologica*, **28**, 259–82.

Goodall, J. (1979) Life and death at Gombe. *National Geographic*, **155**(5) 592–621.

Goodall, J. (1986) *The Chimpanzees of Gombe: Patterns of Behavior*. Harvard University Press, Cambridge, Mass.

Goodall, J., Bandora, A., Bergmann, E., Busse, C., Matama, H. *et al.* (1979) Inter-community Interactions in the Chimpanzee Population of the Gombe National Park, in *The Great Apes* (eds D.A. Hamburg and E.R. McCown). Benjamin/Cummings, Menlo Park, 13–54.

Gorer, G. (1938) *Himalayan Village: An Account of the Lepchas of Sikkim*. M. Joseph, London.

Gotmark, F. and Anderson, M. (1984) Colonial breeding reduces nest predation in the common gull *Larus canus*. *Anim. Behav.*, **32**, 485–92.

Gould, S.J. (1981) *The Mismeasure of Man*. W.W. Norton & Co., New York.

Gouzoules, H., Fedigan, L.M. and Fedigan, L. (1975) Responses of a transplanted troop of Japanese macaques (*Macaca fuscata*) to bobcat (*Lynx lynx*) predation, *Primates*, **16**, 335–49.

Graham, E.E. (1975) Human warfare: an analysis of ecological factors from ethno-historical sources, in *War, Its Causes and Correlates* (eds M.A.Nettleship, R.D. Givens and A. Nettleship). Mouton, The Hague, 451–62.

Grammar, K. (1989) Human courtship behaviour: biological bases and cognitive processing, in *The Sociobiology of Sexual and Reproductive Strategies* (eds O.A.E. Rasa, C. Vogel and E. Voland). Croom Helm, Oxford, 147–69.

Grinnell, G.B. (1910) Coup and scalp among the Plains Indians. *American Anthropologist*, **12**, 296–310.

Grotius, H. (1625) *De jure belli ac pacis libri tres*. Clarendon Press, Oxford, reprinted 1925.

Gudgeon, W.E. (1902–3) The whence of the Maori. *Journal Polynesian Society*, **11–12**.

Guemple, D.L. (1961) *Inuit Spouse-Exchange*. Masters Thesis, Department of Anthropology. University of Chicago.

Guemple, D.L. (1966) *Kinship Reckoning Among the Belcher Island Eskimo*. Doctoral Dissertation, Department of Anthropology. University of Chicago.

Guemple, D.L. (1971) Alliance in Eskimo Society. *Proceedings of the American Ethological Society*, Supplement. University of Washington Press, Washington.

Gumplowicz, L. (1892) *Soziologie und Politik*. Dunker, Leipzig.

Gumplowicz, L. (1883) *Der Rassenkampf.* Wagner, Innsbruck.
Gumplowicz, L. (1885) *Grundriss der Soziologie.* Manz, Wien.
Gumplowicz, L. (1902) *Die Soziologische Staatsidee.* Wagner, Innsbruck.
Haldane, J.B.S. (1932) *The Causes of Evolution.* Longmans, London.
Hall, C.F. (1864) *Life with the Esquimaux.* Low & Marston, London.
Hall, E.T. (1959) *The Silent Language.* Fawcett, Greenwitch, Conn.
Hall, E.T. (1976) The anthropology of space: An organizing model, in *Environmental Psychology, People & their physical settings* (eds H. Proshansky, W. Ittelson, and L. Rivlin). Holt Reinhart and Winston, New York, 158–69.
Hall, K.R.L. and DeVore, I. (1965) Baboon social behavior, in *Primate Behavior: Field Studies of Monkeys and Apes,* (ed. I. DeVore). Holt, Rinehart, and Winston, New York.
Hallpike, C. (1973) Functionalist interpretations of primitive warfare. *Man,* **8**, 451–70.
Hallpike, C. (1979) *The Foundations of Primitive Thought.* Clarendon Press, Oxford.
Halpin, A.W. and Winer, B.J. (1957) A factorial study of the leader behaviour descriptions, in *Leader Behavior: Its Description and Measurement* (eds R.M. Stagdill and A.E. Coons), Bur.Bus.Res.Monogr. 88, Ohio State University, Columbus, Ohio.
Hamburg, D.A. (1963) Emotions in the perspective of human evolution, in *Expression of the Emotions in Man* (ed. P. Knapp), International Universities Press, New York.
Hamburg, D.A. (1970) Recent Research on Hormonal Regulation of Aggressive Behavior, UNESCO Experts' Meeting on the implication of recent scientific research for the understanding of human aggressiveness, Paris, 1970.
Hamburg, D.A. (1973) An evolutionary and developmental approach to human aggressiveness. *Psychoanal. Quart.*, **42**, 185–96.
Hamilton, W.D. (1963) The evolution of altruistic behavior. *American Naturalist,* **97**, 354–6.
Hamilton, W.D. (1964) The genetical evolution of social behaviour I and II. *Journ. of Theor. Biol.*, **7**, 1–52.
Hamilton, W.D. (1971) Selection of selfish and altruistic behavior in some extreme models, in *Man and Beast: Comparative Social Behavior,* (eds J.F. Eisenberg and W.S. Dillon). Smithsonian Institute Press, Washington DC, 57–92.
Hamilton, W.D. (1975) Innate social aptitudes of man: an approach from evolutionary genetics, in *Biosocial Anthropology* (ed. R. Fox) Wiley, New York, 133–55.
Hamilton, W.D. and May, R.M. (1977) Dispersal in stable habitats. *Nature,* **269**, 578–81.
Hampson, N. (1963) *A Social History of the French Revolution.* Routledge and Kegan Paul, London.
Hand, J.L. (1986) Resolution of social conflicts: dominance, egalitarianism, spheres of dominance and game theory. *Quart. Rev. Biol.*, **61**, 667–82.
Hanna, J. (1968) Cold stress and microclimate in the Quechua Indians of Southern Peru, in *High Altitude Adaptation in a Peruvian Community* (eds P.T. Baker *et al.*). University Park, Occasional Papers in Anthropology, No. 1, Pennsylvania State University.
Harcourt, A.H., Harvey, P.H., Larson, S.G. and Short, R.V. (1981) Testis weight, body weight and breeding system in primates. *Nature,* **293**, 441–3.
Harding, R.S.O. (1973) Predation by a troop of olive baboons (*Papio anubis*). *American Journal of Physical Anthropology,* **38**, 578–82.
Harlow, H. (1962) Development of affection in primates, in *Roots of Behavior* (ed. E.J. Bliss). Harper & Row, New York.
Harris, M. (1971)) *Culture, Man, and Nature.* Lowell, New York.
Harris, M. (1972) Warfare old and new. *Natural History,* **81**(3), 18–20.

Harris, M. (1974) *Cows, Pigs, Wars, and Witches: The riddle of culture.* Random House, New York.
Harris, M. (1975) *Culture, Man, and Nature* (2nd edn). Lowell, New York.
Harris, M. (1978) *Cannibals and Kings: The origins of cultures.* Vintage Books, New York.
Harris, M. (1980) *Culture, People, Nature: An introduction to general anthropology.* Harper & Row, New York.
Harris, M. (1984) A cultural materialist theory of band and village warfare: the Yanomamo test, in *Warfare Culture, and Environment* (ed. R.B. Ferguson). Academic Press, New York, 111–40.
Harrison, R. (1973) *Warfare.* Burgess Publication, Minneapolis.
Hart, A. and Lendrem, D.W. (1984) Vigilance and scanning patterns in birds. *Anim. Behav.*, **32**, 1216–24.
Hartmann, L. (1915) *Der Krieg in der Weltgeschichte*, Manz, Wien.
Hartog, H. den (1981) Sociobiologie: Een Oude Hypothese Verpakt als Nieuwe Synthese, in *Sociobiologie ter Discussie* (ed. F.de Waal). Bohn, Scheltema & Holkema, Utrecht, 133–51.
Hausfater, G. and Hrdy, S.B. (eds) (1984) *Infanticide: Comparative and evolutionary perspectives.* Aldine, New York.
Hawkes, E.W. (1916) The Labrador Eskimo. Memoir 91 of the Geological Survey of Canada, *Anthropological series No. 14.* Ottowa.
Healey, M.C. (1967) Aggression and self-regulation of population size in deer-mice, *Ecology*, **48**, 377–92.
Hearne, S. (1795) *A Journey from Prince of Wales's Fort in Hudsons Bay, to the Northern Ocean.* A. Strahan and T. Cadell, London.
Hegner, R.E., Emlen, S.T. and Demong, N.J. (1982) Social organization of the white-fronted bee-eater. *Nature*, **298**(5871), 264–6.
Heinbecker, P. and Irvine-Jones, E.M. (1928) Susceptibility of the Eskimos to the common cold and a study of their natural immunity to diphtheria, scarlet fever and bacterial filtrates. *Journal of Immunology*, **XV**, 395–406.
Hellwald, F. (1883) *Kulturgeschichte in ihrer natürlichen Entwicklung.* Campart, Angsburg.
Helmuth, H. (1973) Cannibalism in paleoanthropology and ethnology, in *Man and Aggression* (2nd edn) (ed. A. Montagu). Oxford University Press, 229–53.
Heymans, G. (1932) *Inleiding tot de Speciale Psychologie.* Bohn, Haarlem.
Himes, J.S. (1980) *Conflict and Conflict Management.* University of Georgia Press, Athens.
Hinde, R.A. (1960) Energy models of motivation. *Symposium Society for Experimental Biology*, **14**, 199–213.
Hinde, R.A. (1966) *Animal Behaviour: A synthesis of ethology and comparative psychology.* McGraw-Hill, London.
Hinde, R.A. (1970) *Animal Behaviour: A synthesis of ethology and comparative psychology.* McGraw-Hill, New York.
Hinde, R.A. (1973) Aggression, in *Biology and the Human Sciences* (ed. J.W.S. Pringle). Clarendon Press, Oxford.
Hinde, R.A. (1974) *Biological Bases of Human Social Behaviour.* McGraw-Hill, New York.
Hinde, R.A. (1979) *Towards Understanding Relationships.* Academic Press, London.
Hinnebusch, R.A. (1982) Libya: personalistic leadership of a populist revolution, in *Political Elites of Arab North Africa: Morocco, Algeria, Tunisia, Libya, and Egypt* (eds I.W. Zartman *et al.*), Longman, New York, 177–222.

Hirschfeld, L., Howe, J. and Levin, B. (1972) Warfare, infanticide, and statistical inference: a comment on Divale and Harris. *Amer. Anthropol.*, **74**, 1318–19.
Hobbes, T. (1651) *Leviathan: Or, the matter, form and power of a commonwealth, ecclesiastical and civil.* (Reprint 1943). Clarendon Press, Oxford.
Hobhouse, L.T. (1924) *Social Development, its nature and conditions.* Allen and Unwin, London.
Hobhouse, L.T., Wheeler, G. and Ginsberg, M. (1915) *The Material Culture and Social Institutions of the Simpler Peoples.* London School of Economics, Series of Studies in Economic and Political Science, Monograph on Sociology, No. 3, London.
Hoffman, S. (1965) *The State of War; Essays in the theory and practice of international politics.* Praeger, New York.
Hoffschulte, B. (1986) The scapegoat theory and sociobiology, paper presented at the 8th meeting of the European Sociobiological Society, July 1986, Bussum, Netherlands.
Hofstadter, R. (1955) *Social Darwinism in American Thought, 1860–1915.* Beacon Press, Boston.
Hold, B.C.L. (1976) Attention structure and rank specific behaviour in preschool children, in *The Social Structure of Attention,* (eds M.C.R. Chance and R.R. Larsen). Wiley, London, 177–201.
Holsti, R. (1913) The relation of war to the origin of the state, *Annales Academiae Scientiarum Fennicae, Helsingfors,* **13**.
Holsti, R. (1912) Some superstitious customs and beliefs in primitive warfare. Festskrift Edvard Westermarck, *Annales Academiae Scientiarum Fennicae,* Helsingfors.
Horowitz, E.L. (1962) Consensus, conflict and cooperation: A sociological inventory. *Social Forces,* **41**, 177–88.
Howard, H.E. (1920) *Territory in Bird Life.* John Murray, London.
Howell, R.W. (1975) Wars without conflict, in *War, Its Causes and Correlates* (eds M.A. Nettleship, R.D. Givens and A. Nettleship). Mouton, The Hague, 675–92.
Hrdy, S.B. (1977) Infanticide as a primate reproductive strategy. *Amer. Sci.*, **65**, 40–9.
Huber, P.B. (1975) Defending the cosmos; Violence and social order among the Anggor of New Guinea, in *War, Its Causes and Correlates* (eds M.A. Nettleship *et al*). Mouton, The Hague, 619–60.
Humphrey, N.K. (1976) The social function of intellect, in *Growing Points in Ethology* (eds P.P.G. Bateson and R.A. Hinde). Cambridge University Press, Cambridge, 303–17.
Huntingford, F.A. and Turner, A. (1987) *Animal Conflict.* Chapman and Hall, London.
Irwin, C. (1981) Inuit ethics and the priority of the future generation. Masters thesis, University of Manitoba.
Irwin, C.J. (1985a) *Sociocultural Biology: Studies in the Evolution of some Netsilingmiut and Other Sociocultural Behaviors.* Doctoral Dissertation, Department of Social Science, Syracuse University.
Irwin, C.J. (1985b) Inuit Navigation, Empirical Reasoning and Survival. *Journal of the Royal Institute of Navigation,* **38**, 2.
Irwin, C.J. (1987) A study in the evolution of ethnocentrism, in *The Sociobiology of Ethnocentrism* (eds V. Reynolds, V. Falger and I. Vine). Croom Helm, London, 131–56.
Irwin, C.J. (1989) The sociocultural biology of Netsilingmiut female infanticide, in *The Sociobiology of Sexual Strategies* (eds A. Rasa, C. Vogel and E. Voland). Chapman and Hall, London, 234–64.

Itani, J. (1982) Intraspecific killing among non-human primates. *Journal of Social and Biological Structures*, **5**(4), 361–8.
Itani, J., Tokuda, K., Furuya, Y., Kano, K. and Shin, Y. (1963) The social construction of natural troops of Japanese monkeys in Takasakiyama, *Primates*, **4**(3), 1–42.
Jacoby, R. (1973) The politics of subjectivity: Slogans of the American New Left. *New Left Review*, **79**, 37–49.
Jaehns, M. (1893) *Über Krieg, Frieden und Kultur.* Allg. Verein f. deutsch Litt. Berlin.
Jaehns, M. (1880) *Geschichte des Kriegswesens von der Urzeit bis zur Renaissance.* Grunow, Leipzig.
James, W. (1890) *Principles of Psychology.* Holt, New York.
James, W. (1910) The moral equivalent of war, in *Memories and Studies.* Longmans, Green & Co., New York, 00
Janis, I.L. (1982) *Groupthink*, Houghton Mifflin, Boston, Mass.
Janson, Ch. H. (1985) Aggressive competition and individual food consumption in wild brown capuchin monkeys (*Cebus apella*). *Behav. Ecol. and Sociobiol.*, **18**, 125–38.
Janson, Ch. H. and van Schaik, C.P. (1988) Recognizing the many faces of primate food competition methods. *Behaviour*, **105**(1–2), 165–86.
Jenness, D. (1922) *Life of the Copper Eskimos.* Report of the Canadian Arctic Expedition, 1913–1918, XII. 1922.
Jensen, A.R. (1982) The debunking of scientific fossils and straw persons. *Contemporary Education Reviews*, Spring.
Jensen, A.R. (1969) How much can we boost IQ and scholastic achievement? *Harvard Educational Review*, **39**, 1–123.
Jerusalem, W. (1915) *Der Krieg im Lichte der Gesellschaftslehre*, Ferdinand Enke, Stuttgart.
Johnson, G.R. (1986) Kin selection, socialization and patriotism: An integrating theory. *Politics and The Life Sciences*, **4**(2), 127–54.
Johnson, R.N. (1972) *Aggression in Man and Animals.* Saunders, Philadelphia.
Jongman, B. and Dennen, J. van der (1988) The Great 'War Figures' Hoax: An investigation in polemomythology. *Bulletin of Peace Proposals*, **19**(2), 197–203.
Jordan, D.S. (1907) *The Human Harvest; a Study of the decay of races through the survival of the unfit.* Amer. Unitarian Assoc., Boston, Mass.
Kallenberg, F. (1892) *Auf dem Kriegspfad gegen die Massai.* Beck, München.
Karli, P. (1956) The Norway rat's killing response to the white mouse; an experimental analysis. *Behaviour*, **10**, 81–103.
Karson, S. and O'Dell, J.W. (1976) A Guide to the Clinical Use of the 16 PF. Institute for Personality and Ability Testing, Champaign.
Kawai, M. (1965) On the system of social ranks in a natural troop of Japanese monkeys; I, basic rank and dependent rank, in *Japanese Monkeys: A Collection of Translations* (eds S.S. Altmann and K. Imanishi). Yerties Regional Primate Center, Atlanta, 66–86.
Kawamura, S. (1965) Matriarchal social ranks in the Minoo-B troop: a study of the rank system of Japanese monkeys, in *Japanese Monkeys: A Collection of Translations* (eds S.S. Altmann and K. Imanishi). Yerties Regional Primate Center, Atlanta, 105–10.
Kawanaka, K. (1973) Intertroop relationships among Japanese monkeys. *Primates*, **14**(2–3), 113–59.
Keith, A. (1947) *A New Theory of Human Evolution.* Peter Smith, Gloucester.
Keller, A.G. (1916) *Societal Evolution.* Macmillan, New York.
Keller, A.G. (1918) *Through War to Peace.* Macmillan, New York.

Kellett, A. (1982) *Combat Motivation: The behavior of soldiers in battle.* Kluwer Nijhoff, The Hague.
Kennedy, J.G. (1971) Ritual and intergroup murder; comments on man, primitive and modern, in *War and the Human Race* (ed. M.N. Walsh). Elsevier, Amsterdam, 40–61.
Kidd, D. (1904) *The Essential Kaffir.* Macmillan, London.
Kipnis, D. (1976) *The Powerholders.* University of Chicago Press, Chicago.
Kirton, M.J. (1961) *Management Initiative.* Acton Society Trust, London.
Kirton, M.J. (1976) Adaptors and innovators: A description and measure, *Journal of Applied Psychology*, **61**(5), 622–9.
Kirton, M.J. (1977) Adaptors and innovators and superior-subordinate identification, *Psychological Reports*, **41**, 289–90.
Kirton, M.J. (1978a) Have adaptors and innovators equal levels of creativity?, *Psychological Reports*, **42**(3), 695–8.
Kirton, M.J. (1978b) Adaptors and innovators in culture clash, *Current Anthropology*, **19**(3), 611–12.
Kirton, M.J. (1984) Adaptors and innovators: Why new initiatives get blocked, *Long Range Planning*, **17**, 137–43; reprinted in Hellriegel & Slocum – *Companion to Organisational Behaviour*, West Publishing and Richards M.D. – *Readings in Management*, South-Western Publishing Co.
Kirton, M.J. (1987a) Adaption-innovation: Problem solvers in organisations, in *Innovation: An Interdisciplinary Perspective* (eds K. Gronhaug and G. Kaufmann). Norwegian Press/ Oxford University Press, London.
Kirton, M.J. (1987b) Adaptors and innovators: Cognitive style and personality, in *Frontiers of Creativity Research: Beyond the Basics* (ed. S.G. Isaksen). Brearly Ltd., Buffalo, New York, 00
Kirton, M.J. (1987c) Kirton Adaption-Innovation Inventory (KAI) – Manual (2nd edn), Occupational Research Centre, Hatfield (only obtainable through M.J. Kirton).
Kirton, M.J. (1989) A theory of cognitive style, in *Adaptors and Innovators: Styles of creativity and problem solving* (ed. M.J. Kirton), Routledge, London, 1–36.
Kirton, M.J. and De Ciantis, S.M. (1986) Cognitive style and personality: The Kirton Adaption-Innovation and Cattell's Sixteen Personality Factor Inventories, *Personality and Individual Differences*, **7**, 141–46.
Kling, A.S. and H.D. Steklis (1976) A neural basis for affiliative behavior in non-human primates. *Brain, Behavior and Evolution*, **13**, 216–38.
Klutschak, H.W. (1881) *Als Eskimo unter den Eskimo.* Hartleben, Wien.
Knabenhans, A. (1917) Der Krieg bei den Naturvölkern. Eine vergleichende Studie über primitive Formen der Kriegführung, in *XVI Jahresbericht d. geogr.-ethnogr,* Gesellschaft Zürich.
Koestler, A. (1967) *The Ghost in the Machine.* Pan, London.
Koestler, A. (1978) *Janus: A Summing Up.* Hutchinson, London.
Konner, M. (1982) *The Tangled Wing.* Holt, New York.
Kortlandt, A. (1965) How do chimpanzees use weapons when fighting leopards. *Yearbook American Philosophical Society*, 327–32.
Kovalevsky, M. (1910) *Sociology* (in Russian).
Krapf-Askari, E. (1972) Women, spears and the scarce good: a comparison of the sociological function of warfare in two central African societies, in *Zande Themes* (eds A. Singer and B.V. Street). Blackwell, Oxford, 19–40.
Krebs, J.R. (1971) Territory and breeding density in the great tit, *Parus major. Ecology*, **52**(1), 2–22.
Krech, D., Crutchfield, R.S. and Ballachey, E.L. (1962) *Individual in Society; A*

Textbook of Social Psychology. McGraw-Hill, New York.
Kriesberg, L. (1973) *The Sociology of Social Conflicts.* Prentice-Hall, Englewood Cliffs, NJ.
Kriesberg, L. (1982) *Social Conflicts.* Prentice-Hall, Englewood Cliffs, NJ.
Krippendorff, E. (1985) *Staat und Krieg; die historische Logik politischer Unvernunft.* Suhrkamp, Frankfurt a. Main.
Kropotkin, P.A. (1902) *Mutual Aid. A factor of evolution.* Heinemann, London.
Kruuk, H. (1972) *The Spotted Hyaena: A Study of Predation and Social Behavior.* University of Chicago Press, Chicago.
Kruuk, H. (1975) Functional aspects of social hunting by carnivores, in *Function and Evolution in Behaviour* (eds G.P. Baerends, C. Beer and A. Manning). Clarendon Press, Oxford, 119–41.
Kühme, W. von (1965) Freilandstudien zur Soziologie des Hyänenhundes (*Lycaon pictus lupinus* Thomas 1902). *Zeitschrift für Tierpsychologie,* **22**, 495–541.
Kuhn, T.S. (1970) The Structure of Scientific Revolutions (2nd edn), University of Chicago Press, Chicago.
Kummer, H. (1968) *Social Organisation of Hamadryas Baboons; A Field Study.* S. Karger, Basel.
Kummer, H. (1971) *Primate Societies: Group Techniques of Ecological Adaptation.* Aldine-Atherton, Chicago.
Kurland, J.A. (1977) Kin selection in the Japanese monkey. *Contributions to Primatology,* Vol. 12, Karger, Basel.
Kuroda, S. (1984) Interaction over food among pygmy chimpanzees, in *The Pygmy Chimpanzee: evolutionary biology and behaviour* (ed. R.L. Susman). Plenum Press, New York, 301–24.
Labrousse, C.B. (1958) The crisis in the French economy at the end of the old regime, in *The Economic Origins of the French Revolution* (ed. Ralph W. Greenlaw). Heath, Boston, Mass., 59–72.
Lack, D. (1966) *Population Studies in Birds.* Clarendon, Oxford.
Lagorgette, J. (1906) *Le rôle de la guerre: Etude de sociologie générale.* Giard, Paris.
Laing, R.D. (1970) *Knots.* Tavistock, London.
Laing, R.D. (1967) *The Politics of Experience,* Penguin, Harmondsworth, Middlesex.
Lancaster. J.B. (1986) Primate social behavior and ostracism, in *Ostracism: a Social and Biological Phenomenon* (eds R.D. Masters and M. Gruter), Elsevier Science Publications Co.Ltd., New York, 215–25.
Lange, F.A. (19866) *Geschichte des Materialismus und Kritik seiner Bedeutung in der Gegenwart.* Baedeker, Iserlohn.
Langton, J. (1979) Darwinism and the behavioral theory of sociocultural evolution: an analysis. *American Journal of Sociology,* **85**, 288–309.
Laqueur, W. (1968) Revolution, *International Encyclopedia of the Social Sciences,* **13**, 501–7.
Lasswell, H.D. (1931) Conflict: Social, in *Encyclopedia of the Social Sciences,* Vol. 4, Macmillan, New York.
Lathrap, D.W. (1968) The hunting economies of the tropical forest zone of South America: An attempt at historical perspective, in *Man the Hunter* (eds R.B. Lee and I. DeVore). Aldine Atherton, Chicago, 23–9.
Laughlin, W.S. (1968) Hunting: An integrating biobehavioral system and its evolutionary importance, in *Man the Hunter* (eds. R.B. Lee and I. DeVore). Aldine Atherton, Chicago, 304–20.
Lavessan, H. De (1918) *La Civilisation et l'Organisation, leur influence sur la guerre.* Alcan, Paris.

Lawick, H. van and Lawick-Goodall, J. van (1971) *Innocent Killers*. Houghton Mifflin, Boston, Mass.

Lawner, R.L. (1954) Social conflict as a subject of investigation in American research from 1919 to 1953. Unpubl. Doct. Dissertation, University of New York.

Leakey, R.E. (1981) *The Making of Mankind*. Dutton, New York.

Leavitt, G. (1977) The frequency of warfare: an evolutionary perspective. *Sociological Inquiry*, **47**, 49–58.

Lee, R.B. (1968) What hunters do for a living, or how to make out on scarce resources, in *Man the Hunter* (eds R.B. Lee and I. DeVore). Aldine Atherton, Chicago, 30–48.

Lee, R.B. and DeVore, I. (eds) (1968) *Man the Hunter*. Aldine Atherton, Chicago.

Leeds, A. (1963) The functions of war, in *Violence and War* (ed. J. Masserman). Grune and Stratton, New York, 69–82.

Leeds, A. and Vayda, A.P. (eds) (1965) *Man, Culture and Animals*. Amer. Assoc. Adv. Sci., Washington DC.

Lefebrvre, G. (1947) *The Coming of the French Revolution*. Princeton University Press, Princeton, NJ.

Lenski, G.E. (1966) *Power and Privilege; A Theory of Social Stratification*. McGraw-Hill, New York.

Lenski, G. and Lenski, J. (1987) *Human Societies; An introduction to macrosociology* (2nd edn). McGraw-Hill, New York.

Lesser, A. (1967) War and the state, in *War* (eds M. Fried, M. Harris and R. Murphy). Natural History Press, Garden City, 92–6.

Letourneau, C. (1895) *La guerre dans les diverses races humaines*. L.Bataille, Paris.

Lévi-Strauss, C. (1949) *Les structures élémentaires de la parenté*. Presses Universitaires de France, Paris.

Lévi-Strauss, C. (1956) The family, in *Man, Culture and Society* (ed. H.L. Shapiro). Oxford University Press, London, 261–85.

Levins, R. (1968) Evolution in changing environments. *Monographs in Population Biology, 2*.

Levins, R. and Lewontin, R. (1985) *The Dialectical Biologist*. Harvard University Press, Cambridge, Mass.

Lewin, K. (1948) *Resolving Social Conflicts*. Harper, New York.

Lewinsohn, R. (1954) *Animals, Men and Myths*. Dutton, New York.

Lewontin, R.C. (1976) Sociobiology – a caricature of Darwinism, in *PSA 1976, 2* (eds F. Suppe and P. Asquith). Philosophy of Science Association, Lansing, MI., 22–31.

Lewontin, R.C. (1979) Sociobiology as an adaptationist program. *Behavioral Science*, **24**, 5–14.

Lifton, R.J. (1986) *The Nazi Doctors*. Basic Books, New York.

Ligon, J.D. and Ligon, S.H. (1982) The cooperative breeding behavior of the green woodhoopoe. *Scientific American*, **247**(1) 126–34.

Lima, S.L. (1987) Vigilance while feeding and its relation to the risk of predation. *J. Theor. Biol.*, **124**, 303–16.

Lippitt, R. and White, R.K. (1958) An experimental study of leadership and group life, in *Readings in Social Psychology* (3rd edn) (eds E.E. Maccoby, T.M. Newcomb and E.L. Hartley). Holt, New York.

Lombroso, C. (1878) *L'Uomo Delinquente*. Bocca, Torino.

Lopreato, J. (1968) Authority relations and class conflict. *Social Forces*, **47**, 70–9.

Lopreato, J. (1981) Toward a theory of genuine altruism in *Homo sapiens*. *Ethology and Sociobiology*, **2**, 113–26.

Lopreato, J. (1984) *Human Nature and Biocultural Evolution*. Allen & Unwin, Boston.

Lopreato, J. and Hazelrigg, L.E. (1972) *Class, Conflict, and Mobility: Theories and*

Studies of Class Structure. Chandler (T.Y. Crowell), San Francisco.
Lopreato, J. and Horton, G. (1987) Revolución en clave sociobiológica, in *La Heterodoxia Recuperada: Entorno a Angel Palerm*, (ed. S. Glantz). Fondo de Cultura Economica, Mexico, 617–44.
Lorenz, K. (1963) *Das sogenannte Böse; zur Naturgeschichte der Aggression*. Borotha-Schoeler, Wien.
Lorenz, K. (1965) *Evolution and Modification of Behavior*. University of Chicago Press, Chicago, Ill.
Lorenz, K. (1966) *On Aggression*. Methuen, London.
Lorenz, K. (1969) *On Aggression*. Bantam Books, New York.
Lotka, A.J. (1922) Contribution to the energetics of evolution. *National Academy of Sciences, Proceedings*, **8**, 147–54.
Lotka, A.J. (1925) *Elements of Physical Biology*. Williams and Wilkins, Baltimore, Md.
Low, H. (1848) *Sarawak, its Inhabitants and Productions*. Bentley, London.
Lubbock, J. (1870) *The Origin of Civilization and the Primitive Condition of Man*. Longmans, London.
Lumsden, C.J. and Wilson, E.O. (1981) *Genes, Mind and Culture, The Coevolutionary Process*. Harvard University Press, Cambridge, Mass.
Lundberg, G.A. (1939) *Foundations of Sociology*. Macmillan, New York.
Lyell, Ch. (1863) *The Geological Evidences of the Antiquity of Man*, Murray, London.
Lyon, G.F. (1824) *The Private Journal of Captain G.F. Lyon of H.M.S. Hecla, During the Recent Voyage of Discovery Under Captain Parry*. Murray, London.
MacArthur, R.H. (1962) Some generalized theoremes of natural selection. *National Academy of Sciences, Proceedings*, **48**, 1893–7.
MacCurdy, J.T. (1918) *The Psychology of War*. Luce, Boston.
MacIver, R.M. (1937) *Society: A textbook of sociology*. Farrar and Rinehart, New York.
MacLeod, W.C. (1931) *The Origin and History of Politics*. Wiley, New York.
Mack, R.W. and Snyder, R.C. (1957) The analysis of social conflict; toward an overview and synthesis. *Conflict Resolution*, **1**, 212–48.
Maringer, J. (1952) *Vorgeschichtliche Kultur*. Benziger Verlag, Berlin.
Maine, H.J. (1888) *International Law*. Murray, London.
Maine, H.J. (1861) *Ancient Law, its connections with the early history of society and its relation to modern ideas*. Murray, London.
Malinowski, B. (1936) The deadly issue. *Atlantic Monthly*, **158**, 659–69.
Malinowski, B. (1941) An anthropological analysis of war. *Amer. Jour. of Sociol.*, **46**, 521–50.
Malmberg, T. (1983) *Human Territoriality: Survey of behavioural territories in man with preliminary analysis and discussion of meaning*. Mouton, New York.
Malthus, T.R. (1798) *An Essay on the Principle of Population*. Johnson, London.
Manning, F.E. (1863) *Old New Zealand, a tale of the good old times; and a history of the war in the North*. Bentley, London.
Mariner, W. (1817) *An Account of the Natives of the Tonga Islands in the South Pacific Ocean* (2 Vols), Murray, London.
Markham, C.R. (1910) A list of the tribes in the valley of the Amazon, including those on the banks of the main stream and of all its tributaries. *J. Anthropol. Inst. of Great Britain and Ireland*, **40**, 73–140.
Marsden, H.M. (1968) Agonistic behavior of young rhesus monkeys after changes induced in social rank of their mothers. *Animal Behaviour*, **16**, 38–44.
Marx, K. (1967) *Capital*. International Publishers, New York.
Marx, K. (1959) Excerpt from A contribution to the critique of political economy, in

Marx and Engels: *Basic Writings on Politics and Philosophy* (ed. L.S. Feuer). Doubleday, Garden City, New York.

Marx, K. and Engels, F. (1947) *The German Ideology.* International Publishers, New York.

Marx, K. and Engels, F. (1955) Wage labour and capital, in *K. Marx and F. Engels, Selected Works in Two Volumes.* Foreign Languages Publishing, Moscow.

Massey, K.A. (1977) Agonistic aids and kinship in a group of pigtail macaques. *Behav. Ecol. and Sociobiol.*, **2**, 31–40.

Masters, R.D. (1982) Is sociobiology reactionary? The political implications of inclusive-fitness theory, *The Quarterly Review of Biology*, **57**, 274–92.

Masters, R.D. (1983) The biological nature of the state. *World Politics*, **35**(2), 161–93.

Mathiassen, T. (1928) *Material Culture of the Iglulik Eskimos.* Report of the Fifth Thule Expedition, 1921–24, VI, No. 1. Gyldendalske Boghandel, Copenhagen.

Maynard Smith, J. (1974) The theory of games and the evolution of animal conflicts. *Journal of Theoretical Biology*, **47**(1), 209–21.

Maynard Smith, J. (1976) Group selection. *Quarterly Rev. of Biol.*, **51**, 277–83.

Maynard Smith, J. (1978) The evolution of behavior. *Sci. Am.*, **239**, 176–92.

Maynard Smith, J. (1982) *Evolution and the Theory of Games.* Cambridge University Press, Cambridge.

Maynard Smith, J. (1984) Game theory and the evolution of behavior. *Behavior and the Brain Sciences*, **7**(1), 95–125.

Maynard Smith, J. (1986) Molecules are not enough. *The London Review of Books*, February.

Maynard Smith, J. and Parker, G.A. (1976) The logic of asymmetric contests. *Animal Behavior*, **24**(1), 159–75.

Maynard Smith, J. and Price, G.R. (1973) The logic of animal conflict. *Nature*, **246**, 15–18.

Mayr, E. (1935) Bernard Altum and the Territory theory. *Proceedings of the Linnaean Society of New York*, **45** and **46**, 24–38.

Mayr, E. (1974) Teleological and teleonomic: A new analysis. *Boston Studies in the Philosophy of Science*, **14**, 19–117.

McClain, E. (1978) Feminists and nonfeminists: contrasting profiles in independence and affiliation, *Psychological Reports*, **43**, 435–441.

McClain, E. (1979) Article on satellizing versus non-satellizing children and the need for affiliation versus the need for independence, *Brain/Mind Bulletin*, **4**(7), 3.

McDougall, W. (1914) *An Introduction to Social Psychology.* (8th edn), Luce, Boston, Mass.

McDougall, W. (1927) *Janus: The conquest of war.* Kegan Paul, Trench, Trubner, London.

McEachron, D.L. and Baer, D. (1982) A review of selected sociobiological principles: application to hominid evolution II. the effects of intergroup conflict. *J. Social and Biol. Structures.* **5**(2), 121–39.

McEnery, J.H. (1985) Toward a new concept of conflict evaluation. *Conflict*, **6**(1), 37–87.

McGrew, W.C. (1979) Evolutionary implications of sex differences in chimpanzee predation and tool use, in *The Great Apes* (eds D.A. Hamburg and E.R. McCown). Benjamin/Cummings, Menlo Park, Cal.

McKenna, J.J. (1982) Primate field studies: the evolution of behavior and its socio-ecology, in *Primate Behavior*, (eds J.L. Forbes and J.E. King). Academic Press, New York, 53–83.

McNeil, E.B. (ed.) (1965) *The Nature of Human Conflict.* Prentice-Hall, Englewood Cliffs, NJ.

McNeill, W.H. (1963) *The Rise of the West; A History of the Human Community.* The University of Chicago Press, Chicago, Ill.
McNeill, W.H. (1982) *The Pursuit of Power.* The University of Chicago Press, Chicago, Ill.
Mech, L.D. (1966) *The Wolves of Isle Royal.* Fauna of National Parks of the US, Vol. 7, US Government Printing Office, Washington DC.
Mech, L.D. (1970) *The Wolf: The Ecology and Behavior of an Endangered Species.* Natural History Press, Garden City.
Mead, M. (1940) Warfare is only an invention – not a biological necessity. *Asia*, **40**, 402–5.
Mead, M. (1962) The psychology of warless man, in *A Warless World* (ed. A. Larson), McGraw-Hill, New York, 131–42.
Mead, M. (1964) Warfare is only an invention – not a biological necessity, in *War. Studies from Psychology, Sociology, Anthropology* (eds L. Bramson and G.W. Goethals). Basic Books, London, 269–75.
Mead, M. (1967) Alternatives to war. *Supplement to Natural History*, **76**, 65–9.
Mehrabian, A. (1972) *Nonverbal Communication.* Aldine, Altherton, Chicago/New York.
Melotti, U. (1979) *L'uomo tra natura e storia. La dialettica delle origini.* Centro Studi Terzo Mondo, Milano.
Melotti, U. (1980) Towards a new theory on the origin of the family: Some hypotheses on monogamy, polygyny, incest taboo, exogamy, and genetic altruism. *The Mankind Quarterly*, **21**(2), 99–133.
Melotti, U. (1981) Towards a new theory of the origin of the family. *Current Anthropology*, **22**(6), 625–38.
Melotti, U. (1982) Oltre la sociobiologia. Verso una nuova scienza unitaria dell'uomo e della società, in *Sociobiologia possibile* (eds M. Ingrosso, S. Manghi, V. Parisi). Angeli, Milano, 35–68.
Melotti, U. (1984a) A sociobiological interpretation of the structures and functions of the human family. *Journal of Human Evolution*, **13**(1), 81–90.
Melotti, U. (1984b) The origin of human aggression: A new evolutionary view. *The Mankind Quarterly*, **24**(4), 379–91.
Melotti, U. (1985a) Competition and cooperation in human evolution. *The Mankind Quarterly*, **25**(4), 323–352.
Melotti, U. (1985b) La sociobiologie comme antidiscipline de l'anthropologie culturelle: une critique de l'interprétation des structures de la parenté chez Lévi-Strauss. *Revue européenne des sciences sociales*, **23**(69), 313–27.
Melotti, U. (1986a) In-group/out-group relations and the issue of group selection, in *The Sociobiology of Ethnocentrism.* (eds V. Reynolds, V. Falger and I. Vine). Croom Helm, London, 94–111.
Melotti, U. (1986b) On the evolution of human aggressiveness, in *Essays in Human Sociobiology* (eds J. Wind and V. Reynolds) Vrije Universiteit, Brussels, 69–81.
Melotti, U. (1986c) Sociobiologia e organizzazione sociale: il caso delle società primitive, in *La Dimensione Bioculturale: Evoluzionismo e scienze dell'uomo oltre la sociobiologia*, (eds S. Manghi and V. Parisi) Centro Studi Terzo Mondo, Milan, 72–102.
Melotti, U. (1987) Prospettive antropologiche della sociobiologia. *Antropologia contemporanea*, **9**(4), 281–95.
Merker, M. (1904) *Die Masai: Ethnographische Monographie eines ostafrikanischen Semitenvolkes.* D. Reimer, Berlin.
Merton, R.K. (1957) *Social Theory and Social Structure.* Free Press, Glencoe.

Meyer, P. (1977) *Kriegs- und Militärsoziologie.* Westdeutscher Verlag. Köln.

Meyer, P. (1981) *Evolution und Gewalt.* Paul Parey, Hamburg.

Michels, R. (1914) *Political Parties.* Free Press, New York.

Milgram, S. (1974) *Obedience to Authority.* Tavistock, London.

Miller, G. (1986) Is scientific thinking different?, *Bulletin of the American Academy of Arts and Sciences,* **36**(5), February.

Miller, R.S. (1967) Pattern and process in competition. *Advances in Ecological Research,* **4**, 1–74.

Mills, C.W. (1956) *The Power Elite.* Oxford University Press, New York.

Mitani, J.C. and Rodman, P.S. (1979) Territoriality: the relation of ranging patterns and home range size to defendability, with an analysis of territoriality among primate species. *Behav. Ecol. and Sociobiol.,* **5**, 241–51.

Moller, A.P. (1982) Coloniality and colony structure in gull-billed terns *Gelochelidon nilotica. J. Ornithol.,* **123**, 41–54.

Montagu. A. (1974) Aggression and the evolution of man, in *The Neuropsychology of Aggression* (ed. R.E. Whalen). Plenum Press, New York, 1–32.

Montagu,. A. (1976) *The Nature of Human Aggression.* Oxford University Press, New York.

Montesquieu, C.L. Baron de (1748) *L'Esprit des Lois.* J.J. Venet, Genève.

Mooney, J. (1894) The Siouan tribes of the East. *Bureau American Ethnology Bulletin,* **22**, whole issue.

Moran, E.F. (1979) *Human Adaptability.* Westview Press, Boulder, Col.

Morey, R.V. and Marwitt, J.P. (1975) Ecology, economy and warfare in Lowland South America, in *War, Its Causes and Correlates* (eds M.A. Nettleship, R.D. Givens and A. Nettleship). Mouton, The Hague, 439–50.

Morgan, L.H. (1851) *League of the Ho-de-no-sau-nee or Iroquois.* Sage, Rochester.

Morren, G.E. (1984) Warfare on the Highland Fringe of New Guinea: the case of the Mountain Ok, in *Warfare, Culture and Environment* (ed. R.B. Ferguson). Academic Press, New York.

Motro, U. (1983) Optimal rates of dispersal. III. Parent-offspring conflict. *Theor. Pop. Biol.,* **23**, 159–68.

Motro, U. and Cohen, D. (1989) A note on vigilance behavior and stability against recognizable social parasites. *J. theor. Biol.,* **136**, 21–25.

Motro, U. and Eshel, I. (1988) The three brothers' problem: kin selection with more than one potential helper. II. The case of delayed help. *Am. Natur.,* **132**, 567–75.

Moyer, K.E. (1968) Kinds of aggression and their physiological basis. *Comm. Behav. Biol.,* **2**, 65–87.

Moyer, K.E. (1976) *The Psychobiology of Aggression.* Harper & Row, New York.

Mühlmann, W.E. (1962) *Homo Creator.* Harrassowitz, Wiesbaden.

Mueller-Lyer, F. (1921) *Phasen der Kultur und Richtungslinien des Fortschritts,* R. Oldenbourg, München.

Mulkay, M.S. (1972) *The Social Process of Innovation,* Macmillan, London.

Mulkay, M. and Gilbert, G.N. (1982) Accounting for error. *Sociology,* **16**, 165–83.

Murdock, G.P. and White, D.R. (1969) Standard cross-cultural sample. *Ethnology,* **8**, 329–69.

Murie, A. (1944) *The Wolves of Mt McKinley.* US Dept. Interior, Fauna Series, no. 5. US Government Printing Office, Washington DC.

Murphy, R.F. (1957) Intergroup hostility and social cohesion. *Amer. Anthropol.,* **59**, 1018–34.

Murphy, R.F. (1958) Reply to H.C. Wilson's reply on the causes of Mundurucu warfare. *Amer. Anthropol.,* **60**, 1196–9.

Murphy, R.F. (1960) *Headhunters' Heritage: Social and Economic Change among the Mundurucu Indians.* University California Press, Berkeley, Cal.

Myer, J.S. and White, R.T. (1965) Aggressive motivation in the rat. *Anim. Behav.*, **13**, 430–3.

Nagel, U. and Kummer, H. (1974) Variation in Cercopithecoid Aggressive Behaviour, in *Primate Aggression, Territoriality, and Xenophobia* (ed. R.L. Holloway). Academic Press, New York, 159–85.

Nansen, F. (1893) *Eskimo Life.* Longmans, Green, London.

Naroll, R. (1964) On ethnic unit classification. *Current Anthropol.*, **5**, 283–312.

Naroll, R. (1970) Cross-cultural sampling, in *A Handbook of Method in Cultural Anthropology* (eds R. Naroll and R. Cohen). Natural History Press, New York, 889–926.

Nasmyth, G. (1916) *Social Progress and the Darwinian Theory.* Putnam, New York.

Nelson, E.W. (1899) *The Eskimo about Bering Strait.* Bureau of American Ethnology, 18th. Annual Report. Washington.

Nelson, S.D. (1974) Nature/Nurture revisited: A review of the biological bases of conflict. *J. Conflict Resolution,* **18**(2), 285–335.

Nettleship, M.A., Givens, R.D. and Nettleship, A.(eds) (1975) *War, its Causes and Correlates.* Mouton, The Hague.

Nettleship, M.A., Givens, R.D. and Nettleship, A. (eds) (1976) *Discussions on War and Aggression.* Mouton, The Hague.

Netto, W.J. and van Hooff, J.A.R.A.M. (1986) Conflict interference and the development of dominance relationships in immature *Macaca fascicularis,* in *Primate Ontogeny, Cognition and Social Behaviour* (eds J. Else and P. Lee). Cambridge University Press, Cambridge, 291–300.

Neumann, G.-H. (1981) *Normatives Verhalten und aggressive Aussenseitereaktionen bei gesellig lebenden Vögeln und Säugern.* Westdeutscher Verlag, Opladen (BRD).

Newcomb, W.W. (1950) A re-examination of the causes of Plains warfare. *Amer. Anthropol.*, **52**, 317–30.

Newcomb, W.W. (1960) Towards an understanding of war, in *Essays in the Science of Culture* (eds G.E. Dole and R.L. Carneiro). Crowell, New York, 317–36.

Nicholson, M.B. (1970) *Conflict Analysis.* English Universities Press, London.

Nicolai, G.F. (1919) *Die Biologie des Krieges.* Orell Füssli, Zürich.

Nishida, T. (1979) The social structure of chimpanzees of the Mahale Mountains, in *The Great Apes* (eds D.A. Hamburg and E.R. McCown). Benjamin/Cummings, Menlo Park, Cal. 72–191.

Nishida, T. (1983) Alpha status and agonistic alliance in wild chimpanzees (*Pan troglodytes schweinfurthii*). *Primates,* **24**, 318–36.

Nobel, G.K. (1939) The role of dominance in the social life of birds. *Auk,* **56**(3), 263–73.

Noë, R. (1986) Lasting alliances among adult male savannah baboons, in *Primate Ontogeny, Cognition and Social Behaviour* (eds. J. Else and P. Lee). Cambridge University Press, Cambridge, 381–92.

Noë, R. (1989) Coalition formation among male baboons. Ph.D. Thesis, University of Utrecht.

Nordau, M. (1889) The philosophy and morals of war. *North Amer. Rev.*, **clxix**, whole issue.

Novikow, J. (1896) *Les luttes entre les sociétés humaines et leur phases successives.* Alcan, Paris.

Novikow, J. (1912) *War and its Alleged Benefits.* Heinemann, London.

Numelin, R.J. (1950) *The Beginnings of Diplomacy: A sociological study of intertribal and*

international relations. Oxford University Press, London.
Oberschall, A. (1973) *Social Conflict and Social Movements*. Prentice-Hall, Englewood Cliffs, NJ.
Oberschall, A. (1978) Theories of social conflict. *Annual Review of Sociology*, **4**, 291–315.
Olson, M. (1982) *The Rise and Decline of Nations (Economic Growth, Stagflation, and Social Rigidities)*. Yale University Press, New Haven.
Ostrogorski, M. (1982) *Democracy and the Organization of Political Parties*, Haskell House, Brooklyn, New York.
Otterbein, K. (1967) The evolution of Zulu warfare, in *Law and Warfare*, (ed. P. Bohannan). The Natural History Press, New York, 351–7.
Otterbein, K.F. (1968) Internal war: a cross-cultural study. *American Anthropol.*, **70**, 277–89.
Otterbein, K.F. (1968) Higi armed combat. *Southwestern J. Anthropol.*, **24**, 195–213.
Otterbein, K.F. (1970) *The Evolution of War: A Cross-Cultural Study*. HRAF Press, New Haven.
Otterbein, K.F. (1970) The anthropology of war, in *Handbook of Social and Cultural Anthropology* (ed. J.J. Honigmann). Rand McNally, New York, 923–58.
Otterbein, K.F. (1976) Warfare, territorial expansion, and cultural evolution. *Amer. Ethnol.*, **3**, 825–7.
Otterbein, K.F. and Otterbein, C. (1965) An eye for an eye, a tooth for a tooth: a cross-cultural study of feuding. *Amer. Anthropol.*, **67**, 1470–82.
Packer, C. (1977) Reciprocal altruism in *Papio anubis*. *Nature*, **265**, 441–3.
Pareto, V. (1916) *A Treatise on General Sociology* (also known as *The Mind and Society*). Dover, New York.
Park, R.E. and Burgess, E.W. (1924) *Introduction to the Science of Sociology*. University of Chicago Press, Chicago, Ill.
Parker, G.A. (1970) Sperm competition and its evolutionary consequences in the insects. *Biological Review*, **45**, 325–67.
Parker, G.A. (1974) Assessment strategy and the evolution of fighting behavior. *Journal of Theoretical Biology*, **47**(1), 223–43.
Parker, G.A. and Hammerstein, P. (1985) Game theory and animal behaviour, in *Evolution: Essays in Honour of John Maynard Smith* (eds P.J. Greenwood, P. Harvey and M. Slatkin). Cambridge University Press, New York.
Parr, A.E. (1966) In search of theory. *Arts and Architecture*, **82**.
Parsons, T. (1951) *The Social System*. Free Press, Glencoe.
Patterson, J.S. (1883) *Conflict in Nature and Life: A study of antagonism in the constitution of things*. Appleton, New York.
Pear, T.H. (1950) *Psychological Factors of Peace and War*. Philosophical Library, New York.
Perry, R.B. (1918) *The Present Conflict of Ideals*. Longmans, Green, New York.
Perry, W.J. (1917a) An ethnological study of warfare, in *Memoirs and Proceedings of the Manchester Literary and Philosophical Society*, **61**(6), 1–16.
Perry, W.J. (1917) The peaceable habits of primitive communities: An anthropological study of the Golden Age. *Hilbert Journal*, **17**, 28–46.
Perry, W.J. (1923) *The Growth of Civilization*, Dutton, New York.
Peters, R. and Mech, L.D. (1975) Behavioral and intellectual adaptations of selected mammalian predators to the problem of hunting large animals, in *Socioecology and Psychology of Primates* (ed. R.H. Tuttle). Mouton, The Hague, 279–300.
Peterson, S.A. and Somit, A. (1978) Sociobiology and politics, in *The Sociobiology Debate* (ed. A. L. Caplan). Harper and Row, New York, 449–61.

Petroff, I. (1884) *Alaska: It's Population, Industries and Resources.* Tenth Census of the United States, VIII.
Pettman, R. (1975) *Human Behaviour and World Politics; A transdisciplinary introduction.* Macmillan, London.
Pfeiffer, J.E. (1972) *The Emergence of Man.* Harper & Row, New York.
Phillips, C.S. (1987) Politics: An aspect of cultural evolution. *Politics and the Life Sciences*, **5**, 2.
Pirsig, R.M. (1974) *Zen and the Art of Motorcycle Maintenance.* Corgi Books, London.
Pitt, R. (1978) Warfare and hominid brain evolution. *J. Theoretical Biol.*, **72**(3), 551–75.
Pittendrigh, C.S. (1958) Adaptation, natural selection and behavior, in *Behavior and Evolution* (eds A. Roe and G.G. Simpson). Yale University Press, New Haven, Conn., 390–416.
Plomin, R. and Rowe, D.C. (1977) A twin study of temperament in young children, *The Journal of Psychology*, **97**, 107–13.
Plomin, R. and Rowe, D.C. (1979) Genetic and environmental etiology of social behaviour in infancy, *Developmental Psychology*, **15**(1), 62–72.
Poirier, F.E. (1970) Dominance structure of the Nilgiri langur (*Presbytis johnii*) of south India. *Folio Primatologica*, **12**, 161–86.
Popper, K.R. (1945) *The Open Society and its Enemies*, Vol. II. Routledge and Kegan Paul, London.
Powell, J.W. (1891) *Indian Linguistic Families North of Mexico.* 7th Ann. Report Bur. Ethnol. Smithsonian Inst., Washington.
Pulliam, H.R. and Dunford, C. (1980) *Programmed to Learn.* Columbia University Press, New York.
Pulliam, H.R., Pyke, G.H. and Caraco, T. (1982) The scanning behaviour of Juncos: a game theoretical approach. *Journal of Theoretical Biology*, **95**, 89–103.
Pusey, A.E. (1979) Intercommunity transfer of chimpanzees in Gombe National Park, in *The Great Apes* (eds D.A. Hamburg and E.R. McCown). Benjamin/Cummings, Menlo Park, Cal., 465–80.
Pusey, A.E. (1980) Inbreeding avoidance in chimpanzees. *Animal Behaviour*, **28**, 543–52.
Rapoport, A. (1965) Is warmaking a characteristic of human beings or of cultures? *Sci. Amer.*, **213**, 115–18.
Rapoport, A. (1960) *Fights, Games and Debates.* University of Michigan Press, Ann Arbor, NJ.
Rapoport, A. (1965) Game theory and human conflict, in *The Nature of Human Conflict* (ed. E. McNeil). Prentice-Hall, Englewood Cliffs, NJ, 195–276.
Rapoport, A. (1966) Models of conflict: cataclysmic and strategic, in *Conflict in Society* (eds A. deReuck and J. Knight). Churchill, London, 259–87.
Rapoport, A. (1974) *Conflict in Man-made Environment.* Penguin Books, New York.
Rasa, O.A.E. (1986) Coordinated vigilance in Dwarf mongoose family groups: The 'Watchman's Song' hypothesis and the costs of guarding. *Ethology*, **71**, 340–4.
Rasa, O.A.E. (1987) The dwarf mongoose: A study of behavior and social structure in relation to ecology, *Advances in the Study of Behavior.* Vol. 17, Academic Press, 121–63.
Rasmussen, K. (1927) *Across Arctic America.* Putnam, London.
Rasmussen, K. (1929) *Intellectual Culture of the Ingloolik Eskimos.* Report of the Fifth Thule Expedition, 1921–1924, VII, No. 1. Gyldendalske Boghandel, Copenhagen.
Rasmussen, K. (1931) *The Netsilik Eskimos: Social Life and Spiritual Culture.* Report of

the Fifth Thule Expedition, 1921–24. Vol. VIII, No. 1–2, Gyldendalske Boghandel, Copenhagen.
Ratzel, F. (1885–1888) *Völkerkunde.* (2 Vols.). Leipzig.
Ratzel, F. (1894) *Völkerkunde.* (2 Vols.) Brockhaus, Leipzig.
Ratzenhofer, G. (1893) *Wesen und Zweck der Politik.* Brockhaus, Leipzig.
Rawling, C.G. (1913) *The Land of the New Guinea Pygmies.* Seeley, Service & Co., London.
Reddin, W.J. (1970) *Managerial Effectiveness.* McGraw-Hill, New York.
Reddin, W.J. (1987) *Effective Management.* McGraw-Hill, New Delhi.
Rensberger, B. (1977) *The Cult of the Wild.* Anchor Press, Doubleday, New York.
Rex, J. (1961) *Key Problems in Sociological Theory.* Routledge & Kegan Paul, London.
Reynolds, V. (1966) Open groups in hominid evolution. *Man,* **1**, 441–52.
Reynolds, V. (1976) *The Biology of Human Action.* Freeman, San Francisco.
Reynolds, V. and Luscombe, G. (1969) Chimpanzee rank order and the function of displays, in *Proceedings of the Second International Congress of Primatology* (ed. C.R. Carpenter). Karger, Basel, 81–6.
Reynolds, V., Falger, V. and Vine, I. (eds) (1987) *The Sociobiology of Ethnocentrism: Evolutionary dimensions of xenophobia, discrimination, racism and nationalism.* Croom Helm, London.
Richardson, L.F. (1960) *Statistics of Deadly Quarrels.* Boxwood Press, Pittsburgh.
Rink, H. (1887) *The Eskimo Tribes. Their Distribution and Characteristics, Especially in Regard to Language; With a Comparative Vocabulary.* Meddelelser om Grönland, XI, Williams & Norgate, London.
Ripley, S. (1967) Intertroop encounters among Ceylon grey langurs (*Presbytis entellus*), in *Social Communication among Primates* (ed. S.A. Altmann). Chicago University Press, Chicago, Ill. 237–54.
Riss, D. and Goodall, J. (1977) The recent rise to the alpha rank in a wild population of free-living chimpanzees. *Folia Primatologica,* **27**, 134–51.
Rivers, W.H. (1922) *History and Ethnology.* Macmillan, London.
Robinson, J.G. (1988) Group size in wedge-capped capuchin monkeys *Cebus olivaceus* and the reproductive success of males and females. *Behav. Ecol. and Sociobiol.*, **23**, 187–97.
Rogers, C.R. (1959) Towards a theory of creativity, in *Creativity and its Cultivation* (ed. H.H. Anderson). Harper, New York.
Rose, S. (ed.) (1982a) *Against Biological Determinism.* Allison & Busby, London/New York.
Rose, S. (1982b) *Towards a Liberatory Biology.* Allison & Busby, London/New York.
Rose, S., Lewontin, R.C. and Kamin, L.J. (1984) *Not in Our Genes. Biology, Ideology and Human Nature.* Penguin Books, Harmondsworth.
Rosenzweig, S. (1981) Toward a comprehensive definition and classification of aggression, in *Multidisciplinary Approaches to Aggression Research* (eds P.F. Brain and D. Benton). Elsevier/North Holland, Amsterdam, 17–22.
Ross, E.A. (1930) *Principles of Sociology.* Century, New York.
Rousseau, J.J. (1762) *Le contrat social ou principes de droit politique.* M.M. Rey, Amsterdam.
Rousseau, J.J. (1755) *Discours sur l'origine et les fondements de l'inégalité parmi les hommes.* M.M. Rey, Amsterdam.
Rowell, T.E. (1972) *The Social Behaviour of Monkeys.* Penguin, Harmondsworth.
Russell, E.W. (1972) Factors of human aggression: a cross-cultural factor analysis of characteristics related to warfare and crime. *Behavior Science Notes,* **7**, 275–312.
Russell, G.W. (1981) Aggression in sport, in *Multidisciplinary Approaches to Aggression*

Research (eds P.F. Brain and D. Benton). Elsevier/North Holland, Amsterdam, 431–46.

Ruyle, E.E. (1973) Genetic and cultural pools: some suggestions for a unified theory of biocultural evolution. *Human Ecology*, **1**, 201–15.

Saabye, H.E. (1818) *Greenland: Being extracts from a journal kept in that country in the years 1770–1778* (Trans. from the German). A.S. Boosey and Sons, London.

Sade, D.S. (1967) Determinance of dominance in a group of free-living rhesus monkeys, in *Social communication among primates* (ed. S. Altmann), Chicago University Press, Chicago.

Sahlins, M.D. (1960) Evolution: specific and general, in *Evolution and Culture* (eds M.D. Sahlins and E. Service). University of Michigan Press, Ann Arbor, Mich., 12–44.

Samuels, A., Silk, J.B. and Rodman, P. (1984) Changes in the dominance rank and reproductive behavior of male bonnet macaques (*Macaca radiata*). *Animal Behavior*, **32**, 994–1003.

Savin-Williams, R.C. (1977a) Dominance in a human adolescent group, *Animal Behaviour*, **25**, 400–6.

Savin-Williams, R.C. (1977b) Dominance-submission behaviour and hierarchies in young adolescents at a summer camp: Predictors, styles, and sex differences, Dissertation, University of Chicago, Ill.

Savin-Williams, R.C. (1979) Dominance hierarchies in groups of early adolescents, *Child Development*, **50**, 923–35.

Savin-Williams, R.C. (1980) Dominance hierarchies in groups of middle to late adolescent males, *Journal of Youth and Adolescence*, **9**(1), 75–85.

Schachter, S. (1951) Deviation, rejection, and communication, *Journal of Abnormal and Social Psychology*, **46**, 190–207.

Schaeffle, A. (1900) *Zur sozialwissenschaftlichen Theorie des Krieges*. Z.f.gesammten Staatswissenschaften, 61.

Scheflen, A. and Scheflen, A. (1972) *Body Language and Social Order*. Prentice Hall, Englewood Cliffs, NJ.

Schellenberg, J.A. (1982) *The Science of Conflict*. Oxford University Press, Oxford.

Schelling, T.C. (1958) The strategy of conflict: prospectus for a reorientation of game theory. *Journal of Conflict Resolution*, **2**, 303–64.

Schelling, T.C. (1960) *The Strategy of Conflict*. Harvard University Press, Cambridge.

Scherer, K.R., Abeles, R.P. and Fischer, C.S. (1975) *Human Aggression and Conflict: Interdisciplinary Perspectives*. Prentice-Hall, Englewood Cliffs, NJ.

Schmitthenner, P. (1930) *Krieg und Kriegführung im Wandel der Weltgeschichte*. Athenaion, Potsdam.

Schultze, F. (1900) *Psychologie der Naturvölker*. Von Veit, Berlin.

Schumacher, E.F. (1975) *Small is Beautiful; Economics as if people mattered*. Harper & Row, New York.

Schumpeter, J.A. (1939) *Business Cycles, a Theoretical, Historical and Statistical Analysis of the Capitalist Process*. McGraw-Hill, New York.

Scott, J.P. (1974) Agonistic behavior of primates: a comparative perspective, in *Primate Aggression, Territoriality, and Xenophobia* (ed. R.L. Holloway). Academic Press, New York, 417–34.

Scott, J.P. (1975) Personal, social and international violence, in *War, Its Causes and Correlates* (eds M.A. Nettleship, R.D. Givens and A. Nettleship). Mouton, The Hague, 173–83.

Scott, J.P. (1980) *The relationship between hunting, predation, and agonistic behavior*. Paper ISRA Meeting, Haren, The Netherlands.

Scott, J.P. (1981) The evolution of function in agonistic behavior, in *Multidisciplinary Approaches to Aggression Research* (eds P. Brain and D. Benton). Elsevier, Amsterdam, 129–57.

Scott, J.P. and Fredericson, E. (1951) The causes of fighting in mice and rats. *Physiological Zoology*, **24**, 273–309.

Scott, J.P. and Fuller, J.L. (1965) *Genetics and the Social Behavior of the Dog.* University of Chicago Press, Chicago, Ill.

Sebeok, T.H. (1987) On a High Horse, *Semiotica*, **67**, 1–2.

Seeck, O. (1910) *Geschichte des Unterganges der antiken Welt.* F. Siemenroth, Berlin.

Segerstråle, U. (1983) Whose truth shall prevail? Moral and scientific interests in the sociobiology controversy. Ph.D. Thesis in sociology. Harvard University (unpublished).

Segerstråle, U. (1986) Colleagues in conflict: An *in vivo* analysis of the sociobiological controversy, *Biology and Philosophy*, **1**, 53–87.

Segerstråle, U. (1987) Scientific controversy as moral/political discourse. *Contemporary Sociology*, **16**(3).

Segerstråle, U. (1989) The (re)colonization of science by the life-world: problems and prospects, in *Social Structure and Culture* (ed. H. Haferkamp). De Gruyter, Berlin and New York.

Service, E.R. (1966) *The Hunters.* Prentice Hall, Englewood Cliffs, NJ.

Service, E.R. (1975) *Origins of the State and Civilization; The process of cultural evolution.* Norton, New York.

Shaw, R.P. (1985) Humanity's propensity for warfare: a sociobiological perspective. *Canad. Rev. Sociol. & Anthropol.*, **22**(2), 158–83.

Shaw, R.P. and Wong, Y. (1987) Ethnic mobilization and the seeds of warfare: an evolutionary perspective, *International Studies Quarterly*, **31**(1), 5–31.

Shaw, R.P. and Wong, Y. (1988) *Genetic Seeds of Warfare; Evolution, Nationalism, and Patriotism.* Unwin Hyman, London.

Shepard, P. (1973) *The Tender Carnivore and the Sacred Game.* Scribner's Sons, New York.

Sherif, M. and Sherif, C.W. (1970) Motivation and intergroup aggression: a persistent problem in levels of analysis, in *Development and Evolution of Behavior* (eds L.R. Aronson, *et al.*) Freeman, San Francisco, 563–79.

Shields, W.M. (1982a) *Philopatry, Inbreeding and the Evolution of Sex.* State University of New York Press, Albany, NY.

Shields, W.M. (1982b) Inbreeding and the paradox of sex: A resolution? *Evolutionary Theory*, **5**, 245–79.

Shirom, A. (1976) On some correlates of combat performance. *Admin. Sci. Quart.*, **21**, 419–32.

Simeons, A.T. (1960) *Man's Presumptuous Brain.* Longmans Green, London.

Simmel, G. (1903) The sociology of conflict. *American Journal of Sociology*, **9**, 490–525, 627–89, 798–84.

Simmel, G. (1955) *Conflict.* Free Press, New York.

Simmel, G. (1966) *Conflict and the Web of Group Affiliations.* Free Press, New York.

Simonds, P.E. (1965) The bonnet macaque in south India, in *Primate Behavior: Field Studies of Monkeys and Apes* (ed. I. DeVore). Holt, Rinehart and Winston, New York, 175–96.

Simpson, G.E. (1937) *Conflict and Community; A study in social theory.* T.S. Simpson, New York.

Simpson, G.G. (1971) *The Meaning of Evolution.* Bantam Books, New York.

Singer, K. (1949a) *The Idea of Conflict.* Melbourne University Press, Melbourne.

Singer, K. (1949b) The meaning of conflict. *Australian Journal of Philosophy*, **27**(3), 145–70.
Skocpol, T. (1979) *States and Social Revolutions.* Harvard University Press, Cambridge, Mass.
Slobodkin, L.B. (1972) On the inconstancy of ecological efficiency and the form of ecological theories. *Conn. Acad. Arts Sci., Trans.*, **44**, 291–305.
Sluckin, A.M. and Smith, P.K. (1977) Two approaches to the concept of dominance in preschool children. *Child Development*, **48**, 917–23.
Smith, C.G. (ed.) *Conflict Resolution: Contributions of the behavioral sciences.* University of Notre Dame Press, London.
Smith, E., Udry, R. and Morris, N. (1985) Pubertal development and friends: A biosocial explanation of adolescent sexual behavior, *J. of Health & Social Behavior*, **26**, 183–92.
Smith, G.E. (1924) *The Evolution of Man.* Oxford University Press, Oxford.
Smith, S.P. (1898) Wars of the northern against the southern tribes of New Zealand. *Journal Polynesian Society*, **7**, 141–234.
Snow, C.P. (1961) Either – Or, *Progressive*, Febr. 1961, 24.
Sociobiology Study Group (1978) Sociobiology: another biological determinism, in *The Sociobiology Debate* (ed. A. Caplan). Harper and Row, New York, 280–90.
Somit, A. (1972) Biopolitics. *Brit. J. Polit. Sci.*, **2**, 209–38.
Sorokin, P.A. (1928) *Contemporary Sociological Theories.* Harper, New York.
Sorokin, P.A. (1937) *Social and Cultural Dynamics.* American Book Company, New York.
Sorokin, P.A. (1947) *Society, Culture and Personality.* Harper, New York.
Sorokin, P.A. (1957) *Social Change and Cultural Dynamics.* Sargent, Boston, Mass.
Sorokin, P.A. (1966) *Sociological Theories of Today.* Harper & Row, New York.
Southwick, C.H. (1969) Aggressive behaviour of rhesus monkeys in natural and captive groups, in *Aggressive Behaviour* (eds S. Garattini and E.B. Sigg). Excerpta Medica, Amsterdam, 32–43.
Southwick, C.H., Beg, M.A. and Siddiqi, M.R. (1965) Rhesus monkeys in north India, in *Primate Behavior* (ed. I. DeVore). Treubner King, New York.
Southwick, C.H., Siddiqi, M.F., Farooqi, M.Y. and Pal, B.C. (1974) Xenophobia among free-ranging rhesus groups in India, in *Primate Aggression, Territoriality, and Xenophobia* (ed. R.L. Holloway). Academic Press, New York, 185–210.
Speer, A. (1969) *Inside the Third Reich.* Macmillan, New York.
Speier, H. (1941) The social types of war. *American Journal of Sociology*, **46**, 445–54.
Spencer, H. (1873) *The Study of Sociology.* Appleton, New York.
Spencer, H. (1885) *The Principles of Sociology* (3 Vols). Williams & Norgate, London.
Spencer, H. (1892) *The Principles of Ethics* (2 Vols). Williams & Norgate, London.
Spencer, H. (1896) *Principles of Sociology* (2 vols). Appleton, New York.
Spencer, H. (1897) *Social Statics.* Appleton, New York.
Spencer, W.B. and Gillen, F.J. (1912) *Across Australia* (2 Vols). Macmillan, London.
Spengler, O. (1918) *Der Untergang des Abendlandes.* Wilhelm Braunmüller, Vienna/Leipzig.
Spits, F.C. (1977) Hobbes' views on war and peace, in *Declarations on Principles* (eds R.J. Akkerman *et al*). Sythoff, Leiden, 101–19.
Srb, A.M., Owen, R.D. and Edgar, R.S. (1965) *General Genetics* (2nd edn). Freeman & Co., San Francisco/London.
Stagner, R. (ed.) (1967) *The Dimensions of Human Conflict.* Wayne State University Press, Detroit.
Stamps, J.A. and Metcalf, R.A. (1980) Parent-offspring conflict, in *Sociobiology:*

Beyond nature/nurture? (eds G. Barlow and J. Silverberg). Westview Press, Boulder, 589–618.
Stanley, S.M. (1981) *The New Evolutionary Timetable.* Basic Books, New York.
Stea, D. (1965) Space, territory and human movements. *Landscape*, **15**, 13–16.
Stea, D. (1970) Home range and use of space, in *Spatial behavior of older people* (eds L.A. Pastalan and D.H. Carson). University of Michigan Press, Ann Arbor, Mich.
Steenhoven, G. van den (1959) *Legal Concepts Among the Netsilik Eskimos of Pelly Bay, N. W.T. Canada.* Department of Northern Affairs, N.C.R.C. Report 59–3.
Stefansson, V. (1914) The Stefansson-Anderson Arctic Expedition of the American Museum: Preliminary Ethnological Report. *Anthropological Papers of the American Museum of Natural History*, **XIV**(1).
Stefansson, V. (1921) *The Friendly Arctic.* Macmillan, New York.
Steinmetz, S.R. (1892) *Ethnologische Studien zur ersten Entwicklung der Strafe.* J.A. Barth, Leipzig.
Steinmetz, S.R. (1896) Endokannibalismus. *Mitteilungen der anthropologischen Gesellschaft in Wien*, **26**, 1–60.
Steinmetz, S.R. (1907) *Philosophie des Krieges.* J.A. Barth, Leipzig.
Steinmetz, S.R. (1929) *Soziologie des Krieges.* J.A. Barth, Leipzig.
Steklis, H.D. (1990) Men, women and evolution, in *Male and Female: The Child and the Adult* (ed. Anne Campbell). Andromeda, Oxford (in press).
Stempel, J.D. (1981) *Inside the Iranian Revolution.* Indiana University Press, Bloomington, Ind.
Stinchcombe, A.L. (1968) *Construction Social Theories.* Harcourt Brace, New York.
Stouffer, S.A. *et al.* (1949) *Studies in Social Psychology in World War II, Vol. 2 The American Soldier.* Princeton University Press, Princeton.
Strachey, A. (1957) *The Unconscious Motives of War: A psychoanalytic contribution.* Allen & Unwin, London.
Strayer, F.F. and Strayer, J. (1976) An ethological analysis of social agonism and dominance relations among preschool children, *Child Development*, **47**, 980–89.
Strelau, J. (1974a) Temperament as an expression of energy level and temporal features of behaviour. *Polish Psychological Bulletin*, **5**, 119–27.
Strelau, J. (1974b) *Experimental Investigations of the Relations between Reactivity as a Temperament Trait and Human Action*, paper presented at the International Conference on 'Temperament and Personality', October 1974, Warsaw, Poland.
Struhsaker, T.T. and Leland, L. (1987) Colobines: infanticide by adult males, in *Primate Societies* (eds B.B. Smuts *et al.*) University of Chicago Press, Chicago, Ill., 83–97.
Strum, S.C. (1975) Primate predation: interim report on the development of a tradition in a troop of olive baboons. *Science*, **56**, 44–68.
Sumner, W.G. (1906) *Folkways.* Ginn, New York.
Sumner, W.G. (1911) *War and Other Essays.* Yale University Press, New Haven.
Sumner, W.G. (1913) *Earth-hunger and Other Essays.* Yale University Press, New Haven.
Sumner, W.G. and Keller, A.G. (1927) *The Science of Society.* Yale University Press, New Haven.
Swatez, G.M. (1970) The social organization of a university laboratory, *Minerva: A Review of Science Learning and Policy*, **8**, 36–58.
Symons, D. (1979) *The evolution of human sexuality*, Oxford University Press, Oxford.
Szyliowicz, J.S. (1975) Elites and modernization in Turkey, in *Political Elites and Political Development in the Middle East* (ed. F Tachau). Schenkman, Cambridge, Mass., 23–66.

Tajfel, H. (1979) Human intergroup conflict: useful and less useful forms of analysis, in *Human Ethology, Claims and Limits of a New Discipline* (eds M. von Cranach, K. Foppa, W. Lepenies and D. Ploog). Cambridge University Press, Cambridge, 396–434.

Talmon, J.L. (1980) *The Myth of the Nation and the Vision of Revolution, The Origins of Ideological Polarization in the 20th Century*, Oxford University Press, London.

Tarde, G. (1897) *L'Opposition Universelle.* Alcan, Paris.

Tarde, G. (1899) *Social Laws.* Macmillan, New York.

Tedeschi, J.T., Melburg, V. and Rosenfeld, P. (1981) Is the concept of aggression useful? in *Multidisciplinary Approaches to Aggression Research* (eds P.F. Brain and D. Benton). Elsevier/North Holland, Amsterdam, 23–37.

Teleki, G. (1973) *The Predatory Behavior of Wild Chimpanzees.* Bucknell University Press, Lewisburg.

Teleki, G. (1975) Primate subsistence patterns: collector-predators and gatherer-hunters. *Journal of Human Evolution*, **4**, 125–84.

Terrace, H. (1984) *Nim.* Random House, New York.

Terrace, H. (1985) In the beginning was the 'Name'. *American Psychologist*, **40**(9).

Thibaut, J.W. and Kelley, H.H. (1959) *The Social Psychology of Groups.* Wiley, New York.

Thom, R. and Zeeman, E.C. (1974) Catastrophe theory: Its present state and future perspectives, in *Dynamical systems: proceedings of a symposium held at the University of Warwick 1973/1974* (ed. A. Manning). Springer-Verlag, New York.

Thomas, L. (1981) Debating the unknowable. *The Atlantic Monthly*, July, 49–54.

Thomson, J. (1885) *Through Massailand.* Cass Library of African Studies, No. 46, London.

Thomson, B.H. (1908) *The Fijians: A study of the decay of custom.* Heinemann, London.

Thomson, D. (1980) Adaptors and innovators: A replication study on managers in Singapore and Malaysia. *Psychological Reports*, **47**, 383–7.

Thorndike, E.L. (1913) *Original Nature of Man.* Columbia University Press, New York.

Thurlings, J.M. (1965) The dynamic function of conflict. *Sociologia Neerlandica*, **2**(2), 142–60.

Tiger, L. (1985) Ideology as brain disease. *ZYGON*, **20**(1), 31–9.

Tiger, L. (1987) *The Manufacture of Evil: Ethics, Evolution, and the Industrial System.* Bessie Books/Harper & Row, New York.

Tiger, L. and Fox, R. (1971) *The Imperial Animal.* Holt, Rinehart and Winston, New York; (1989) 2nd edn, Henry Holt, New York.

Tinbergen, N. (1953) *Social behaviour in animals.* Methuen, London.

Tinbergen, N. (1963) *The Herring Gull's World.* Collins, London.

Tinbergen, N. (1967) Adaptive features of the black-headed gull *Larus ridibundus* L. *Proceedings of XIV International Ornithology Congress*, 43–59.

Tinbergen, N. (1968) On war and peace in animals and man. *Science*, **160**, 1411–18.

Tinbergen, N. (1976) Ethology in a changing world, in *Growing Points in Ethology* (eds P.G. Bateson and R.A. Hinde). Cambridge University Press, Cambridge, 507–27.

Tinbergen, N. (1981) On the history of war, in *Aggression and Violence* (eds L. Valzelli and L. Morgese). Edizioni Centro Culturale, Saint Vincent, 31–8.

Tindale, N.B. (1974) *Aboriginal Tribes of Australia; their terrain, environmental controls, distribution, limits, and proper names.* University of California Press, Berkeley, Cal.

Tobach, E. (1978) The methodology of sociobiology from the viewpoint of a comparative psychologist, in *The Sociobiology Debate* (ed. A.L. Caplan). Harper and Row, New York, 411–23.

Tooby, J. and Cosmides, L. (1988) The evolution of war and its cognitive foundations. Proceedings of the Institute of Evolutionary Studies 88, Ann Arbor, 1–15.
Topinard, P. (1900) *L'Anthropologie et la Science Sociale*. Babé et Lecroshier, Paris.
Torrance, E.P. and Horng, R.Y. (1980) Creativity and style of learning and thinking characteristics of adaptors and innovators, *The Creative Child and Adult Quarterly*, **5**(2), 80–5.
Toynbee, A.J. (1950) *War and Civilization*. Oxford University Press, New York.
Toynbee, A.J. (1972) *A Study of History*. Oxford University Press, London.
Trivers, R.L. (1971) The evolution of reciprocal altruism. *Quarterly Review of Biology*, **46**(4), 35–57.
Trivers, R.L. (1974) Parent-offspring conflict. *American Zoologist*, **14**(1), 249–64.
Trivers, R.L. (1985) *Social Evolution*. Benjamin/Cummings, Menlo Park, Cal.
Trotter, W. (1916) *Instincts of the Herd in Peace and War*. Fisher Unwin, London.
Turnbull, C.M. (1965) *Wayward Servants: The Two Worlds of the African Pygmies*. Eyre & Spottiswood, New York.
Turner, L.M. (1887) On the Indians and Eskimos of the Ungava District, Labrador. *Transactions of the Royal Society of Canada*, **V**, sec. 2.
Turner, L.M. (1894) *Ethnology of the Ungava District, Hudson Bay Territory*. Bureau of American Ethnology, 11th Annual Report, 1889–90.
Turney-High, H.H. (1949) *Primitive War: Its practice and concepts*. University of South Carolina Press, Columbia.
Turney-High, H.H. (1971) *Primitive War: Its practice and concepts*. University of South Carolina Press, Columbia.
Tutin, C.E.G. (1979) Mating patterns and reproductive strategies in a community of wild chimpanzees (*Pan troglodytes schweinfurthii*). *Behav. Ecol. and Sociobiol.*, **6**. 29–38.
Tylor, E.B. (1874) *Primitive Culture*. Appleton, New York.
Tylor, E.B. (1889) On a method of investigating the development of institutions, applied to the laws of marriage and descent. *Journal of the Anthropological Institute*, **18**, 245–72.
Vaccaro, M. (1886) *La lotta per l'esistenzia e suoi effetti nell'humanita*. Setth, Roma.
Vaccaro, M. (1898) *Les Bases Sociologiques du Droit et de l'Etat*. Alcan, Paris.
van Valen, L. (1976) Energy and evolution. *Evolutionary Theory*, Vol. 1, 179–86.
van de Velde, F. (1956) Les rigles du portage des phogues pris par la chasse aux aglus. *Anthropologica*, **3**, 5–15.
Vandenberg, S.G. (1967) Hereditary factors in normal personality traits (As measured by inventories), in *Recent Advances in Biological Psychiatry*, Vol. 9, Plenum Press, New York.
Vandenbergh, J.G. (1966) Rhesus monkey bands. *Natural History*, **75**, 22–27.
Van Den Berghe, P.L. (1978) Bridging the paradigms: biology and the social sciences, in *Sociobiology and Human Nature* (eds M.S. Gregory , A. Silvers and D. Sutch). Jossey-Bass, San Francisco, 33–52.
van der Dennen, J.M.G. (1977) *De Apologeten van de Oorlog*. Polemologisch Instituut, Groningen.
van der Dennen, J.M.G. (1981) On War: Concepts, definitions, research & data – A literature review and bibliography, in *UNESCO Yearbook on Peace and Conflict Studies*, Paris, 128–89.
van der Dennen, J.M.G. (1984a) Ontstaan en Evolutie van de 'Primitieve' Oorlog. *Transaktie*, **13**(4) 321–45.
van der Dennen, J.M.G. (1984b) Source materials for the study of 'primitive' war. A bibliography containing some 5,500 entries on warfare, feuding and intratribal

violence in preliterate societies. Polemological Institute, University of Groningen.

van der Dennen, J.M.G. (1986) Four fatal fallacies in defense of a myth: the aggression-warfare linkage, in *Essays in Human Sociobiology* (eds J. Wind and V. Reynolds) Vol. 2, V.U.B. Study Series, 26, Brussels, 43–68.

van der Dennen, J.M.G. (1987) Ethnocentrism and ingroup/outgroup differentiation, in *The Sociobiology of Ethnocentrism* (eds V. Reynolds, V. Falger and I. Vine), Croom-Helm, London/Sydney, 1–47.

van der Dennen, J.M.G. (1988) The Ethnological Inventory Project; Containing the band-level and tribal societies, ethnies and ethno-linguistic groups of the world and in the world's history. 5 Vols., 3rd rev. ed., Polemological Institute, State University of Groningen.

van der Dennen, J.M.G. (1990) A comparison of Hobhouse *et al.*'s and Q. Wright's list of primitive peoples and warfare.

van der Dennen, J.M.G. (1990) De Primitieve Oorlog in Evolutionair Perspectief, in Oorlog: multidiscipline beschonwingen (eds J. van Hoof *et al.*), Stirbeg, Hoogezand, 274–93.

van der Molen, P.P. (1981) Self-will and population cycles; the concept of a genetically-determined behavioural trait 'Thing-oriented and self-willed versus social and compliant' and its implications on the level of population dynamics, *Genetical Society Newsletter*, **7**, 24.

van der Molen, P.P. (1989) Adaption – innovation and changes in social structure: on the anatomy of catastrophe, in *Adaptors and Innovators: styles of creativity and problem solving* (ed. M.J. Kirton). Routledge, London, 158–98.

van der Molen, P.P. (1987) Social role blindness and selection on the self-will versus compliance trait; their evolutionary stability and the way they cause cyclic social changes and catastrophes, *Heymans Bulletin* HB-87-870-EX, Dept. of Psychology, Rijks Universiteit Groningen (RUG).

van Hooff, J.A.R.A.M. (1977) De adaptieve betekenis van agressief gedrag, in *Agressief Gedrag; Oorzaken en Functies*, (eds P.R. Wiepkema and J.A.R.A.M. van Hooff). Bohn, Scheltema & Holkema, Utrecht, 5–27.

van Hooff, J.A.R.A.M. (1982) Coalitions and positions of influence in a chimpanzee community, in *The Biology of Primate Sociopolitical Behavior* (eds G. Schubert and A. Somit). Northern Illinois University Press, DeKalb, 2–15.

van Hooff, J.A.R.A.M. (1988) Sociality in primates, a compromise of ecological and social adaptation strategies, in *Perspectives in the Study of Primates* (eds A.Tartabini and M.L. Genta) De Rose, Cosenza, 9–23.

van Hooff, J.A.R.A.M. and Netto, W.J. (in press) Conflict interference and the development of dominance relationships in immature *Macaca fascicularis*. *International Journal of Primatology.*

van Hooff, J.A.R.A.M. and de Waal, F.B.M. (1975) Aspects of an ethological analysis of polyadic agonistic interactions in a captive group of *Macaca fascicularis*, in *Contemporary Primatology* (eds S. Kondo, M. Kawai and A. Ehara). Karger, Basel, 269–74.

van Noordwijk, M.A. (1985) The Socio-Ecology of Sumatran Long-tailed Macaques (*Macaca fascicularis*): II. The Behaviour of Individuals. Dissertation, University of Utrecht.

van Rhijn, J.G. (1981) Units of behaviour in the black-headed gull *Larus rudibundus* L. *Animal Behaviour*, **29**(2), 586–97.

van Schaik, C.P. (1983) Why are diurnal primates living in groups? *Behaviour*, **87**, 120–44.

van Schaik, C.P. (1985) The Socio-ecology of Sumatran Long-tailed Macaques; I.

Costs and Benefits of Group Living. Dissertation, University of Utrecht, Netherlands.
van Schaik, C.P. (1989) The ecology of social relationships amongst female primates, in *Comparative Socio-Ecology of Mammals and Humans*, (eds V. Standon and R. Foley). Blackwell, Oxford, 195–217.
van Schaik, C.P. and van Hooff, J.A.R.A.M. (1983) On the ultimate causes of primate social systems. *Behaviour*, **85**, 1–2, 91–117.
van Sommers, P. (1972) *The Biology of Behaviour*. Wiley, New York.
Vayda, A.P. (1960) Maori warfare. *Polynesian Society Maori Monograph*, **2**, Wellington.
Vayda, A.P. (1961) Expansion and warfare among swidden agriculturists. *American Anthropologist*, **63**, 346–58.
Vayda, A.P. (1967) Research on the functions of primitive war. *Peace Res. Soc. Internat. Papers*, **7**, 133–8.
Vayda, A.P. (1968) Hypotheses about functions of war, in *War: The Anthropology of Armed Conflict and Aggression* (eds M. Harris and R. Murphy). The Natural History Press, New York, 85–91.
Vayda, A.P. (1970) Maoris and muskets in New Zealand: Disruption of a war system. *Political Science Quarterly*, **85**, 560–84.
Vayda, A.P. (1971) Phases of the process of war and peace among the Marings of New Guinea. *Oceania*, **42**, 1–24.
Vayda, A.P. (1974) Warfare in ecological perspective. *Ann. Rev. Ecol. and Systematics*, **5**, 183–93.
Vayda, A.P. (1976) *War in Ecological Perspective: Persistence, Change and Adaptive Processes in Three Oceanian Societies*. Pergamon, New York.
Vayda, A.P. and Leeds, A. (1961) Anthropology and the study of war. *Anthropologica*, **3**(2), 131–4.
Veblen, T. (1928) *The Theory of the Leisure Class*. Vanguard Press, New York.
Vehrencamp, S.L. (1983) A model for the evolution of despotic versus egalitarian societies. *Animal Behaviour*, **31**, 667–82.
Vierkandt, A. (1896) *Naturvölker und Kulturvölker*. Ferdinand Enke, Stuttgart.
Vine, I. (1973) Social spacing in animals and man. *Social Science Information*, **12**(5), 7–50.
Volgyes, I. (1978) Modernization, stratification, and elite development in Hungary. *Social Forces*, **57**, 500–21.
Volterra, V. (1928) Variations and fluctuations of the number of individuals in animal species living together. *J. Cons. Int. Explor. Mer.*, **3**, 3–51.
Voorzanger, B. (1987) *Woorden, Waarden en de Evolutie van Gedrag*. Vrije Universiteit, Amsterdam.
Voslensky, M. (1984) *Nomenklatura: The Society Ruling Class*. Doubleday, Garden City, New York.
Vree, J.K. de (1982) *Foundations of Social and Political Processes*. Prime Press, Bilthoven.
Waitz, F.T. (1859–62) *Anthropologie der Naturvölker*. Fleischer, Leipzig.
Walker, R.B. (1987) Realism, change and international political theory. *International Studies Quarterly*, **31**(1), 65–86.
Walters, J.R. and Seyfarth, R.M. (1986) Conflict and cooperation, in *Primate Societies* (eds B.B. Smuts *et al.*), University of Chicago Press, Chicago, Ill., 306–17.
Warner, W.L. (1931) Murngin warfare. *Oceania*, **2**, 457–94.
Washburn, S.L. (1959) Speculations on the inter-relations of the history of tools and biological evolution, in *The Evolution of Man's Capacity for Culture* (ed. J.N. Spuhler). Wayne State University Press, Detroit, Ill.

Washburn, S.L. and Avis, V. (1958) Evolution of human behavior, in *Behavior and Evolution* (eds A. Roe and G.G. Simpson). Yale University Press, New Haven, 421–36.

Washburn, S.L. and DeVore, I. (1961) The social life of baboons. *Scient. Amer.*, **204**, 62–71.

Washburn, S.L. and Hamburg, D.A. (1968) Aggressive behavior in Old World monkeys and apes, in *Primates* (ed. P.C. Jay). Holt, Rinehart & Winston, New York, 458–78.

Washburn, S.L. and Howell, F.C. (1960) Human evolution and culture, in *The Evolution of Man* (ed. S. Tax). University of Chicago Press, Chicago, 33–56.

Washburn, S.L. and Lancaster, C.S. (1968) The evolution of hunting, in *Man the Hunter* (eds R.B. Lee and I. DeVore). Aldine Atherton, Chicago, 293–303.

Watanabe, K. (1979) Alliance formation in a free-ranging troop of Japanese macaques. *Primates*, **20**, 459–74.

Watts, C.R. and Stokes, A.W. (1971) The social order of turkeys. *Sci. Amer.*, **224**(6), 112–18.

Watzlawick, J.H. and Fish, R. (1973) *Change.* Palo Alto, Cal.

Weber, M. (1948) *From Max Weber: Essays in sociology,* translated, edited, and with an introduction by H.H. Gerth and C. Wright Mills. Routledge and Kegan Paul, London.

Weber, M. (1964) *Wirtschaft und Gesellschaft* (2 Vols). Mohr & Siebeck, Tübingen.

Wedgwood, C.H. (1930) Some aspects of warfare in Melanesia. *Oceania,* **1**, 5–33.

Weick, E.I. (1969) *The Social Psychology of Organizing.* Addison-Wesley, Reading, Mass.

Weisfeld, G.E. (1980) Social dominance and human motivation, in *Dominance Relations: An Ethological Perspective on Human Conflict,* (eds Donald R. Omark, F.F. Strayer and D.G. Freedman). Garland, STPM Press, New York, 273–86.

West Eberhard, M.J. (1975) The evolution of social behaviour by kin selection. *Quart. Rev. Biol.*, **50**, 1–33.

Westermarck, E.A. (1889) *The History of Human Marriage.* Allerton Press, New York.

Westermarck, E.A. (1907–9) *Ursprung und Entwicklung der Moralbegriffe.* Klinkhardt, Leipzig.

Weule, K. (1916) *Der Krieg in den Tiefen der Menschheit.* Kosmos, Stuttgart.

Weyer, E.M. (1932) *The Eskimos: Their Environment and Folkways.* Yale University Press, New Haven, Conn.

Wheeler, V. (1974) Drums and Guns: A Cross-Cultural Study of the Nature of War. Ph.D. Dissertation, University of Oregon.

Wheeler, W.M. (1928) *Emergent Evolution and the Development of Societies.* Norton, New York.

White, L.A. (1949) *The Science of Culture: A Study of Man and Civilization.* Farrar, Straus & Cudahy, New York.

White, R.K. and Lippitt, R. (1960) *Autocracy and Democracy; An experimental inquiry.* Harper & Brothers, New York.

Whiting, J.W.M. (1944) The frustration complex in Kwoma society. *Man,* **44**, 140–4.

Whyte, W.H. (1957) *The Organization Man.* Cape, London.

Wickler, W. (1985) Coordination of vigilance in bird groups. The 'Watchman's Song' hypothesis. *Z. Tierpsychol.*, **69**, 250–3.

Wiepkema, P.R. (1977a) Oorzaken van agressief gedrag, in *Agressief Gedrag; Oorzaken en functies,* (eds P.R. Wiepkema & J.A.R.A.M. van Hooff). Bohn, Scheltema & Holkema, Utrecht, 28–44.

Wiepkema, P.R. (1977b) Agressief Gedrag als Regelsysteem, in *Agressief Gedrag;*

Oorzaken en Functies, (eds P.R. Wiepkema and J.A.R.A.M. van Hooff). Bohn, Scheltema & Holkema, Utrecht, 69–78.
Wiese, L. von and Becker, H. (1932) *Systematic Sociology*. Wiley, New York.
Wilber, K. (1983) Kierkegaard's passion, *ReVision*, **6**(1), 81–5, excerpt from *A History of Western Psychology* (in progress).
Wilkinson, D. (1980) *Deadly Quarrels*. University of California Press, Los Angeles, Cal.
Willhoite, F.H. (1980) Evolutionary biology and political authority. *Human Ethology Newsletter*, **31**, 10–24.
Williams, G.C. (1966) *Adaptation and Natural Selection*. Princeton University Press, Princeton, NJ.
Williams, T. and Calvert, J. (1858) *Fiji and the Fijians* (2 Vols). Heglin, London.
Wilmsen, E.N. (1973) Interaction, spacing behavior, and the organization of hunting bands. *J. Anthropol. Res.*, **19**(1), 1–31.
Wilson, E.O. (1970) Competitive and aggressive behavior. *Social Science Information*, **9**(6), 123–54.
Wilson, E.O. (1975a) *Sociobiology: The new synthesis*. Harvard University Press, Cambridge, Mass.
Wilson, E.O. (1975b) Human Decency is Animal. *New York Magazine*, October.
Wilson, E.O. (1978) *On Human Nature*. Harvard University Press, Cambridge, Mass.
Wind, J. (1984) Sociobiology and the human sciences: An introduction. *Journal of Human Evolution*, **13**, 3–24.
Woodcock, A. and Davis, M. (1978) *Catastrophe Theory*. Avon, New York.
Woods, F.A. and Baltzly, A. (1915) *Is War Diminishing?* Houghton Mifflin, Boston.
Woolfenden, G.E. and Fitzpatrick, J.W. (1984) *The Florida Scrub Jay: Demography of a Cooperative-breeding Bird*. Princeton University Press, Princeton, NJ.
Woolrych, A. (1973) The English revolution: An introduction, in *Revolutions: A Comparative Study*, (ed. L. Kaplan). Vintage Books, New York, 77–111.
Wrangham, R.W. (1979) On the evolution of ape social systems. *Social Science Information*, **18**, 335–68.
Wrangham, R.W. (1980) An ecological model of female-bonded primate groups. *Behaviour*, **75**, 262–300.
Wrangham, R.W. (1982) Mutualism, Kinship and Social Evolution, in *King's College Sociobiology Group*, (eds) 269–89.
Wright, Q. (1942) *A Study of War*. University of Chicago Press, Chicago, Ill.
Wuketits, F.M. (1985) *Zustand und Bewusstsein. Leben als biophilosophische Synthese*. Hoffmann & Campe, Hamburg.
Wynne-Edwards, V.C. (1962) *Animal Dispersion in Relation to Social Behaviour*. Oliver & Boyd, Edinburgh.
Yamada, M. (1966) Five natural troops of Japanese monkeys in Shodishama Island. *Primates*, **7**, 313–62.
Yoshiba, K. (1968) Local and intertroop variability in ecology and social behavior of common Indian langurs, in *Primates*, (ed. P. Jay). Holt, Rinehart & Winston, New York, 217–42.
Young, C.W. (1975) An evolutionary theory of the causes of war, in *War, Its Causes and Correlates*, (eds M.A. Nettleship, R.D. Givens and A. Nettleship). Mouton, The Hague, 199–208.
Zahavi, A. (1977) Reliability in communication systems and the evolution of altruism, in *Evolutionary Ecology*, (eds B. Stonehouse and C. Perrins). Macmillan, London.
Zartman, I.W. (1975) Algeria: a post-revolutionary elite, in *Political Elites and Political Development in the Middle East*, (ed. F. Tachau). Schenkman, Cambridge, Mass., 255–92.

Zeeman, E.C. (1976) Catastrophe theory. *Scientific American*, **234**(4).
Zimen, E. (1978) *Der Wolf, Mythos und Verhalten.* Meyster, München.
Zimmerman, J.L. (1971) The territory and its density dependent effect in *Spiza americana. Auk*, **88**, 591–612.
Zuckerman, M. (1974) The sensation seeking motive, in *Progress in Experimental Personality Research*, (ed. B.A. Maher), Vol. 7. Academic Press, New York.

Author index

Author index

Subject index

Subject index

9 780412 337703

SPRINGER NATURE

GPSR Compliance

The European Union's (EU) General Product Safety Regulation (GPSR) is a set of rules that requires consumer products to be safe and our obligations to ensure this.

If you have any concerns about our products, you can contact us on ProductSafety@springernature.com

In case Publisher is established outside the EU, the EU authorized representative is:

Springer Nature Customer Service Center GmbH
Europaplatz 3
69115 Heidelberg, Germany

The manufacturer's authorised representative in the EU is Springer Nature Customer Service Centre GmbH, Europaplatz 3, 69115 Heidelberg, Germany. If you have any concerns regarding our products, please contact ProductSafety@springernature.com

Printed and bound by CPI Group (UK) Ltd, Croydon, CR0 4YY
30/09/2026
02225436-0009